D0022563

Virus Structure and Assembly

Edited by
Sherwood Casjens
University of Utah Medical Center

Jones and Bartlett Publishers, Inc.
Boston / Portola Valley

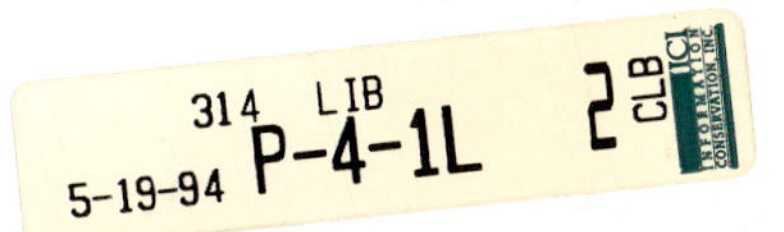

Editorial offices: Jones and Bartlett Publishers, Inc., 30 Granada Court, Portola Valley, California 94025

Sales and customer service offices: Jones and Bartlett Publishers, Inc., 20 Park Plaza, Boston, Massachusetts 02116

Library of Congress Cataloging in Publication Data

Virus structure and assembly.

1. Viruses—Morphology. I. Casjens, Sherwood.
[DNLM: 1. Viral Proteins—metabolism. 2. Viruses.
QW 160 V82118]
QR450.V57 1985 576′.64 85-12719
ISBN 0-86720-044-8

Cover Photo: Electron micrograph of phage T4, tobacco mosaic virus and turnip yellow mosaic virus. Magnification, 115,000. Micrograph taken by J. T. Finch.

Printer/Binder: Alpine Press

Design/Production: Unicorn Production Services, Inc.

Printed in the United States of America
Printing Number (last digit) 10 9 8 7 6 5 4 3 2 1

Preface

The elucidation of the structures and mechanisms of assembly of viruses has been the subject of a very large number of investigations, both because of the intrinsic scientific significance of the topic and because of the importance of viruses in human health and well being. In addition, viruses are often studied as apparently simpler versions of features of and processes that occur in the host cell. From the latter viewpoint, viruses have been exceedingly valuable in the study of macromolecular assembly reactions, macromolecular superstructure, membrane proteins, and many other areas. Although research problems in the different virus systems under study often have common themes, so many experimental systems are now progressing at a rapid pace that it is virtually impossible for even the researcher within the field to stay abreast of new developments in all subareas. The last review to cover the structure and assembly of viruses as a group was in 1975. Since then, particular aspects of the topics discussed in this volume have been reviewed, but the reviews are invariably specific to one or a few experimental systems. In addition, variation among this myriad of different "organisms" tends to obscure underlying similarities and to inhibit researchers who study other macromolecular structures from understanding the topic of virus assembly as a whole. For this reason, when possible, chapters in this volume deal with timely biochemical and biophysical problems rather than individual experimental systems.

I undertook to assemble this book with several goals in mind. I believe that workers within the field of virus assembly will profit by having current information and ideas from other related systems compiled in a single volume. Thus, effort has been made to have the chapters as up to date as possible and to deal with current aspects of the problems under discussion. However, in addition and perhaps more important, I hope that by organizing the volume as much as possible around problems rather than individual viruses, the information will often be presented in a novel manner and be more comprehensible and useful to students and researchers in related fields. To this end, an introductory chapter (Chapter 1) has been included to provide interested persons access to the current ideas, goals, and status of experimentation in the field of virus structure and assembly. Finally, in the past, researchers in this field have been extraordinarily open and noncombative. I hope that this volume will stimulate productive scientific contact between experimental systems and help keep it a rewarding field in which to work.

In the chapter on structure and assembly of icosahedral shells (Chapter 2) protein-protein interactions in such shells are discussed, with emphasis on the recently derived atomic-resolution structures of several simple RNA viruses. Since it has not been the focus of a previous review, the chapter on nucleic acid packaging (Chapter 3) is broad in scope, covering topics from the assembly of simple viruses such as tobacco mosaic virus to the complex packaging reactions of the dsDNA bacteriophages. This chapter also serves as a general review of the various virus assembly systems under study, with emphasis on nucleic acid-protein interactions. Several chapters, those concerning the nature of assembly pathways (Chapter 4), spatial control of assembly (Chapter 5), and the role of the host in assembly (Chapter 8), concentrate on the dsDNA bacteriophages, because most information on these subjects has been obtained from these experimental systems. These three chapters also emphasize the nature of protein-protein interactions in viruses. Two chapters, covering the assembly of the budding viruses (Chapter 6) and the filamentous bacteriophages (Chapter 7) address the role of lipid membranes in virus assembly.

Some aspects of virus assembly are not focused upon in this volume; for example, the physical chemistry of assembly, enzymes assembled into virus particles, covalent protein modification during assembly, regulation of the expression of the genes coding for virus structural proteins, *de novo* synthesis of lipid membranes during virus assembly, and functions of virions such as nucleic acid delivery into cells. These topics, though not unimportant, are either not well enough understood or do not currently appear to have sufficient generality to warrant such treatment in the limited space available here. They are dealt with as individual cases when appropriate.

I thank the authors and David Freifelder for their efforts toward achieving the goals of this volume, and John Finch for supplying the cover micrograph.

Salt Lake City,
March, 1985

Sherwood Casjens

Contents

List of Contributors

Peter B. Berget, Department of Biochemistry, University of Texas Health Center; Houston, Texas.

Sherwood Casjens, Department of Cellular, Viral and Molecular Biology, University of Utah Medical Center; Salt Lake City, Utah.

John W. Erickson, Department of Biological Sciences, Purdue University; West Lafayette, Indiana. Present Address: Abbott Laboratories, Department of Pharmaceutical Discovery; Abbott Park, Illinois.

Costa Georgopolous, Department of Cellular, Viral, and Molecular Biology, University of Utah College of Medicine; Salt Lake City, Utah.

Roger W. Hendrix, Department of Biological Sciences, University of Pittsburgh; Pittsburgh, Pennsylvania.

Javier Lopez, Department of Biochemistry, Duke University Medical Center; Durham, North Carolina.

Michael G. Rossmann, Department of Biological Sciences, Purdue University; West Lafayette, Indiana.

Ellen G. Strauss, Department of Biology, California Institute of Technology; Pasadena, California.

James H. Strauss, Department of Biology, California Institute of Technology; Pasadena, California.

Kit Tilly, Department of Biochemistry and Molecular Biology, Harvard University; Cambridge, Massachusetts. Present Address: Department of Physiological Chemistry, University of Wisconsin; Madison, Wisconsin.

Robert E. Webster, Department of Biochemistry, Duke University Medical Center; Durham, North Carolina.

1
An Introduction to Virus Structure and Assembly

Sherwood Casjens
University of Utah Medical Center

It is after all, not man but the universe that is subtle.

Barry Lopez

HISTORICAL ASPECTS OF VIRUS STRUCTURE

Near the end of the nineteenth century scientists in Holland, Russia and Germany discovered that several disease agents could be transmitted by fluids that had been passed through filters thought to block the passage of all bacteria. These agents were called "filterable viruses" to distinguish them from the more general class of disease-causing agents, including bacteria, that were then encompassed by the term "virus." Other workers soon found intracellular changes in "infected" cells, even though the agents causing them were too small to be seen themselves, and the belief that filterable viruses were intracellular parasites became prevalent (see Luria et al., 1978). Understanding of the structure of these causative agents, now simply called "viruses," had to wait some years for technological advances to make more detailed analyses possible. The next advance came when Elford and Andrewes (1932) used filtration methods to determine the "characteristic diameter" of different viruses. These and subsequent studies indicated that virus diameters varied from roughly 25 to 120 nm. At this point progress on understanding virus structure began an acceleration that has yet to slow. Schlesinger (1934) showed that a bacteriophage was composed mainly of protein and nucleic acid. Stanley (1935) was able to crystallize tobacco mosaic virus, and Bernal and Fankuchen (1941) found that virus crystals had well-behaved x-ray diffraction patterns. These observations, along with early electron microscopic visualizations (Kausche et al., 1939), confirmed that viruses were structured, rather than amorphous entities. This conclusion was extended by the finding that each of several types of viruses each contained a single major type of protein molecule present in many copies per virus particle or virion (Harris and Knight, 1955; Niu and Fraenkel-Conrat, 1955; Schramm et al., 1955). At about the same time, it became evident that the protein forms the exterior surface of virus particles (virions), and the nucleic acid is internal (Schmidt et al., 1954; Franklin et al., 1957).

SYMMETRY AND VIRUS STRUCTURE

Crick and Watson (1956) used the information just presented to make the seminal observation that viruses must be constructed with protein building blocks in arrays that have either helical or cubic symmetry (see below). The *basic premise* of their argument was that such arrays must contain protein subunits arranged in such a manner that all protein subunits occupy identical positions. That is, each subunit should have *identical bonding contacts with its neighbors*, a presumably desirable property for identical proteins.

Helical structures are "open" in the sense that they can enclose any volume by simply varying their length. It is quite easy to visualize how protein molecules can be arranged helically (i.e., along a screw axis of symmetry) such that each has identical bonding properties (Figure 1). This idea was first stated in a less rigorous way by Crane (1950). Klug et al. (1958) have summarized the symmetry elements that must define such structures. The only exceptional subunits in such an arrangement are the ones at the ends, which do not have all of their bonding sites occupied. All of the rod-shaped viruses of known structure, do in fact have structural subunits in helical arrays (see for example, Stubbs et al., 1977).

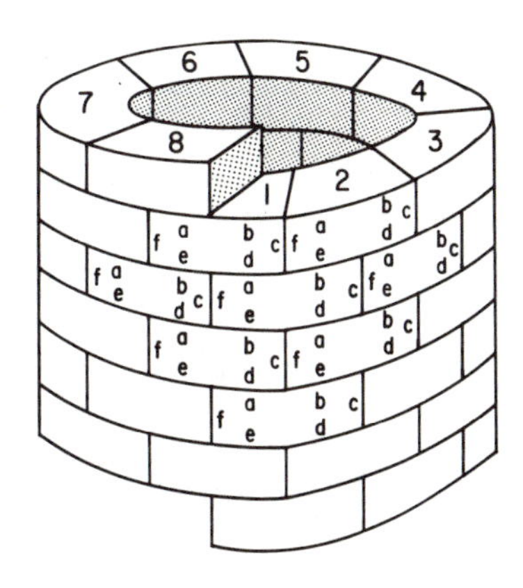

FIGURE 1 Identical intersubunit bonding between subunits in a helical array. A one-start helix with 7-1/2 subunits per turn is shown. Note that a-d, b-e, and c-f contacts are made in identical fashion between all pairs of adjacent subunits.

The viruses that have closed protein shells are geometrically more complex. How can asymmetric protein subunits be placed in space such that they build a closed structure in which all subunits have identical bonding properties? Mathematically this is considered by the utilization of symmetry axes. If all of the protein subunits have identical positions with respect to their neighbors, one should be able to perform proper rotations about the symmetry axes that define the structure, such that any chosen subunit or subunit contact can be perfectly superimposed on any other subunit or subunit contact (i.e., the rotated structure should be indistinguishable from the structure before the rotation). Clearly, if this series of operations is possible, all subunits occupy identical positions in the structure. Alternatively one can imagine the virus structure to be generated from a single subunit placed in space, by creating n new subunits through rotation of the first about the n-fold symmetry axes of a point group. (An arrangement of symmetry axes in space is called a point group.) Since proteins are asymmetric, there can be no mirror planes in the appropriate point groups.

There are only three point groups that can give rise to structures satisfying these requirements. A conceptual argument for this limit, not presented in detail by Crick and Watson (1956), proceeds as follows (adapted in part from Holden, 1971): The asymmetric protein subunits can be thought of as being placed on the surface of a

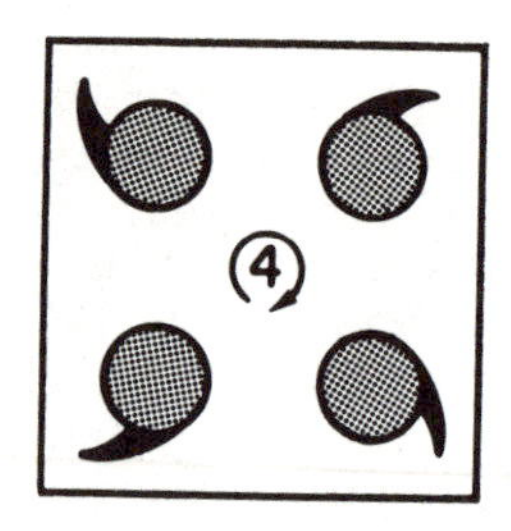

FIGURE 2 Identical asymmetric subunits in equivalent positions on an equilateral triangle and square. These subunits can be superimposed by rotations of 360/*n* degrees, where the *n*-fold rotational symmetry axes are shown within the circular arrows. They therefore occupy identical positions on these regular polygons. The positioning of subunits is arbitrary and diagrammatic. Any arrangement with *n*-fold symmetry is acceptable. If a shell without holes were being built the subunits would have to abut each other and fill the faces.

polyhedral structure such that they can be rotated into each other by proper rotations about the axes defined by the shape of the polyhedron (or point group that defines the polyhedron). Imagine placing a subunit on the surface of such a polyhedron, and creating *n* identical subunits by rotation about the *n*-fold symmetry axis in the center of the face on which it lies. Such a symmetry axis *must* exist if we are able to place subunits on a face so that they can be rotated into each other, and their spatial relationship to each other and to the face edges (subunits on adjacent faces) is constant. Thus, an equilateral triangular face will contain three subunits, a square face four, etc. (Figure 2). If we can then use this face to generate a closed structure in which all faces, edges and corners (relationships between faces) are indistinguishable, we will have generated a closed shell in which *all* subunits are indistinguishable and therefore have identical bonds with their neighbors. Plato, in the fourth century B.C., was already aware that structures could exist that contain identical faces and corners, and described them in his dialogue *Timeaus*. A rigorous proof that there are only three spatial arrangements of symmetry axes (point groups) that always result in the superposition of a closed structure onto itself can be found in Weyl (1952). A conceptual rationalization that there are only three such point groups follows: Only regular polygons can be used in the construction of a protein shell with the symmetry properties outlined above, since there must be a rotational symmetry axis in the center of each face (above). Furthermore, in building such regular polyhedra these regular polygons (faces) must be joined at only one type of simple convex corner, since the subunits must all bond equivalently about these corners. Figure 3 shows that a triangular face can join in three ways to make simple convex corners, and squares and pentagons can do so in only one way. All other ways of joining like polygons fail to give simple convex corners and hence cannot be used

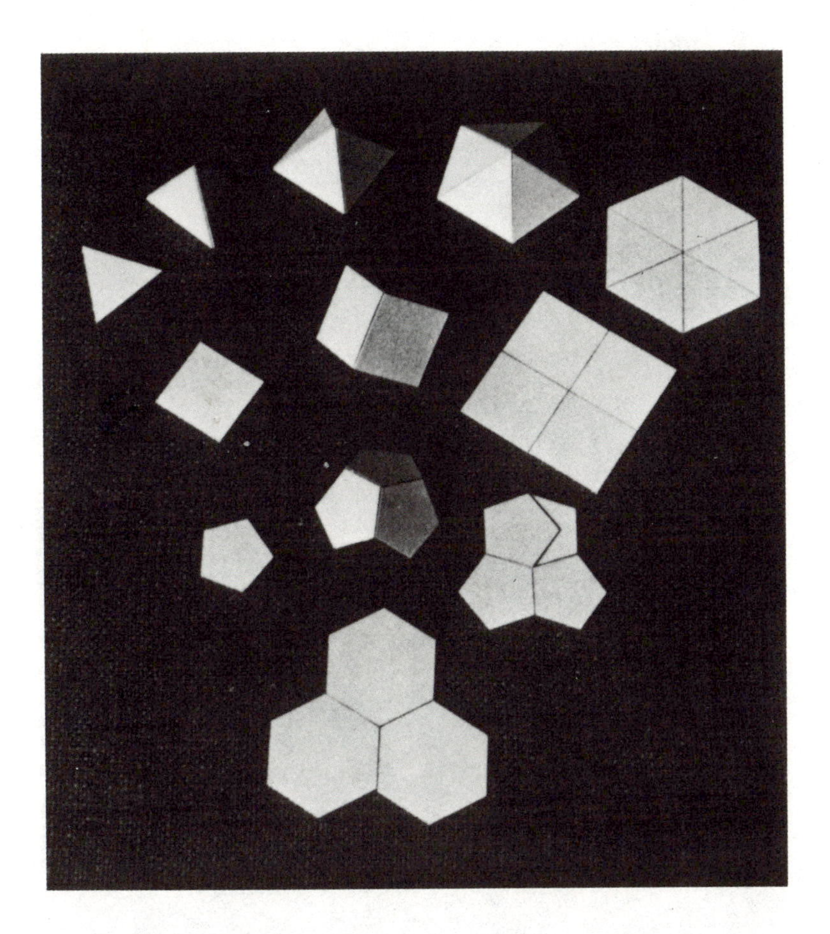

FIGURE 3 Simple convex corners that can be constructed from one type of regular polygon. At least three sides must be joined to make a simple convex corner. Three equilateral triangles, squares, or pentagons can be joined to make such a corner. Four and five equilateral triangles will also make convex corners. Six equilateral triangles, four squares and three hexagons join to form flat surfaces. Any larger number of these three polygons, as well as three or more of all other regular polygons, cannot form simple convex corners since the sum of their internal angles is greater than 360 degrees. Reproduced from Holden (1971) with the permission of Columbia University Press.

to assemble regular polyhedra. Note that joining three or more regular polyhedra with more that five sides cannot yield such corners, since the sum of their inside angles is not less than 360 degrees. By inspection, it is clear that each of these five types of corners defines one and only one regular solid in which all faces and corners are equivalent.

Table 1 Platonic solids

Polyhedron	Faces	Corners	Edges	Rotational axes			
				2-fold	3-fold	4-fold	5-fold
Tetrahedron	4	4	6	3	4	0	0
Cube	6	8	12	6	4	3	0
Octahedron	8	6	12	6	4	3	0
Dodecahedron	12	20	30	15	10	0	6
Icosahedron	20	12	30	15	10	0	6

The values in the table are the number of the indicated feature that are present in the polyhedra.

FIGURE 4 The Platonic solids. Solid and empty versions of (from left to right) tetrahedron, cube, octahedron, dodecahedron and icosahedron are shown. All faces, edges, and corners are indistinguishable within each of these regular polyhedrons. Reproduced from Holden (1971) with the permission of Columbia University Press.

We now know these five regular polyhedra, the tetrahedron, cube, octahedron, dodecahedron, and icosahedron, as the Platonic solids (Figure 4; Table 1). They fall into only three point groups, since the cube and octahedron, and the dodecahedron and icosahedron are "dual" pairs. That is, they have identical arrangements of symmetry axes in space. The two different shapes in the dual pairs can be thought of as arising from the two possible choices for the type of symmetry axis at the center of the faces (shown in Figure 5 for the dodecahedron and icosahedron; see comments on shape below). Thus, there are only three point groups that can be utilized to generate closed protein shells in which all proteins occupy identical positions.

FIGURE 5 The icosahedron and dodecahedron are dual structures. The icosahedron (left) and dodecahedron (right) have identical symmetries (i.e., they contain an equal number of symmetry axes identically oriented in space). One each of the 2-, 3-, and 5-fold axes are indicated by a square, triangle, and pentagon, respectively. Reproduced from Holden (1971) with the permission of Columbia University Press.

These three point groups are said to have "cubic" symmetry after a representative member of the class. Since faces with n-fold symmetry contain n subunits, tetrahedral, octahedral, and icosahedral structures can only be built from 12, 24 and 60 asymmetric identical units, respectively (Crick and Watson, 1956).

Soon after Crick and Watson advanced their postulates, it was found that several viruses do in fact have icosahedral shape (Williams and Smith, 1958; Horne et al., 1959), and that the subunits are arranged with icosahedral symmetry (Finch and Klug, 1959; Horne et al., 1959; Klug and Finch, 1960). In addition, it was found that many viruses must contain more than 60 subunits (Horne and Wildy, 1961). Caspar and Klug (1962) developed an insightful theory specifying the possible arrangement of protein subunits in structures with more than 60 subunits. The essence of their idea was that these proteins might occupy *quasi-equivalent positions* in the structure. Thus, the subunits could retain their the basic bonding properties, but slight deformations would allow them to occupy slightly different environments. This is perhaps best shown by example. Consider the "face" of an icosahedral virus shell. Figure 6 shows a face with three equivalent subunits. A larger face can be assembled from the same subunits by combining several of these faces, now called "*facets*." Caspar and Klug called this process "triangulation," because it is equivalent to subdividing the original face into smaller triangular units, and they defined the *triangulation number T* as the number of facets per face. Figure 6 shows four facets per face, $T = 4$, as an example. Such triangulation can be performed in many different ways, giving many different arrangements of facets with respect to the icosahedral face. If facets are allowed to bend, there is no geometric reason why facet edges must be congruent with the face edges. However, face corners must coincide with facet corners (see below). Subunits now occupy positions on the *facets* analogous to those occupied on the face in our previous

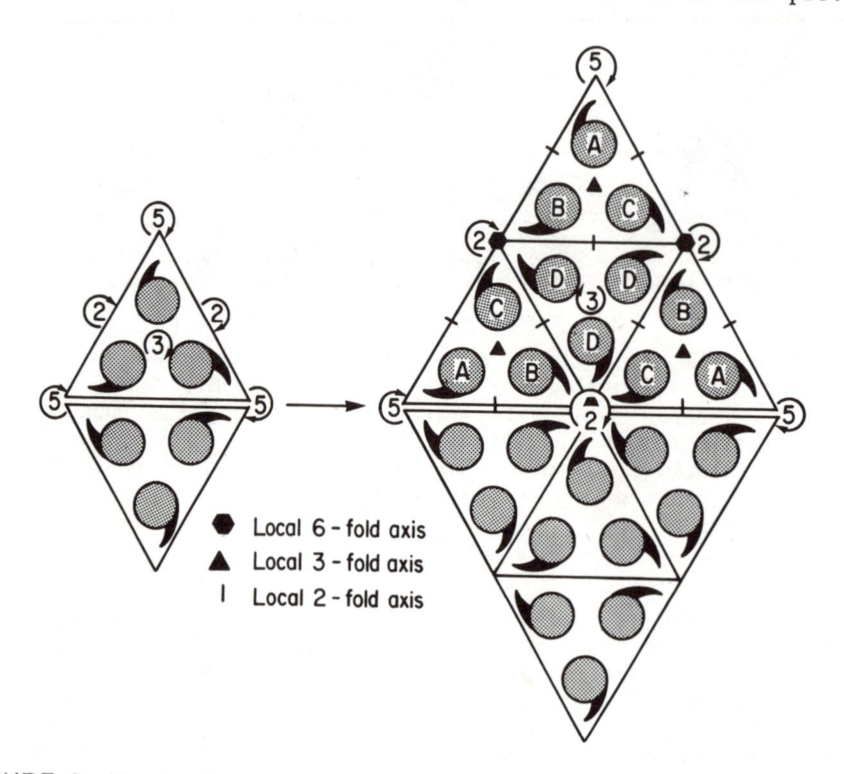

FIGURE 6 $T = 4$ triangulation of an equilateral triangular face. Shown on the left are two faces of a $T = 1$ icosahedron, flattened into a single plane. The "true" or icosahedral rotational symmetry axes are shown on the upper face by circular arrows. Note that the 5-fold axes are the corners of the icosahedron and that all of the asymmetric subunits can be superimposed by rotations of 360/n degrees about the *n*-fold symmetry axes. Shown on the right are two $T = 4$ faces flattened into a single plane (made from the same subunits as on the left). The true symmetry axes for the whole icosahedron are again indicated by circular arrows on the upper face. These faces have true symmetry axes at the same relative positions as the $T = 1$ faces. In addition, they have axes about which there is approximate local or quasi-symmetry. These local 2-, 3-, and 6-fold axes are shown as solid lines, triangles, and hexagons, respectively. There are no local 5-fold axes; the only five fold axes in the structure pass through the corners. In the $T = 4$ subunit distribution, the true 2-fold axes coincide with local 6-fold axes. This is not a universal feature of triangulated faces.

$T = 1$ example. Bonding among subunits on a given facet (around *quasi* or *local* 3-fold symmetry axes; see Figures 6 and 7) is similar to bonding around the true 3-fold axis on the original face. Bonding between subunits at the face corners (around true 5-fold axes) likewise retains $T = 1$ characteristics. However, bonding between subunits where other facet corners meet (quasi or local 6-fold axes) must be somewhat different than the other intersubunit contacts. Most notably, all subunits in the larger (triangulated) structure can be viewed as participating in rings of five subunits (pentamers) around each corner of the icosahedron and rings of six subunits (hexamers) elsewhere, whereas the original contains only rings of five. Thus, in the mental triangulation process, some subunits and/or subunit contacts must have been deformed to yield hexamers rather than pentamers; these hexamers have been inserted into the $T = 1$ structure between the pentamers in such a manner that the hexamer-pentamer contacts are quasi-equivalent to the pentamer-pentamer contacts in the $T = 1$ structure. No new pentamers are created in such a mental triangulation. In this way, Caspar and Klug envisioned viruses larger than those built from 60 subunits. The $T = 4$ example in Figures 6 and 7 would be built from 240 subunits.

Only certain strictly defined numbers of subunits can be placed in quasi-equivalent positions to make triangulated icosahedrons (Caspar and Klug, 1962). Caspar (1965) has presented a simple mathematical derivation of an expression for all possible ways to perform triangulation, expressed in terms of the triangulation number. To obtain this expression, we ask how many different sizes of equilateral triangles (which represent the faces of different-sized icosahedrons) can be properly superimposed on a planar net of facets (Figure 8). We can then imagine cutting out each of these triangles, duplicating each into 20 copies, and joining them together to form all possible sizes of icosahedral shells. The only constraint on this process is that the corners of the triangle representing the face (which will join with other identical corners at the 5-fold vertices of the icosahedron) must always fall on intersection points (6-fold axes) on the facet net. The reason for this constraint is probably best appreciated by actually drawing a facet net and attempting to construct icosahedra from it; in essence, if the face corners do not fall on facet corners, the resulting structure is biologically nonsensical, with arbitrarily truncated protein subunits and nonequivalent interactions. Since the face size is determined by the length of its side S, the problem reduces to asking how many different lengths of lines (S) can be drawn between pairs of intersection points on the facet net. We can specify a path between a given point (0,0 in Figure 8) and any other point (H,K) in terms of steps along the net in the directions defined by the unit vectors h and k. It is sufficient to consider a 1/6 sector since there is 6-fold symmetry about the point (0,0) in an infinite net. The line, S, between (0,0) and (H,K) is the hypotenuse of a right triangle with sides $H + K/2$ and $(\sqrt{3}/2)K$ (see Figure 8), and its length can be given in terms of H and K. By Pythagoras:

$$S^2 = (H + K/2)^2 + (\sqrt{3}K/2)^2$$

$$S^2 = H^2 + HK + K^2$$

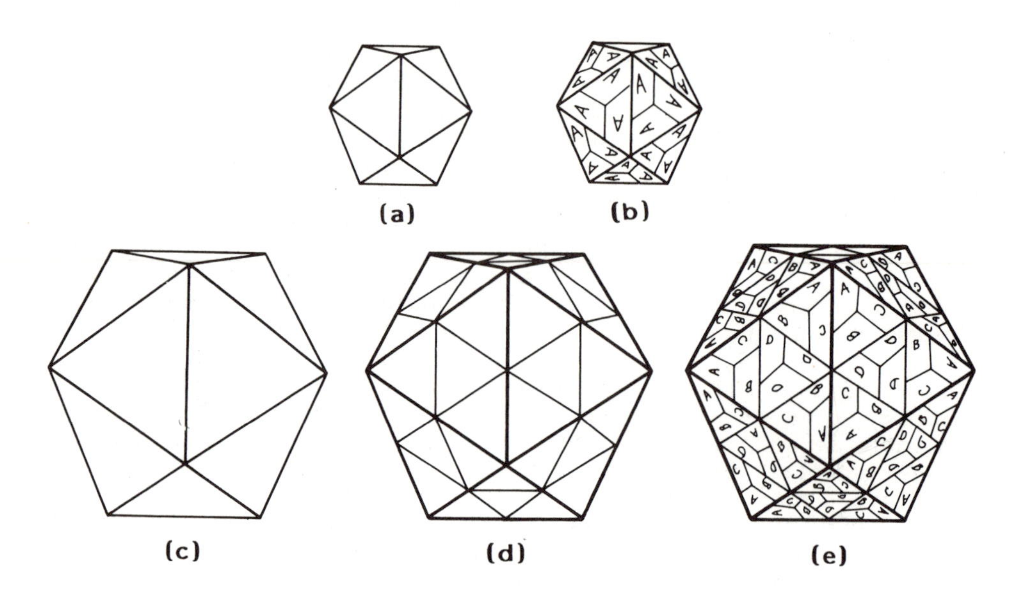

FIGURE 7 Coat protein environments in a $T = 1$ and a $T = 4$ capsid built with the same protein subunit. The faces shown in Figure 6 are assembled into complete icosahedral capsids. Panels (a) and (b) of the diagram show an icosahedron and 60 protein subunits placed in identical environments on its surface. This is a $T = 1$ structure. If we wish to assemble a triangulated icosahedron from these subunits, its facets must be the same size as the $T = 1$ faces (see Figure 6 and text). Panels (c) and (d) of the figure show such a larger icosahedron (c) with each face assembled from 4 such facets; therefore, it is a $T = 4$ structure. Panel (e) shows 240 identical protein subunits placed on the surface of the $T = 4$ icosahedron so that the relationship between subunits within each facet is similar to the $T = 1$ face. A $T = 4$ structure must have four different quasi-equivalent subunit environments; these are indicated by A, B, C, and D. Refer to Figure 6 and note that if the structure were rotated 120 degrees about a local 3-fold axis so that subunit A were superimposed upon subunit C, for example, the rest of the structure would not be superimposed upon itself. However, a rotation of 120 degrees about the true 3-fold axis results in a perfect superposition of the entire structure upon itself. The A subunits bond between faces and between facets in an identical manner (rings of 5 or pentamers) in the two structures. B, C, and D subunits participate in rings of 6 (hexamers) with bonds that are "quasi- equivalent" to those between subunits within pentamers (i.e., the same surfaces participate in analogous bonds). In this case, where the capsid has icosahedral shape, it is easy to visualize that the A, B, C, and D subunits occupy different environments. A subunits make two contacts out of the facet plane. B and C subunits make only one contact out the the facet plane (on different subunit surfaces for the two types of environment), and D subunits make no contacts out of the facet

(continued)

FIGURE 7 (continued)
plane. Also, note that the arrangement of subunits in different quasi-equivalent environments around local rotational axes does not precisely reflect the symmetry implied by those axes, showing that the local symmetry is necessarily imprecise (e.g., one B and one D subunit are related to each other by a local 2-fold axis).

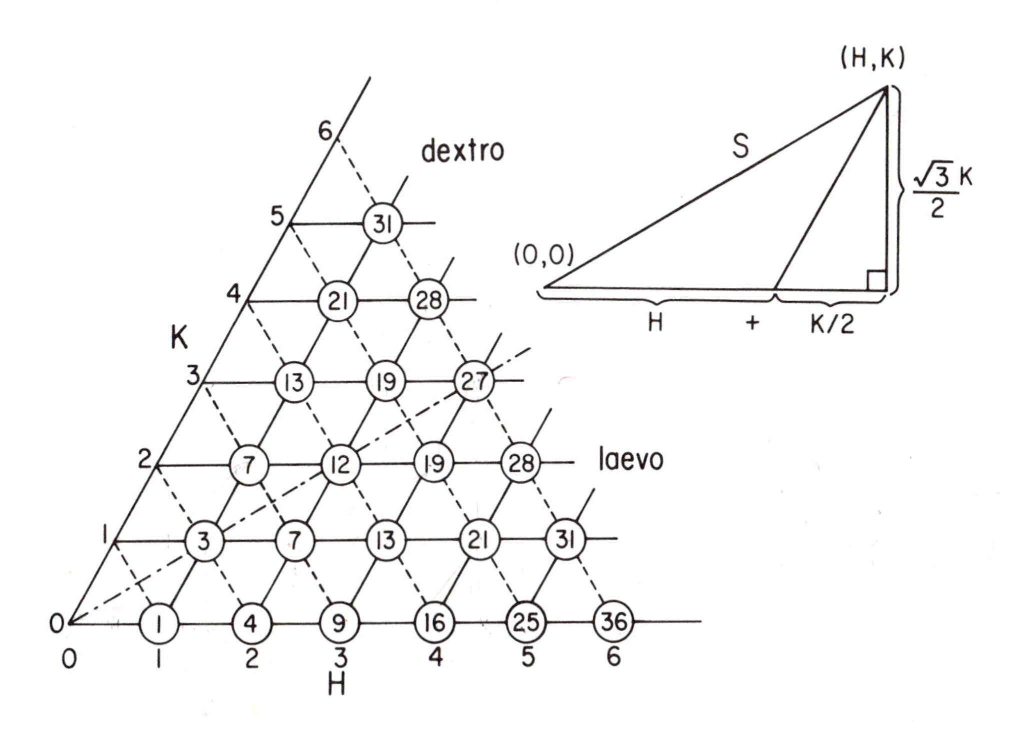

FIGURE 8 Triangulation numbers on an equilateral triangular net. The left part of the figure shows a 1/6 sector of an equilateral triangular net, with each triangle representing a facet. The indices H and K, each representing all integers greater than or equal to 0, define points of 6-fold symmetry, and therefore potential 5-fold symmetry by quasi-equivalence (see text). Thus, all vectors S (0,0 to H,K) define possible face sides, both with respect to length and spatial relationship to facets, thereby describing all possible triangulated surface lattices (after Caspar, 1965). The right part of the figure shows diagrammatically that by the theorem of Pythagoras the length of vector S is $[(H + K/2)^2 + (\sqrt{3}K/2)^2]^{1/2}$. Therefore, the area of an equilateral triangular face with sides of length S is $\sqrt{3}/4(H^2 + HK + K^2)$, in which H and K are the index values of the S vector.

In addition, since the area of an equilateral triangle is $(\sqrt{3}/4)S^2$ and the vectors h and k are unit length, the triangulation number T defined above as (area of a face)/(area of a facet), equals

$$T = (\sqrt{3}/4)S^2/(\sqrt{3}/4) = S^2$$

so

$$T = H^2 + HK + K^2$$

in which H and K can be any integer greater than or equal to 0. These pairs of integers describe all possible surface lattices for triangulated deltahedrons (polyhedra with equilateral triangular faces, including icosahedrons). Note that all face edge lengths S not coincident with sector (facet) edges or the line bisecting the sector appear twice with mirror symmetry about the bisect. Thus, there are two enantiomorphic forms of $T = 7, 13, 19, 21, \ldots.$ surface lattices. Where $H > K > 0$ the surface lattice is said to be left-handed; and where $K > H > 0$, it is right-handed (Klug and Finch, 1965). For some values of T, such as 91 and other larger values, more than one combination of H and K give the same value of T. However, for values of T of 84 or less, the range of values for all well-understood viruses (see however Carrascosa et al., 1984), the surface lattice is unambiguously defined by the triangulation number and when appropriate, the handedness. Thus, the structure of icosahedral viruses is usually discussed in terms of T numbers.

Some workers refer to the P number of a virus structure. If H and K have a common factor f such that $H = fa$ and $K = fb$, then

$$T = (a^2 + ab + b^2)(f^2) = P(f^2)$$

in which a and b are all possible integers greater than or equal to 0 with no common factor, and f is any integer > 0. Thus, each P number ($a^2 + ab + b^2 = 1, 3, 7, 13, \ldots$) defines a subset of the T numbers when combined with all integers f in the above expression. All surface lattices defined by a given P (e.g., $P = 3$, $T = 3, 12, 27, \ldots.$) are said to belong to the same "P class." P classes have geometric significance in that all members of a given class have the same angle between facet and face edges (see Figure 8). The biological significance, if any, of this subclassification is not known.

The possible ways of triangulating an icosahedron are defined by an infinite series of T numbers (1, 3, 4, 7, 9, 12, 13, ...). Viruses have been found that have structures compatible with triangulation numbers varying from 1 to 25 (the iridoviruses may have T numbers as large as 147, though their structure is not yet well understood; Wrigley, 1970). A triangulated icosahedral virus has $20T$ facets and three subunits per facet, so the number of protein subunits required to assemble a structure is $60T$. Thus, according to Caspar and Klug's (1962) theory, icosahedral structures should be built only from 60, 180, 240, 420, 540, 720, ... asymmetric protein subunits. These subunits occupy T distinguishable sets of quasi-equivalent positions. This can be understood as follows: there is a true 3-fold axis in the center of each *face* of a triangulated icosahedron, so the subunits within a 1/3 sector of each face are *identical* to those in the other two sectors. Rotation about the true icosahedral symmetry axes cannot rotate any of these subunits upon each other, since the triangulation process retains all of the original true symmetry axes, but does not generate any new ones. Triangulation generates only quasi-symmetry axes, which are only approximately true for even the immediate area (see Figures 6 and 7). Thus, each subunit within the 1/3 sector occupies a different position in that each has a different spatial relationship to the 5-fold axes. Since there are $3T$ subunits per face, there are T distinct positions on any face. All faces are equivalent, by virtue of the (true) 5-fold axes, so there are exactly T different quasi-equivalent subunit environments, each repeated 60 times on such an icosahedrally symmetric surface.

For convenience, all of the above arguments have been made with reference to an icosahedron. However, it must be emphasized that icosahedral *shape* is not required for structures with icosahedral *symmetry*. All known viruses have only convex corners, and convex structures with icosahedral symmetry can vary in shape from an icosahedron to a sphere to a dodecahedron, depending upon the bond angles between subunits in the different quasi-equivalent positions. Thus, many actual shapes are possible (see Holden, 1971). For example, on tomato bushy stunt virus ($T = 3$) pairs of facets are flattened into diamond-shaped surfaces. The resulting shape is a polyhedron with 30 diamond faces and icosahedral symmetry, called a rhombic triacontahedron (Winkler et al., 1978). Since many viruses are approximately spherical, or have unknown precise shape, viruses not obviously icosahedral in shape are often referred to as "spherical."

Icosahedral symmetry is not unique in its ability to generate closed structures that can be triangulated, since both tetrahedrons and octahedrons have equilateral triangular faces (and a cube can be "quadrangulated"). All viruses with closed shells, which have been analyzed to date, are icosahedral or have icosahedral caps separated by a cylinder of facets. Why should this be true? There are at least two possible explanations. (1) All closed-shell viruses could be descendants of a single ancestor, which happened upon icosahedral symmetry by chance, or (2) as was pointed out by Caspar and Klug (1962) the differences among the various quasi-equivalent positions on an icosahedron are smaller than the differences required on a structure with tetrahedral or octahedral symmetry. The subunit rings on these last structures can be viewed as being composed of trimers and tetramers, or tetramers and hexamers. Thus, if it was advantageous for viruses to be able to build shells larger than those made of 60 subunits, it might be have been easier to evolve a subunit that could form pentamers and hexamers than the less similar oligomers required to assemble tetrahedral or octahedral shells.

Electron micrographs of icosahedral virus particles often show considerable surface structure, and the *morphological units* or *capsomeres* seen on the surface rarely correspond to individual protein subunits (Horne and Wildy, 1961; Caspar and Klug, 1962). This was simply explained by Caspar and Klug, through the suggestion that an outward projecting portion of the protein subunit is asymmetrically distributed with respect to the portion that forms the closed shell. If outward projecting domains of neighboring subunits are close together, they can, when analyzed by low-resolution techniques, merge to form a "cluster" (Figure 9). A protuberance near the local and

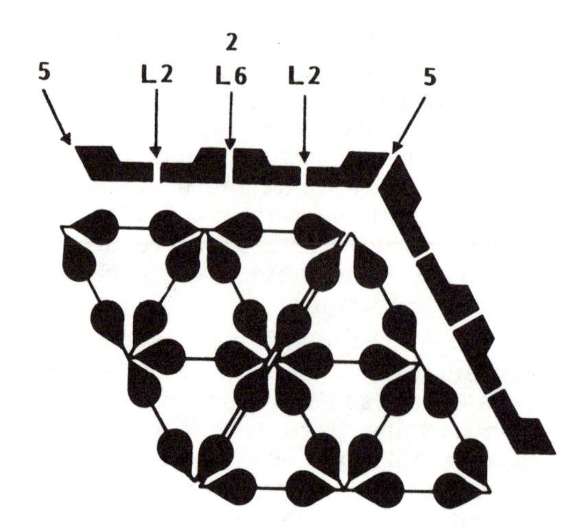

FIGURE 9 Hexamer-pentamer clustering of capsomeres or morphological units. Two faces of a $T = 4$ icosahedron are shown diagrammatically. Subunits are arranged so that the outward protruding portions, shown in black, lie on the facet boundaries (cf., Figure 6), near the facet corners. Thus, when viewed by techniques such as negative staining or heavy metal shadowing, which emphasize surface topology, the subunits appear to be clustered as pentamers at the corners of the icosahedron (5-fold axes) and as hexamers at the local 6-fold axes. Above and to the right is shown a diagrammatic vertical section along the face sides.

true 2-fold axes (facet edges) is said to be "dimer clustered," one near the local and true 3-fold axes (facet centers) is "trimer clustered," and one near the local 6-fold and true 5-fold axes (facet corners) is "hexamer-pentamer clustered." Not surprisingly, examples of each of these types of clustering have been observed in nature, as well as ones that are apparently unclustered and one that has two outward projecting domains with both trimer and hexamer-pentamer clustering (Williams and Fisher, 1974; Aebi et al., 1974; Casjens and King, 1975; Casjens, 1979). These clusters or capsomeres may or may not reflect strength of intersubunit bonds or oligomers active in the assembly of the structure.

X-ray structural analyses of simple $T = 1$ and $T = 3$ viruses have clearly borne out the prediction that subunits would be positioned on the surface according to the rules of triangulation (Olson et al., 1983; Rossmann et al., 1983; Liljas et al., 1982; Fukuama et al., 1983; Harrison, 1983). However, Caspar and Klug (1962) predicted small subunit and bond deformations to give "quasi-equivalent" subunit contacts, whereas atomic resolution studies have found fairly constant subunit structure and *substantially different* types of intersubunit bonds at different "quasi-equivalent" positions (see Chapter 2, this volume). Curiously, in the only more complex ($T = 7$) virus to be analyzed by crystallographic methods, Rayment et al. (1982) found that the local 6-fold axes of polyoma virus shells are occupied by pentamers! This means that the bonding between subunits is not really quasi-equivalent at all, but there appear to be three very different, nonequivalent types of bonds between subunits (see chapter 2, this volume). It is not clear how the spatial arrangement of these different bonds is controlled during assembly, or if this could be a general property of more complex shells. Electron micrographic analyses of the surface structure of adenovirus, phages T4, T7, and P22 (T = 25, 13, 7, and 7, respectively) are difficult to reconcile with this particular "nonequivalent" type of intersubunit bonding (Crowther and Franklin, 1972; Aebi et al., 1974; Casjens, 1979; Steven et al., 1983). Caspar and Klug (1962) used quasi-equivalent intersubunit bonding a *basic premise* in deriving the notion of triangulation. Since nonequivalence now seems a *possible* rule rather than the exception, their premise *may* not have been valid (in spite of the fact that the correct conclusion was reached, namely that subunits occupy positions dictated by triangulation)! If this is true, the ideas of quasi-equivalence may have to be reconsidered. See chapter 2 of this volume and Olson et al. (1983) for a more detailed discussion of this point. No general rationalization of nonequivalent subunit arrangement in virus shells has been put forward to replace Caspar and Klug's 1962 theory. An opposite view is that it is likely that some viruses do in fact form icosahedral shells by quasi-equivalent bonding (e.g. phage T7, see Steven et al., 1983), and the nonequivalences seen might be evolutionary drift from true quasi-equivalence (Olson et al., 1983). In addition to the historical importance of the idea of quasi-equivalence, the above detailed discussion of quasi-equivalence is justified by this later view. At present, it seems most useful to view these nonequivalences as variations on the quasi-equivalence theme until more data accumulates and an alternate theory is presented (but with awareness that these ideas are currently under renewed discussion).

Many more complex icosahedral viruses clearly utilize the basic idea of triangulation or quasi-symmetry (disregarding the details of subunit bonding) in subunit placement, but in addition embellish and modify the structure in various ways. For example, numerous virus shells have asymmetric protein units that contain multiple polypeptides (the consequences of which were discussed by Dunker, 1974). In others, the pentamers at the icosahedral 5-fold axes are made from polypeptides that are different from those in the hexamers on the faces (Ginsberg, 1979; Black and Showe, 1983); in this case, as with the simpler viruses discussed above, the hexameric subunits should still occupy $T - 1$ different quasi-equivalent positions. One virus has physical trimers occupying the local 6-fold positions (Periera and Wrigly, 1974; Crowther and Franklin, 1972). Some viruses have elongated the icosahedron by assembling extra "belts" of facets between icosahedral caps (Aebi et al., 1974; Eiserling, 1979; Dreidonks et al., 1976), and some contain what appear to be double icosahedral shells (Silverstein et al., 1976). In many bacteriophages one corner (point of 5-fold symmetry) is modified extensively by the attachment of adsorption apparatuses (tails) and nucleic acid packaging machinery (Earnshaw and Casjens, 1980) (see Chapters 3, 4 and 5 this volume for further details). These last modifications destroy the strict icosahedral symmetry of the protein shells. It is not yet certain how only one corner is chosen for this modification, but it is likely that the proteins at this unique corner serve as nucleation points for the assembly of the shell (Murialdo and Becker, 1978). A number of icosahedral viruses have evolved mechanisms for enclosing the structure that contains the nucleic acid with a lipid "envelope" (see Chapter 6, this volume). These envelopes are made from typical lipid bilayers and virus-coded proteins; the lipid bilayer may be assembled *de novo* or "stolen" from the host cell membranes. Helical virus protein components can also be more complex than the simple helix shown in Figure 1. Some have special proteins bound to each end, bound replication enzymes and/or envelopes (see Chapters 3, 6, and 7, this volume).

VIRUS CLASSIFICATION

This aspect of virology is important to the study of the structure of viruses for a number of reasons. It is certainly of interest to compare the structures of the many different types of viruses, both similar and different morphologically, in the pursuit of an overall understanding of how these parasites have solved the problems of intracellular replication and spread from cell to cell. Often such comparisons lead to the understanding of underlying general principles. Perhaps a more practical reason to have some knowledge of classification is that there is an extremely large number of very diverse viruses under study. Grouping viruses into families with similar properties allows consolidation of information within each group, thereby lessening the total number of facts that need to be remembered. Since many readers (even virologists) may not be familiar with all virus groups, Table 2 summarizes the major groups of viruses classified with respect to structural criteria that should be useful in the remainder of this volume. The mode of classification shown has the disadvantage of separating the enveloped viruses, even though they have this common feature, and separating viruses such as retroviruses and hepadnaviruses which are likely to utilize similar replication strategies, but which encapsidate different replication intermediates. For further details concerning the classification of individual viruses or the detailed biochemical properties of the virus groups, the reader is advised to consult Fraenkel-Conrat and Wagner (1974), Joklik (1980), and Mathews (1981 and 1982).

VIRUS ASSEMBLY

The study of virus assembly began when Fraenkel-Conrat and Williams (1955) found that they could *reconstitute* tobacco mosaic virus from separated RNA and protein components derived from virus particles. Since that time a number of simple viruses have been reconstituted *in vitro*. More complex viruses such as the dsDNA phages can also be assembled *in vitro*, but in most cases one must begin from "assembly-naive" components isolated from infected cells. This second approach, which in general requires mutations of the virus that block

Table 2 Structural classification of viruses

ssRNA					Typical Member
Icosahedral-naked					
	Single chromosome				
		$T = 1$	Picornaviruses		Poliovirus
			Caliciviruses		Vesicular exanthema
			Plant viruses		Satellite tobacco necrosis
		$T = 3$	Leviviruses		R17, MS2
			Plant viruses		
				Tombusviruses	Tomato bushy stunt
				Sobemoviruses	Southern bean mosaic
				Tymoviruses	Turnip yellow mosaic
				Luteoviruses(?)	Barley yellow dwarf
		$T = 4$	Insect viruses		Nudaurelia virus β
	Multiple chromosome				
		$T = 1$	Comoviruses		Cowpea mosaic
		$T = 1$ (elongated)			Alfalfa mosaic
		$T = 3$	Plant viruses		
				Bromoviruses	Brome mosaic
				Cucumoviruses	Cucumber mosaic
		$T = ?$*	Plant viruses		
				Ilarviruses	Tobacco streak
				Nepoviruses	Tobacco ringspot
			Nodaviruses		Black beetle virus
Icosahedral-enveloped					
	Single chromosome				
			Togaviruses		
				Alphaviruses	Sindbis
				Flaviviruses	St. Louis encephalitis
	Single chromosome-diploid				
			Retroviruses		Avian sarcoma
Helical-rigid-naked					
	Single chromosome				
			Tobamoviruses		Tobacco mosaic
	Multiple chromosomes				
		Plant viruses			
			Hordeiviruses		Barley stripe
			Tobraviruses		Tobacco rattle
Helical-flexuous-naked					
	Single linear chromosome				
		Plant viruses			
			Potexviruses		Potato X
			Potyviruses		Potato Y
			Closteroviruses		Beet yellows
			Carlaviruses		Carnation latent
Helical-flexuous-enveloped					
	Single linear chromosome				
		(+) strand in virion			
			Coronaviruses		Avian infectious bronchitis

(continued)

(Table 2, continued)

Classification	Example
Helical-flexuous-enveloped (cont.)	
(-) strand in virion	
Rhabdoviruses	Vesicular stomatitis
Paramyxoviruses	Sendai
Arenaviruses	Lymphocytic choriomeningitis
Multiple linear chromosomes	
(-) strand in virion	
Orthomyxoviruses	Influenza A
Bunyaviruses	La Crosse virus
Unknown structure	
Multiple chromosomes, enveloped	Tomato spotted wilt
dsRNA	
Icosahedral	
Naked	
Multiple linear chromosomes	
Reoviruses	Reovirus
Birnaviruses	Pancreatic necrosis
Fungal viruses	Pc virus
Lipid-containing (not acquired by budding)	
Multiple linear chromosomes	
Cystoviruses	Phage ϕ6
ssDNA	
Icosahedral-naked	
Isometric	
Single circular chromosome	
Microviruses	ϕX174
Single linear chromosome	
Parvoviruses	H-1
Elongated	
Multiple circular chromosomes	
Geminiviruses	Cassava latent
Helical-naked	
Single circular chromosome	
Flexuous	
Inoviruses	Filamentous phages
Rigid(?)	
Plectoviruses	Phage MV-L1
dsDNA	
Icosahedral	
Naked	
Single linear chromosome	
Myoviruses	Tailed phages, cyanophages
Adenoviruses	Adenovirus

(Table 2, continued)

Icosahedral (cont.)	
Single circular chromosome	
Papovaviruses	SV-40
Calimoviruses•	Cauliflower mosaic
Enveloped	
Single linear chromosome	
Herpesviruses	Herpes simplex-1
Hepadnaviruses•	Hepatitis B
Single circular chromosome	
Plasmaviruses▼	Phage MV-L2
Lipid-containing (not acquired by budding)	
Single linear chromosome	
Iridoviruses	Frog virus-3
Tectiviruses	Phage PRD1
Poxviruses▼	Vaccinia
Single circular chromosome	
Corticoviruses	Phage PM2
Baculoviruses▼	Nuclear polyhedrosis

Viruses are grouped according to symmetry of the virion, type of nucleic acid in the virion, the presence of an envelope, and other structural features (after Williams and Fisher, 1974; Casjens and King, 1975; Matthews, 1982). The International Committee on the Taxonomy of Viruses nomenclature has been followed (Matthews, 1982). The indicated characteristics are not meant to be definitive, but only denote salient structural differences between groups. The right column indicates a group member whose assembly and/or structure has been studied.

The term "naked" indicates that no lipid is present in the virion; "enveloped" indicates the presence of a lipid bilayer surrounding the particle, which is obtained by budding through a cellular membrane (see Chapter 6, this volume); (+) and (-) refer to the message-sense and anti-message strand, respectively.

*These viruses are not yet well understood; the ilarvirus particles may be variable in T number.

•The chromosomes of these two virus groups contain specific single-strand breaks or gaps.

▼These viruses, although generally "spherical" or "bacilliform" in shape, have structures that are not yet understood and may not contain icosahedral elements.

assembly and accumulate assembly precursors or intermediates, was pioneered most elegantly by workers studying phage T4 (Epstein et al., 1963; Edgar and Wood, 1966; King, 1968; Kikuchi and King, 1975; Plishker et al., 1983). A large body of work on the molecular biology of these phages and some other viruses has shown that assembly almost invariably follows very specific pathways, and mechanisms are present that control the mode of assembly (shape) as well as the order and rate of the steps in the assembly pathways (discussed in Chapters 4 and 5, this volume; Caspar, 1980; King, 1980; Wood, 1980). The assembly steps often consist of controlled, specific noncovalent binding of one protein to another. Such binding reactions usually occur spontaneously, but they are occasionally catalyzed by other proteins (King and Casjens, 1974; Wood and Conley, 1979). In addition covalent modification of proteins is also common in these pathways. Covalent modifications known to accompany virus assembly include phosphorylation (Erikson et al., 1977), glycosylation (Chapter 6, this volume), proteolytic cleavage (Hershko and Fry, 1975), nucleic acid cleavage (Chapter 3, this volume), and possibly covalent joining of proteins (Hendrix and Casjens, 1974; Hendrix, personal communication; Etchison and Walter, 1977). Although virus particles are usually built exclusively from virus-encoded proteins, proper assembly often requires the participation of host proteins that are not components of the completed virion (see Chapters 6 and 8, this volume). Much of the current work in this field is focused on obtaining detailed chemical and physical understanding of the nature and control of the individual steps in pathways for virus assembly. Subsequent chapters of this volume deal with particular structural and assembly problems in the most active areas in this field.

The ultimate goal of the study of virus assembly is to understand the way in which molecular structure and energetics determine the dynamics of the assembly process. Thus, both kinetic and equilibrium (thermodynamic) aspects are of interest, in addition to atomic-level structural detail of the components. Only a few virus assembly steps have been analyzed with respect to their detailed kinetic or thermodynamic aspects. The physical analysis of phage T4 head-tail joining and tail-fiber attachment have recently been reviewed by Bloomfield (1983). These studies have begun to examine the fraction of encounters between reaction components that give rise to productive assembly, the ways in which proteins catalyzing these reactions might function (in this case, T4 whiskers aiding tail-fiber attachment), and the activation energies of what are believed to be "relatively simple" assembly reactions.

The thermodynamics of TMV assembly is also under study. Early examination of TMV assembly was important in understanding the contribution of solvent-ordering changes during protein-association reactions (Lauffer and Stevens, 1968). The negative free energy change that drives the association of two protein molecules ($-\Delta G$) is composed of two parts, the enthalpy (ΔH) and entropy (ΔS) of association, since $\Delta G = \Delta H - T\Delta S$. The two components that drive the reaction forward, negative enthalpy change (heat of bond formation) and positive entropy change. The first is generally thought to be derived from protonation and increased hydrogen bond formation during association, and the second is considered to be a result of the release of ordered solvent molecules during the formation of hydrophobic and electrostatic interactions (Chothia and Janin, 1975; Osawa and Asakura, 1975; McCarthy and Allewelle, 1983). Lauffer and coworkers have found that the TMV coat-protein polymerization reaction is strongly endothermic (has a positive ΔH) and has a large positive entropy change (Lauffer and Stevens, 1968; Sturtevant et al., 1981). The magnitude of ΔS for TMV-protein assembly is among the highest for any protein-associating system studied, reflecting its tight self-association. Thus, the TMV assembly reaction is apparently driven not by chemical bond formation between subunits, but by solvent disordering resulting from hydrophobic and electrostatic intersubunit contacts. It is not yet known if most or all virus-assembly reactions are entropically driven. The analysis of other nonviral protein-associating systems has shown that some are entropically driven and some are enthalpically driven (Ross and Subramanian, 1980; McCarthy and Allewelle, 1983). It is not yet clear how applicable these systems are to virus assembly, but they do point out that it is not necessary for such reactions to be entropically driven. Analysis of the intersubunit contacts in viruses whose x-ray diffraction-derived structures are known, generally shows that there are hydrophobic contacts as well as a considerable number of hydrogen bonds and electrostatic interactions between subunits (Bloomer et al., 1978; Harrison, 1980; Rossmann et al., 1983; Hopper et al., 1984; Jones and Liljas, 1984; Chapter 2 this volume). The way in which virus-capsid subunits are currently envisioned to bind one another is by hydrophobic bonding and electrostatic interactions, possibly with protonation and increased hydrogen bonding playing a role as well (with the relative contributions of these components being unclear). Three-dimensional fit, resulting in van der Waals interactions, correct hydrogen-bonding, and electrostatic interaction, is a clear requirement for successful hydrophobic bonding as well (see Chothia and Janin, 1975). Thus, although it may not be the driving force for association, fit determines whether or not a successful interaction can occur (i.e., the specificity of interaction). The ultimate goal of understanding the detailed steps of virus assembly, and in particular the control of these steps will certainly be challenging, and the continued pursuit of this goal should continue to yield interesting scientific findings.

VIRUS ASSEMBLY AND MACROMOLECULAR SYNTHESIS

The time of assembly of virus particles after infection of susceptible cells is regulated at the level of expression of the viral structural-protein genes, and the assembly process itself can in turn signal changes in macromolecular metabolism. Study of these aspects of virus assembly are currently in their infancy, so few general rules for virus biochemistry have yet emerged; however, a number of interesting systems are under study. In most of the more complex bacterial and animal viruses, expression of the virion structural-protein genes is regulated in such a manner that these proteins are synthesized only late in infection, at a time after replication of the nucleic acid has begun. This control is apparently exerted by a variety of mechanisms (too numerous to review here) in the different virus families. However, in all cases studied, blocks of structural genes are coordinately regulated. The consequence of this arrangement is that the order of the virus-assembly steps are controlled by the properties of the proteins involved rather than by the temporal synthesis of the assembly components, (see Chapter 4, this volume). In many cases the relative rates of synthesis of the virion structural proteins reflect the relative numbers of the different protein molecules required to build the virion. This is particularly clear in the dsDNA bacteriophages, where in some cases alteration of the relative rates of synthesis causes failure in the proper assembly process (Floor, 1970; Showe and Onorato, 1978). In other cases such differential rates of synthesis may simply reflect evolutionary pressure to economize overall viral protein synthesis.

Currently one well-documented case (phage P22) is known in which the status of the assembly process exerts control on the expression of one of the genes involved in the assembly. With this phage, expression of the scaffolding protein gene responds to one aspect of the assembly process, namely, the relative rates of coat-protein synthesis and of DNA packaging. The scaffolding protein is the major internal component of virion precursor particles. It leaves these particles when DNA is packaged, and it is reused in the assembly of additional precursor structures with newly-made coat-protein molecules. Thus, it acts catalytically in the assembly of phage particles. The scaffolding protein fills the reservoir of available binding sites on the major coat protein during the assembly of the precursor particles, and any excess unassembled scaffolding protein specifically depresses the rate of synthesis of additional scaffolding protein (at a posttranscriptional level). In this way, under a variety of assembly conditions the amount of scaffolding protein synthesized is just sufficient to associate with the available coat protein (King et al., 1978; Casjens and Adams, 1985).

There are several examples of control of nucleic acid synthesis by some feature of the assembly process. The best-understood of these is the translational repression of the bacteriophage R17 RNA replicase gene by the coat protein (Steitz, 1979; Carey et al., 1983). Late in infection, as R17 coat protein accumulates (and particle assembly begins), expression of the replicase gene is depressed. This decrease in gene expression may help to avoid attempts at packaging chromosomes actively engaged in macromolecular synthesis or to help attain the proper ratio of nucleic acid to coat protein required for assembly. Less well-understood are the observations that DNA replication with microviruses (e.g., bacteriophage øX-174), parvoviruses, and adenoviruses appears to be coupled to successful virus assembly (see Chapter 3, this volume), and two major virion proteins of an enveloped virus have been implicated in the control of transcription and replication (Blumberg et al., 1981; Pinney and Emerson, 1982; Perrault et al., 1983; Patton et al., 1984). These types of findings point out that virus structural proteins may be multifunctional and have roles beyond participation in the actual assembly of virions and delivery of virion nucleic acid into cells.

OTHER USES OF THE STUDY OF VIRUS STRUCTURE AND ASSEMBLY

Viral assembly systems, because of their apparent simplicity, have often been used as model systems for studying other biological phenomena, such as macromolecular superstructure and assembly, and numerous aspects of the biochemistry and biophysics of macromolecules.

A particularly productive approach has been to make use of the detailed genetic analyses of phages lambda and P22 that have been carried out over a period of many years. The assembly systems of these phages have been used to test the validity of the analysis of "second-site revertants" (mutations in one protein that correct a genetic defect in a second protein) for determining where protein-protein contacts occur (Georgopoulos et al., 1973; Jarvik and Botstein, 1975; Sternberg, 1976). The reasoning employed in such analyses is that if a mutation in one gene is compensated for by a mutation in a second gene, these gene products probably interact as subunits of a multisubunit protein with the mutations in the binding regions of each polypeptide chain. Indeed studies of these model systems have demonstrated that second-site revertants are usually located in genes encoding proteins that interact physically with the protein harboring the initial mutation. In another type of genetic experiment Jarvik and Botstein (1973) successfully used the P22 system to test a method utilizing temperature-sensitive and cold-sensitive mutations to determine the temporal sequence of action of two gene products that act in the same pathway. Casjens and Hendrix (1974) and Katsura (1981) used the lambda assembly system to examine genetic recombination and complementation between analogous genes in closely related species. Genetic relations between the structural genes of lambda have also been instrumental in understanding the evolution of viruses in terms of the exchange of "modules" of genes (Echols and Murialdo, 1978; Campbell and Botstein, 1983).

In structural studies Katsura (1980) and King and co-workers (Yu and King, 1984) have used the major coat protein of lambda and the tail protein of P22 as model structural proteins for studying various ways by which a missense mutation can affect protein function. These systems have also been used to examine the different routes by which a polypeptide chain can fold to form a finished protein. As was alluded to above, proteins involved in virus assembly have been extremely useful in studying the biochemistry of protein-protein interactions. For example, particular virus proteins are unique in their ability to catalyze other protein-assembly reactions (Wood and Conley, 1979) and perhaps to catalyze the proper folding of another protein (Wood, 1979). Some enveloped animal virus systems have been important in understanding the anchoring of proteins to membranes and in elucidating the mechanism by which carbohydrates are added to proteins (Wickner, 1979; Garoff et al., 1982; Chapters 6 and 7, this volume). In addition, virus particles and related structures have been very useful in developing and testing methods for the determination of macromolecular structure, for example, in x-ray diffraction analysis (Harrison, 1969) and in obtaining high resolution and three-dimensional information from electron micrographs (reviewed by Steven, 1981).

Recently, virus assembly has taken on new importance as viruses have found increased use as vectors for the cloning, isolation, and characterization of nucleic acid sequences and the delivery of genetically engineered nucleic acid sequences into various cells (see discussion, Chapter 3, this volume). Viruses have also achieved new medical relevance as designed particles carrying a variety of antigens have been used in the development of multipurpose and easily produced vaccines. Viruses will undoubtedly continue to be useful in the analysis of many complex biological phenomena, in addition to being studied for their intrinsic interest.

ACKNOWLEDGEMENTS

I thank Victor Bloomfield, Roger Hendrix, and Michael Rossmann for their critical and thoughtful comments on this chapter.

REFERENCES

Aebi, U., R. Bijlenga, J. van der Broek, R. van der Broek, F. Eiserling, C. Kellenberger, E. Kellenberger, V. Mesyanzhinov, L. Muller, M. Showe, R. Smith, and A. Steven. 1974. *J. Supramol. Struct.*, 2, 253.

Bernal, J. and I. Fankuchen. 1941. *J. Gen. Physiol.*, 25, 111.

Bloomer, A., J. Champness, G. Bricogne, R. Staden, and A. Klug. 1978. *Nature*, 276, 362.

Bloomfield, V. 1983. In *Bacteriophage T4*. C. Mathews, E. Kutter, G. Mosig, and P. Berget, eds. p. 270. Amer. Society for Microbiology.

Blumberg, B. M. Leppert, and D. Kolakofsky. 1981. *Cell*, 23, 837.

Campbell, A. and D. Botstein. 1983. In *Lambda II*. R. Hendrix, J. Roberts, F. Stahl, and R. Weisberg, eds. Cold Spring Harbor. p. 365.

Carey, J., T. Lowery, and O. Uhlenbeck. 1983. *Biochemistry*, 22, 4723.

Carrascosa, J., J. Carazo, A. Carrascosa, N. Garcia, A. Santisteban, and E. Vinuela. 1984. *Virology*, 132, 160.

Casjens, S. 1979. *J. Mol. Biol.*, 131, 1.

Casjens, S. and M. Adams. 1985. *J. Virol.*, 53, 185.

Casjens, S. and R. Hendrix. 1974. *J. Mol. Biol.*, 90, 20.

Casjens, S., and J. King. 1975. *Ann. Rev. Biochem.*, 44, 555.

Caspar, D. 1980. *Biophys. J.*, 32, 103.

Caspar, D. 1965. In *Viral and Rickettsial Infections of Man,* 4th Edition. P. Horsfall and I. Tamm, eds. p. 51. Lippincott.

Caspar, D. and A. Klug. 1962. *Cold Spring Harb. Symp. Quant. Biol.,* 27, 1.

Chothia, C., and J. Janin. 1975. *Nature,* 256, 705.

Crane, H. 1950. *The Scientific Man.,* 70, 376.

Crick, F.H.C. and J.D. Watson. 1956. *Nature,* 177, 473.

Crowther, A., and R. Franklin. 1972. *J. Mol. Biol.,* 68, 181.

Driedonks, R., P. Krigsman, and J. Mellema. 1976. *Phil. Trans. Soc. Lond. B.,* 276, 131.

Dunker, A. 1974. *J. Virol.,* 14, 878.

Echols, H. and H. Murialdo. 1978. *Microbiol. Rev.,* 42, 577.

Edgar, R., and W. Wood. 1966. *Proc. Nat. Acad. Sci. USA,* 55, 498.

Eiserling, F. 1979. In *Comprehensive Virology.* Vol. 13. H. Fraenkel-Conrat and R. Wagner, eds. p.543. Plenum Press.

Elford, W., and C. Andrewes. 1932. *Brit. J. Exp. Pathol.,* 13, 446.

Epstein, R., A. Bolle, C. Steinberg, G. Kellenberger, E. Boy de la Tour, R. Chevalley, R. Edgar, M. Susman, G. Denhardt, and I. Lielausis. 1963. *Cold Spring Harbor Symp. Quant. Biol.,* 28, 375.

Erikson, E., J. Brugge, and R. Erikson. 1977. *Virology,* 80, 177.

Etchison, D., and G. Walter. 1977. *Virology,* 77, 783.

Finch, J., and A. Klug. 1959. *Nature,* 183, 1709.

Floor, E. 1978. *J. Mol. Bio.,* 47, 293.

Fraenkel-Conrat, H., and R. Wagner. *Comprehensive Virology.* Vol 1. Plenum Press.

Fraenkel-Conrat, H. and R.C. Williams. 1955. *Proc. Nat. Acad. Sci. USA,* 41, 690.

Franklin, R., A. Klug, and K. Holmes. 1957. *Ciba Foundation Symp. on the Nature of Viruses,* p.39.

Fukuyama, K., S. Abdel-Meguid, J. Johnson, and M. Rossmann. 1983. *J. Mol. Biol.,* 167, 873.

Garoff, H., C. Kondor-Koch, and H. Riedel. 1982. 1982. *Curr. Topics.*

Microbiol. Immunol., 99, 1.

Georgopoulos, C., R. Hendrix, S. Casjens, and A.D. Kaiser. 1973. *J. Mol. Biol.*, 76, 45.

Harris, J. and C.A. Knight. 1955. *J. Biol. Chem.*, 214, 215.

Harrison, S. 1969. *J. Mol. Biol.*, 42, 457.

Harrison, S. 1980. *Biophys. J.*, 32, 139.

Harrison. S. 1983. *Adv. Virus Res.*, 28, 175.

Hendrix, R., and S. Casjens. 1974. *Proc. Nat. Acad. Sci. USA*, 71, 1451.

Hershko, A., and M. Fry. 1975. *Ann. Rev. Biochem.*, 44, 775.

Holden, A. 1971. *Shapes, Space, and Symmetry*. Columbia University Press.

Hopper, P., S. Harrison, and R. Sauer. 1984. *J. Mol. Biol.*, 177, 701.

Horne, R., G. Russell, and A. Trim. 1959. *J. Mol. Biol.*, 1, 234.

Horne, R., and P. Wildy. 1961. *Virology*, 15, 348.

Jarvick, J. and D. Bottstein. 1973. *Proc. Nat. Acad. Sci. USA*, 70, 2046.

Jarvick, J. and D. Bottstein. 1975. *Proc. Nat. Acad. Sci. USA*, 72, 2738.

Joklik, W. 1980. *Principles of Animal Virology*. Appleton-Century-Crofts.

Jones, T. and L. Liljas. 1984. *J. Mol. Biol.*, 177, 735.

Katsura, I. 1980. *J. Mol. Biol.*, 142, 387.

Katsura, I. 1981. *J. Theor. Biol.*, 88, 503.

Kausche, V., E. Pfankuch, and H. Ruska. 1939. *Naturwiss.*, 27, 292.

Kikuchi, Y., and J. King. 1975. *J. Mol. Biol.*, 99, 695.

King, J., 1968. *J. Mol. Biol.*, 32, 231.

King, J., 1980. In *Biological Regulation and Development*, Vol. 3. R. Goldberger, ed. p. 101. Plenum Press.

King, J. and S. Casjens. 1974. *Nature*, 251, 112.

King, J., C. Hall, and S. Casjens. 1978. *Cell*, 15, 551.

Klug, A., F. Crick, and H. Wyckoff. 1958. *Acta Cryst.*, 11, 199.

Klug, A., and J. Finch. 1960. *J. Mol. Biol.*, 2, 201.

Klug, A., and J. Finch. 1965. *J. Mol. Biol.*, 11, 403.

Lauffer, M., and C. Stevens. 1968. *Adv. Virus Res.*, 13, 1.

Liljas, L., T. Unge, T. Jones, K. Fridborg, S. Lövgren, U. Soglund, and B. Strandberg. 1982. *J. Mol. Biol.*, 159, 93.

Luria, S., J. Darnell, D. Baltimore, and A. Campbell. 1978. *General Virology*, 3rd Ed., Wiley.

Matthews, R. 1981. *Plant Virology,et al., 2nd Ed., Academic Press.*

Matthews, R. 1982. Intervirology, 17, 1.

McCarthy, M., and N. Allewelle. 1983. *Proc. Nat. Acad. Sci. USA*, 80, 6824.

Murialdo, H. and M. Becker. 1978. *Microbiol.*, 43, 529.

Niu, C., and H. Fraenkel-Conrat. 1955. *Biochim. Biophys. Acta.*, 16, 597.

Olson, A., G. Bricogne, and S. Harrison. 1983. *J. Mol. Biol.*, 171, 61.

Osawa, F., and S. Asakura. 1975. *Thermodynamics of the Polymerization of Protein.* Academic Press.

Patton, J., N. Davis, and G. Wertz. 1984. *J. Virol.*, 49, 303.

Pereira, H., and N. Wrigly. 1974. *J. Mol. Biol.*, 85, 617.

Perrault, J., G. Clinton, and M. McClure. 1983. *Cell*, 35, 175.

Pinney, D. and S. Emerson. 1982. *J. Virol.*, 42, 897.

Plishker, M., M. Chidambaram, and P. Berget. 1983. *J. Mol. Biol.*, 170, 119.

Rayment, I., T. Baker, D. Caspar, and W. Murakami. 1982. *Nature*, 295, 110.

Ross, P., and S. Subramanian. 1980. *Biophys. J.*, 32, 79.

Rossmann, M., C. Abad-Zapatero, M. Hermodson, and J. Erickson. 1983. *J. Mol. Biol.*, 166, 37.

Schlesinger, M. 1934. *Biochem. Zeit.*, 264, 306.

Schramm, G. , G. Braunizer, and J. Schneider. 1955. *Nature*, 176, 456.

Schmidt, P., P. Kaesberg, and W. Beeman. 1954. *Biochim. Biophys. Acta*, 14, 1.

Showe, M. and L. Onorato. 1978. *Proc. Nat. Acad. Sci. USA*, 75, 4165.

Silverstein, S., J. Christman, and G. Acs. 1976. *Ann. Rev. Biochem.*, 45, 375.

Stanley, W. 1935. *Science*, 81, 644.

Steitz, J. 1979. In *Biological Regulation and Development*. R. Goldberger, ed. Vol. 1, p. 349. Plenum.

Sternberg, N. 1976. *Virology*, 71, 568.

Steven, A. 1981. *Meth. in Cell Biol.*, 22, 297.

Steven, A., P. Serwer, M. Bisher, and B. Trus. 1983. *Virology*, 124, 109.

Stubbs, G., S. Warren, and K. Holmes. 1977. *Nature*, 267, 216.

Sturtevant, J., G. Velicelebi, R. Jaenicke, and M. Lauffer. 1981. *Biochemistry*, 20, 3792.

Weyl, H. 1952. *Symmetry*. Princeton University Press.

Wickner, W. 1979. *Ann. Rev. Biochem.*, 48, 23.

Williams, R., and H. Fisher. 1974. *An Electron Micrographic Atlas of Viruses*. Charles C. Thomas.

Williams, R., and K. Smith. 1958. *Biochim. Biophys. Acta*, 28, 464.

Winkler, F., C. Schutt, and S. Harrison. 1978. *Nature*, 265, 509.

Wood, W. 1979. *Harvey Lect.*, 73, 203.

Wood, W. 1980. *Quart. Rev. Biol.*, 55, 353.

Wood, W. and M. Conley. 1979. *J. Mol. Biol.*, 127, 15.

Wrigley, N. 1970. *J. Gen. Virol.*, 6, 179.

Yu, M. and J. King. 1984. *Proc. Nat. Acad. Sci. USA*, 81, 6584.

2
Structure and Assembly of Icosahedral Shells

Michael G. Rossmann and John W. Erickson
Purdue University

INTRODUCTION

Virus assembly is a problem of quaternary structure. Somewhere a molecular "blueprint" must reside for specifying the various unique interactions required for correct assembly. Some coat proteins have the capacity for regulating their own assembly, as demonstrated by the formation of empty capsids from purified coat protein in the absence of nucleic acid (Bancroft, 1970). Other viruses have the information for assembly dispersed over several accessory proteins, which serve as a temporary scaffold to direct the formation of new virions, sometimes in concert with the viral nucleic acid, (e.g., øX174; Hayashi, 1978). In other cases, for example, RNA bacteriophages, interactions with minor capsid proteins participate in a maturation event (Hung, 1976). This chapter will deal primarily with those icosahedral viruses possessing only a single type of subunit, which includes most of the simple RNA plant and bacterial viruses. Features of the multisubunit small isometric phages and RNA animal viruses that are relevant to assembly will also be discussed.

GENERAL PRINCIPLES

Polyhedral Viruses

Crick and Watson (1956) proposed that the protein-coat structures of most simple spherical viruses, whether animal, plant, or bacterial, were based on the design of regular polyhedra. They recognized (Crick and Watson, 1957) that the information in the viral nucleic acid would be sufficient to code for a protein whose molecular weight was only a fraction of that of the intact virus capsid. Thus, they inferred that the nucleic acid is encapsulated by multiple copies of identical protein subunits, necessitating identical environments and hence limiting the possible assembly to the Platonic solids.

It soon became apparent that the simple viruses invariably had an icosahedral structure. For instance, by using X-ray diffraction techniques Caspar (1956) demonstrated that tomato bushy stunt virus (TBSV) was icosahedral; later, Huxley and Zubay (1960) and Nixon and Gibbs (1960) observed that turnip yellow mosaic virus (TYMV) was also icosahedral. However, the principles laid down by Crick and Watson required modification, because the number of subunits per virus was usually larger than the 60 permitted icosahedral symmetry. This observation led Caspar and Klug (1962) to develop the concepts of quasi-symmetry that have been the foundation for the structural classification of viruses for more than two decades. High resolution structures have now become available and show that quasi-equivalence of subunit contacts is frequently violated. Hence, the very foundations of which Crick, Watson, Caspar, and Klug built the geometrical perception of self-assembly have been undermined. Nevertheless, the overall structural organization of protein subunits in viral coats follows the anticipated organizational patterns predicted by Caspar and Klug (see Chapter 1). Thus, the paradoxical situation exists that essentially correct surface lattices had been predicted from a basic hypothesis that have been applied too rigorously. An understanding of quasi-symmetry is still essential to the analysis of virus structure and assembly, if only to recognize where there are significant departures from the theory. Indeed, the concept of the triangulation number, *T*

Table 1 The symmetry of oligomeric proteins

Point group*	Number of subunits	Examples	References to examples
2	2	Malate dehydrogenase	10
		Liver alcohol dehydrogenase	4
		Hemoglobin	7
3	3	Bacterial aldolase	13
		Bacterial chlorophyll protein	12
		Hemagglutinin spike of influenza virus	17
222	4	Lactate dehydrogenase	1
		Yeast alcohol dehydrogenase	8
		Glyceraldehyde-3-phosphate dehydrogenase	5
4	4	Neuraminidase spike of influenza virus	16
32	6	Insulin	11
		Aspartate transcarbamylase	14
		Hemocyanin	15
622	12	Glutamine synthetase	9
17	17	TMV disk protein	3
23?	24	Ferritin	2
432	24	Dihydrolipoyl transsuccinylase	6
532	60	Spherical viral coats of SBMV, TBSV, and STNV	18

*For a definition and explanation of point groups and the Hermann-Mauguin notation used here, see Cotton (1971).

References: (1) Adams et al., 1970; (2) Banyard et al., 1978; (3) Bloomer et al., 1978; (4) Brändén et al., 1975; (5) Buehner et al., 1974; (6) DeRosier et al., 1971; (7) Dickerson and Geis, 1983; (8) Eklund et al., 1976; (9) Heidner et al., 1978; (10) Hill et al., 1972; (11) Hodgkin, 1974; (12) Matthews et al., 1979; (13) Mavridis et al., 1982; (14) Monaco et al., 1978; (15) van Schaick et al.,1982; (16) Varghese et al., 1983; (17) Wilson et al., 1981; (18) this chapter.

is generally applicable to all viruses with closed protein shells even when these shells might themselves be covered by a lipid coat or by another protein shell.

Symmetry of the Coat Protein

Why do all known simple virus coats exhibit icosahedral symmetry, while other closed point groups occur frequently in many biological assemblies (Table 1)? Caspar and Klug (1962) pointed out that icosahedral symmetry would permit the RNA to be covered by the largest number of subunits, that is, by subunits with the smallest size. However, some constraints may also be imposed by the requirements of quasi-symmetry. For instance, an octahedron demands the quasi-equivalence of a tetramer with fourfold symmetry and a hexamer with sixfold symmetry, whereas an icosahedron requires only the quasi-equivalence of a tetramer with fourfold symmetry to a hexamer with sixfold symmetry, whereas an icosahedron requires only the quasi-equivalence of a fivefold pentamer to a sixfold hexamer. In other words, the aggregation of more than 60 subunits into quasi-equivalent positions imposes least strain for an icosahedron, though the absence of quasi-symmetrical subunit contacts in those viruses examined at higher resolution does raise problems with this line of reasoning.

Slight corruption of symmetry, as in hexokinase (Steitz et al., 1977), or alteration of the symmetrical contacts, as in hemoglobin (Baldwin and Chothia, 1979), have functional benefits. The tendency of slight departures from symmetry for particular functional advantages is also mimicked in viral assemblies.

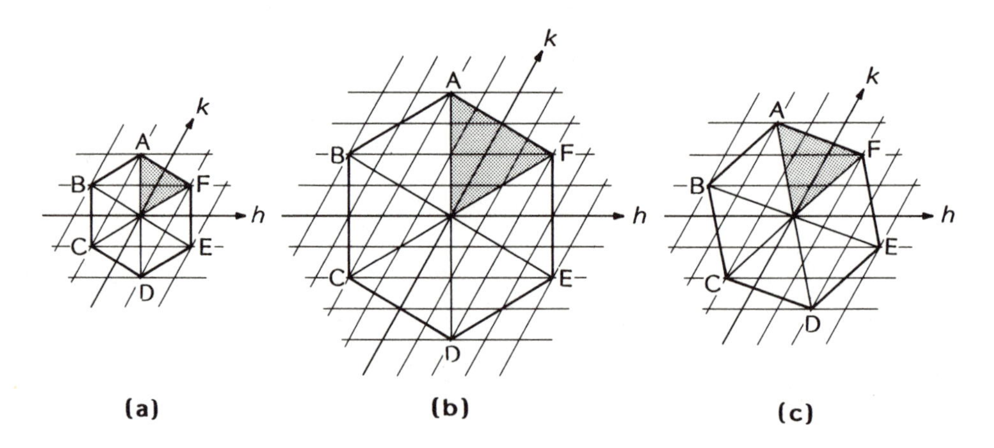

FIGURE 1 Symmetrical positions of fivefold vertices A, B, C, D, E surrounding a fivefold vertex at the origin on a hexagonal grid with axes h,k at (fh,fk). The position of the fivefold vertices satisfies the condition that T is a constant where $T= f^2P$ and $P = h^2 + hk + k^2$. The shaded area is the part that must be excised, so that F is made to coincide with A. (a) $T = 3$ with $P = 3$, $f = 1$. (b) $T = 4$ with $P = 1$, $f = 2$. (c) $T = 7$ with $P = 7$, $f = 1$.

Quasi-Symmetry

Caspar and Klug (1962) considered the folding of two-dimensional hexagonal nets to simulate quasi-equivalence. Structures differed only by the selection of positions of the icosahedral fivefold axes. Thus, the major breakdown of equivalence between subunits is in the difference between pentamers and hexamers. The only exact symmetry is that due to the icosahedron, while all other symmetries are local.

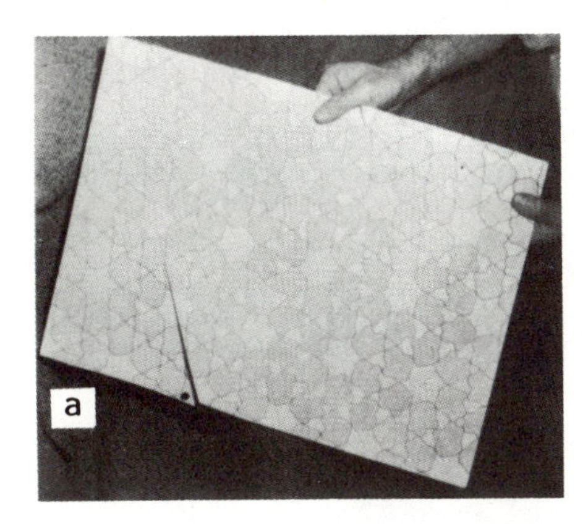

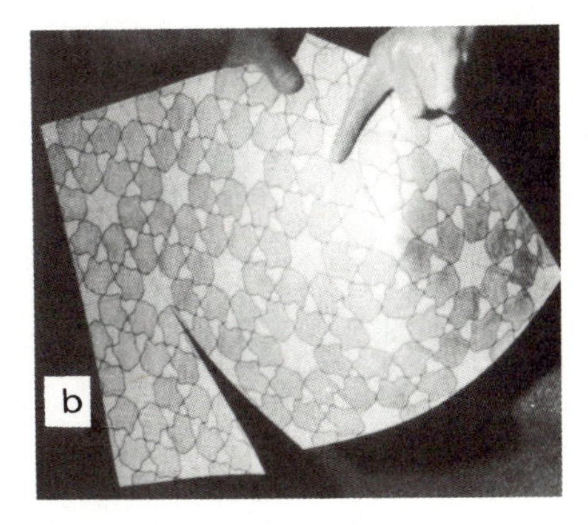

FIGURE 2 The folding of the hexagonal net into a closed surface. (a) Slitting along a line joining two lattice points. (b) Forming a cone by transforming a 6-vertex into a 5-vertex. Note that the bonding pattern of the net is preserved locally. (c) Forming another 5-vertex has produced part of a closed surface. In this example, the disposition of 5-vertices has been chosen in such a way that a complete surface with icosahedral symmetry would have the triangulation number T = 4. (From Caspar and Klug, 1962.)

The position of fivefold vertices surrounding any selected fivefold vertex within the hexagonal net is given by (fh, fk), in which f, h, and K are integers (Figure 1). The values of h and k must be confined to a given integer constant P such that $P = h^2 + hk + k^2$. Caspar and Klug demonstrated the construction of hollow figures of selected quasi-symmetries by making scissors cuts in a hexagonal net and folding one-sixth of the paper behind its neighbor at each pentameric vertex (Figure 2).

The triangulation number, T, is defined as the number of quasi-equivalent units on a surface and is given by $T = Pf^2$. The number of subunits in the complete icosahedral surface is $60T$. The simplest assembly is $T = 1$ with all subunits having identical environments. $T = 2$ cannot be formed from any combination of the integers h, k, and f. A $T = 3$ icosahedron requires 180 subunits in sets of three (referred to as A, B, and C) (Figure 3). These three covalently identical subunits would require three quasi-equivalent conformations and environments. For those triangulation numbers based on a P value requiring $h \neq k$ ("skew classes"), there will be two possible surface lattices related by a mirror plane. Hence, it is necessary to designate such lattices as dextro or laevo, as is done with the $T = 7$ (dextro)

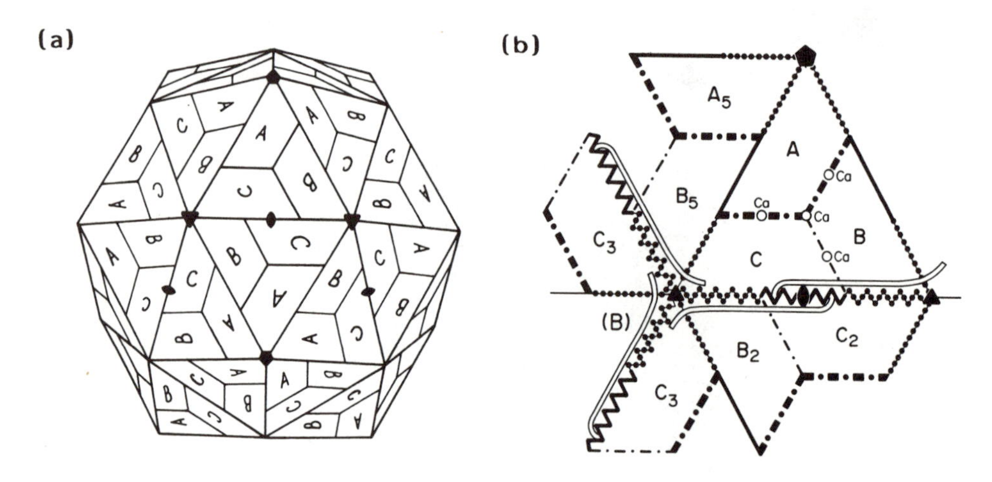

FIGURE 3 (a) Diagrammatic representation of the spatial distribution of the three quasi-equivalent subunits A, B, and C in the $T = 3$ surface lattice of TBSV and SBMV. (b) The position of the Ca^{2+} sites and the additional ordered arms and β annuli of the C subunits are indicated in greater detail than in (a). Notice that all A subunits are identical, as are B and C subunits. The subscripts denote the type of symmetry operation (rotation) that relates a given subunit to a standard asymmetric unit (subunits A, B, and C). The different types of subunit contacts (indicated by the form of the subunit borders) are also shown in (b). For instance, the diagram indicates that the AA_5 and CB_5 interactions are identical. However, the CB subunit contacts are only quasi-equivalent to the BA and AC contacts. Note that in a $T = 3$ structure the quasi-sixfold axis is coincident with the icosahedral threefold axis shown as ▲.

viruses (h = 1, k = 2, f = 1), such as the papilloma group (Klug and Finch, 1965).

A table of some spherical viruses, each categorized by its triangulation number, is given in Table 2, which shows that as the triangulation number increases, the viral diameter d also increases. The consequence of putting an ever-increasing number of roughly equally sized subunits into the surface of the virus suggests that $dT^{-1/2}$ is approximately constant (Table 2). Viruses of greater complexity may have more than one type of protein in the coat, often as a result of post-translational cleavage. Their icosahedral asymmetric unit is the sum of the structures of all such proteins. However, it would not be surprising if some aspects of quasi-symmetry were

Table 2 Correlation of viral size with triangulation number

Virus	Diameter, Å	T	Mol wt. of subunit	Diameter/ $\sqrt{T}$	Reference
STNV	190	1	21,500	190	11
AMV protein aggregate	195	1	24,250	195	6
SBMV protein aggregate	175	1	28,500	175	5
SBMV	300	3	28,500	173	1
CCMV	266	3	19,600	154	13
R17	250	3	14,000	144	2
TRV	280	4?	13,000	140	9
P4	400	4	36,500	200	7
Sindbis core	390	4	30,000	195	10
Polyoma	500	7d	42,000	189	14
P22	610	7	50,000	231	3
Lambda	640	7 l	37,500	241	17
T7	600*	7 l	38,000	227	16
P2	580	9	36,500	193	7
Reovirus	830	13 l	3 proteins	230	12
T4 (short)	850	13 l	46,000	249	4
SP01	870*	16	45,000	217	4
Herpes simplex-1	900*	16	?	225	15
Adenovirus	750*	25	105,000	150	8

*Determined by electron microscopy.

References: (1) Abad-Zapatero et al., 1980; (2) Boedtker and Gesteland, 1975; (3) Casjens, 1979; (4) Eiserling, 1979; (5) Erickson and Rossmann, 1982; (6) Fukuyama et al., 1983; (7) Geisselsoder et al., 1982; (8) Ginsberg, 1979; (9) Heuss et al., 1981; (10) Kääriäinen and Söderlund, 1978; (11) Liljas et al., 1982; (12) Metcalf, 1982; (13) Rayment et al., 1977; (14) Rayment et al., 1982; (15) Roizman and Furlong, 1974; (16) Steven et al., 1983; (17) Williams and Richards, 1974.

retained among such protein aggregates. For instance, the $T = 1$ asymmetric unit of cowpea mosaic virus (CpMV) might have three similar domains, two in the large protein and one in the small protein, which pack as in a $T = 3$ virus (Johnson and Hollingshead, 1981).

Rigidity of Protein Structure

Caspar and Klug (1962) assumed that proteins could be *elastically* distorted, at least to a limited extent. They stated (although never proved) that the local distortions necessary to accommodate quasi-symmetry would be less than 0.5 A in juxtaposed atoms and 5° in bond direction. Furthermore, these considerations were dependent on two-dimensional surfaces, whereas simple viruses usually have a protein shell of about 40 A thickness. If, then, the real thickness of the coat protein is taken into account, the *absolute* change in *overall* subunit dimension for, say, southern bean mosaic virus (SBMV) would have to be 15 A in some places in order to maintain the same subunit-subunit contact areas in a pentamer as opposed to a hexamer. Nevertheless, the local distortions would still be within the limits set by Caspar and Klug.

A large number of protein structures have been solved to atomic resolution since 1962. Many of these have similar polypeptide backbone folds. Whenever domain structures have similar topology, then these are exceedingly rigid and suffer no distortion, even when amino acid sequences differ quite radically. A typical example derives directly from the observation of virus structure. The three different, quasi-equivalent subunits of the $T = 3$ virus, SBMV, can be superimposed, atom by atom, with a r.m.s. deviation of less than 1.4 A or, taking only the C_α atoms, a r.m.s. deviation of 1.2 A (Rossmann et al., 1983). Similar figures for the superposition of the SBMV subunit and the shell domain of the related virus, TBSV, show a r.m.s. deviation of 2.2 A for equivalenced C_α atoms (Rossmann et al., 1983b). Indeed, there are numerous examples of the rigidity of protein structures (cf. Rossmann et al., 1974; Matthews and Rossmann, 1983). The origin of this rigidity resides in the intercalation of residues into gaps between residues on opposing β sheets, as in a β barrel or between a β sheet and α helices (cf. Lesk and Chothia, 1982; Chothia and Janin, 1981, 1982; Cohen et al., 1980). In other words, the overall rigidity of a fold is maintained not by the local environment of each atom or atomic bond, but by the interlocking of secondary structural features in widely separated portions of the polypeptide. Thus, observation of protein folding shows that structural domains (see Rossmann and Argos, 1981, for definition of domain) do not permit distortion to nearly the extent required by the quasi-equivalent concepts of Caspar and Klug.

Another way of assessing the exceptional rigidity of protein domains is by observing the constancy of the number of subunits in, for instance, tobacco mosaic virus (TMV) disk protein. There are always 17 subunits in one ring of the disk (Bloomer et al., 1978), although there are 16-2/3 subunits per turn in the intact virus. However, it should be far easier to form a ring of 16 subunits instead of 17 than to form a pentamer cluster instead of a hexamer cluster.

Table 3 Examples of protein flexibility

Type	Example	Reference
A1. Order-disorder	a. βA arm in SBMV and TBSV	1,7
	b. Activation domain of trypsinogen	5
	c. RNA binding loop of TMV	3,14
A2. Hinge movement	a. S-P domain rotation	7,11
	b. Hexokinase	2
	c. Phosphoglycerate kinase	2
	d. Immunoglobulins	8
A3. Conformational changes of loops and β bends	a. Lactate dehydrogenase	12
	b. Triose phosphate isomerase	10
	c. Phosphorylase	9
B. Alteration of binding surface between one subunit and another	a. Subunit associations in SBMV and TBSV	13
	b. Hemoglobin	6
	c. TMV, disk and virus subunit	4

References: (1) Abad-Zapatero et al., 1980; (2) Anderson et al., 1979; (3) Bloomer et al., 1978; (4) Champness et al., 1976; (5) Fehlhammer et al., 1977; (6) Fermi and Perutz, 1981; (7) Harrison et al., 1978; (8) Huber et al., 1976; (9) Kasvinsky et al., 1978; (10) Phillips et al., 1977; (11) Robinson and Harrison, 1982; (12) Rossmann et al., 1972; (13) Rossmann et al., 1983a; (14) Stubbs et al., 1977.

The extreme persistence for maintaining the particular oligomeric aggregates clearly demonstrates the rigid properties of most protein subunits.

Flexibility of Protein Structure

Considerable flexibility has been observed in protein structures (Huber, 1979; Huber and Bennett, 1983), but not as a consequence of elastic deformation of domains. Protein flexibility can be attained by tertiary and quaternary structural changes (Table 3), in the following ways:

1. Order-disorder. A typical example is the ordering of the "βA arm" in the C subunit of SBMV or TBSV, as shown in Figure 3 (Abad-Zapatero et al., 1978). In A- or B-type quasi-equivalent subunits, these 20-or-so residues are disordered. The βA arm is ordered only in the C subunit, where it is inserted into a hydrophobic cleft

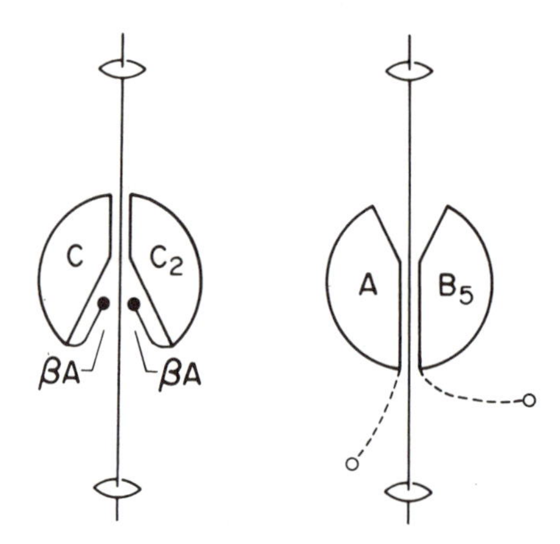

FIGURE 4 Diagrammatic representation of the altered binding mode between two monomers related by a twofold axis. The subunits contact one another on different surfaces in the two types of dimers. The CC_2 dimer straddles the icosahedral twofold axis in SBMV, whereas the AB_5 dimer straddles a quasi-twofold axis. See Figure 3 for subunit nomenclature and the location of these dimers in the capsid. The inside of the capsid is below the subunits as they are drawn. The disordered N-terminal βA arm is shown as dashed lines in the AB_5 dimer. (From Rossmann, 1983.)

formed between adjacent B and C subunits (Figures 4 and 5).

2. Hinge movement. A typical example is the hinge between the two domains S and P of TBSV (Harrison et al., 1978), which is different for the AB_5 dimers on the quasi-twofold axes and the CC_2 dimers on the icosahedral twofold axes. Removal of Ca^{2+} ions from TBSV causes the P domains to rotate by about 100° in the C subunits and by about 35° in the A and B subunits (Robinson and Harrison, 1982).

3. Conformational changes in loop and β-bend regions. The conformation of a few residues can be altered in some enzymes by the binding of substrate or other ligands.

4. Alteration of binding surface between one subunit and another. This is caused by the binding of substrate or ligands. The difference between the CC_2 and AB_5 dimers in SBMV, caused by binding Ca^{2+} ions and RNA, is an excellent example (Figure 4). The best-known example occurs in hemoglobin, in which the $\alpha_1\beta_1$ dimer alters its relation to the $\alpha_2\beta_2$ dimer in the relaxed state (oxygenated) and tensed state (without oxygen).

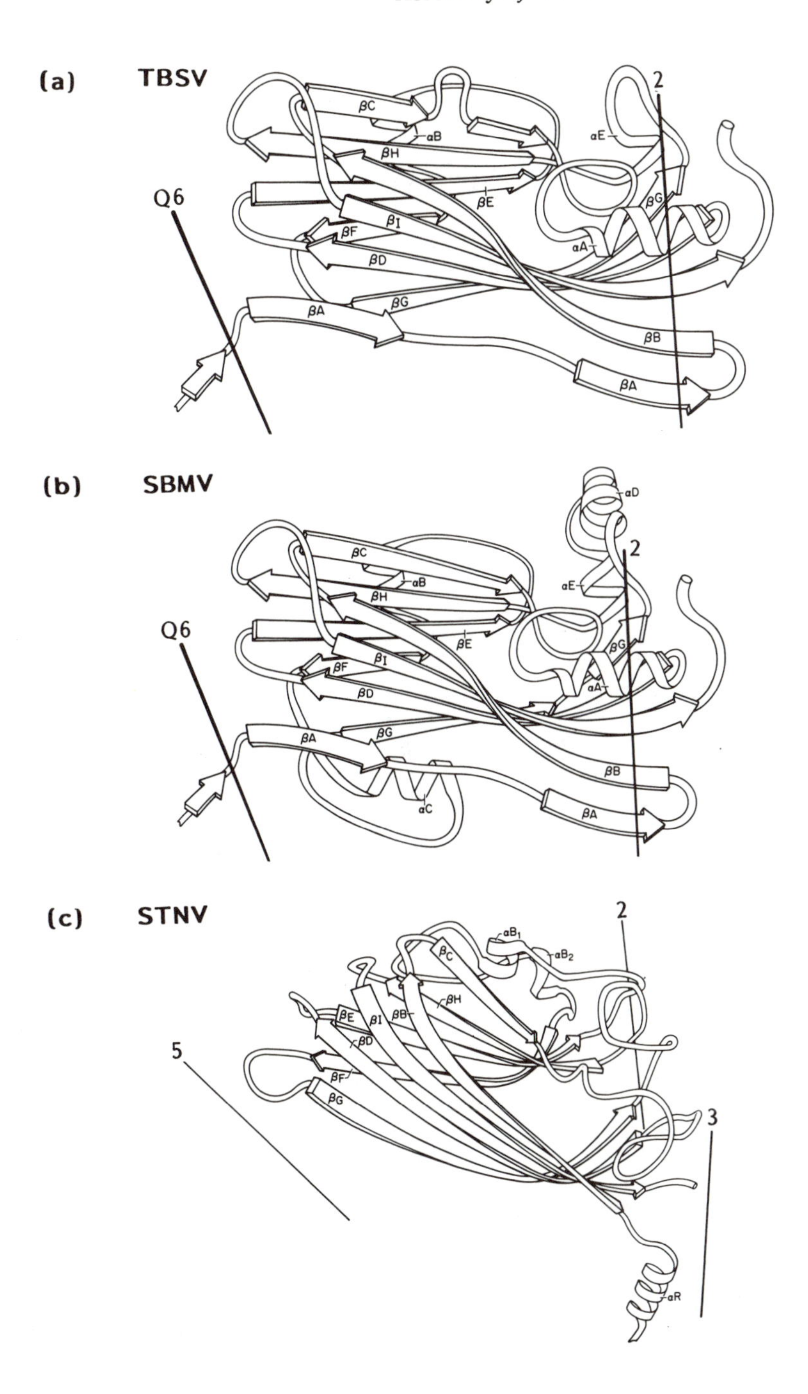

FIGURE 5 Backbone folding diagrams for the ordered portions of (a) TBSV (without its P domain), (b) SBMV, and (c) STNV coat proteins, shown in roughly comparable orientations. The C subunit (see Figure 3) with the ordered βA arm is shown for TBSV and SBMV. (From Rossmann et al., 1983b.)

In summary, the elastic deformations required by the Caspar and Klug theory of quasi-symmetry do not occur in proteins. However, there are alternative possibilities that depend on the environment of the protein and that can affect binding surfaces. Thus, the assembly of a virus must be viewed as a cooperative set of events, in which the association of a limited number of subunits alters their relationship to subsequent subunit additions. In short, Caspar and Klug considered unchangeable binding surfaces, implying path-independent assembly, whereas further knowledge of protein structure shows that binding surfaces do change in a path-dependent assembly process.

Similar concepts have been proposed by Caspar (1980) and by Fuller and King (1980). Caspar coined the word "autostery" to describe self-controlled switching from an unsociable to an associable conformation. He suggests that, "The energy to drive the change from the unsociable to the intrinsically less stable associable conformation is provided by the intersubunit bonds. Thus, the growing structure acts as the autosteric effector to control its own assembly".

Assembly pathways

An analysis of the various oligomeric units from which an icosahedron might be assembled is presented in Table 4 for viruses with $T \leq 12$. For instance, a simple $T = 1$ particle could be built from 60 monomeric subunits, 30 dimers, 20 trimers, or 12 pentamers, depending on the preferred subassembly aggregates of the subunit. The precise symmetry of these oligomers will vary according to the surface lattice. For example, the hexamers in a $T = 3$ lattice must have threefold symmetry and, therefore, can be described as $(BC)_3$. In this nomenclature, B and C are subunits with quasi-equivalent environments and conformations. The hexamers in a $T = 7$ lattice need not have perfect symmetry and may therefore be described as (BCDEFG).

Thus, assembly of icosahedra with $T > 1$ is likely to require the preassembly of oligomers whose association (but not symmetry) can be altered by the binding to other preassembled oligomers, to RNA, or to ligands such as cations. For instance, a $T = 3$ icosahedron can be constructed from two different types of dimers (AB and CC_2), the only oligomer that can exist in more than one state in the $T = 3$ particle. Indeed, many $T = 3$ virus subunits have a strong tendency to dimerize (Table 5). A $T = 7$ particle could be constructed from a 3-state trimer or 4-state dimer; alternatively, it could be formed from different types of pentamers in which pentamers would replace the onefold hexamers, as is the case polyoma virus (Rayment et al., 1982).

There must be switches that control the assembly pathway (see also Chapter 4, this volume). This is shown vividly in the structure of alfalfa mosaic virus (AMV) and the closely related ilarviruses (Matthews, 1979). The former are bacillus-shaped particles with hemispherical ends. Each hemisphere has an icosahedral $T = 1$ structure (Abdel-Meguid et al., 1982; Driedonks et al., 1977). The diameter of the cylindrical portion connecting the two ends is about 190 A, consistent with the diameter of AMV $T = 1$ particles. The protein subunits are arranged on a hexagonal surface lattice in the cylindrical portion of the virus (Gibbs et al., 1963; Hull et al., 1969;

Table 4 Number of oligomeric building blocks required to assemble various icosahedral surface lattices

The structure is built of:	Number of subunits	Value of T					Formula
		1	3	4	7	12	
Pentamers	60	12	12	12	12	12	1
+ 3-fold hexamers	$20h_A$	—	20	—	—	20	
+ 2-fold hexamers	$30h_B$	—	—	30	—	30	
+ asymm.* hexamers	$60h_C$	—	—	—	60	60	
3-fold trimers	$20t_A$	20	—	20	20	—	2
+ asymm. trimers	$60t_B$	—	60	60	2•60	4•60	
2-fold dimers	$30d_A$	30	30	—	30	—	3
+ asymm. dimers	$60d_B$	—	60	2•60	3•60	6•60	
Monomers	$60T$	60	180	240	420	720	$60T$

Formulas:

1. $1 + 2h_A + 3h_B + 6h_C = T$, where $h_A = 0$ or 1, $h_B = 0$ or 1, $h_C = 0, 1, 2, \ldots.$
2. $t_A + 3t_B = T$, where $t_A = 0$ or 1, and $t_B = 0, 1, 2, \ldots.$
3. $d_A + 2d_B = T$, where $d_A = 0$ or 1, and $d_B = 0, 1, 2, \ldots.$

*asymm. = asymmetric.

Table 5 Oligomers used for capsid assembly

Virus	Triangulation number, T	Probable oligomer size	References
SBMV	3	2	4,8
TCV	3	2	5
AMV	Bacilliform with $T = 1$ ends, cylindrical hexagonal net	2	3
BMV	3	2	2,6
Qβ	3	5,6	9
Polyoma	7	5	1,7

References: (1) Baker et al., 1983; (2) Cuillel et al., 1983a,b; (3) Driedonks et al., 1976,1977; (4) Erickson and Verduin, to be published; (5) Golden and Harrison, 1982; (6) Pfeiffer and Hirth, 1974; (7) Rayment et al., 1982; (8) Sehgal and Hsu, 1977; (9) Takamatsu and Iso, 1982.

Mellema and van den Berg, 1974). The virus itself has a divided genome with the larger RNA components encapsulated in longer particles. Investigation of the self-assembly of several strains of AMV in the presence of various nucleic acids has shown that the nucleic acid controls the elongation of the virus particles (Driedonks et al., 1978a,b). Therefore, information is derived from the nucleic acid when to build hexamers for the cylindrical body or when to build pentamers for the end surfaces. A mutant of AMV, which forms tubes of variable lengths without ends, apparently lacks the switch for making pentamers (Cremers et al., 1981). The situation for ilarviruses is intermediate to icosahedral and bacillus-shaped viruses where the pentamer and hexamer formation is not rigidly controlled (van Volten-Doting et al., 1981; Cusack et al., 1983). These irregularly shaped particles appear to have randomly distributed pentameric and hexameric clusters on their surface.

STRUCTURE

Virus Structures at Atomic Resolution

At the time of writing, TBSV (T = 3), the related turnip crinkle virus (TCV) (T = 3), SBMV (T = 3), and satellite tobacco necrosis virus (STNV) (T = 1) are the only spherical viruses whose structures have been determined to better than 3.5 A resolution (Table 6). This resolution is sufficient to trace the polypeptide chain and to identify amino acids. The analysis of the empty T = 1 protein particle of AMV has progressed to 4.5 A resolution. This sparsity of information is due to the magnitude of the crystallographic problem, though a variety of other plant and animal virus structures may be solved within the next several years.

The RNA region of the electron density maps resulting from the X-ray diffraction investigations has not shown any extensive ordered structure. Most of the nucleic acid must be packed fairly tightly and somewhat ordered, as rapid conformational changes are not observed (McCain et al., 1982a,b; Munowitz et al., 1980). The protein subunits of TBSV, TCV, SBMV, STNV, and possibly AMV have essentially the same polypeptide topology (Figure 5). The coat protein of these viruses contains a very basic amino-terminal domain (R) which is random in the structure. It lacks icosahedral symmetry and may be different in each virion and is intertwined with the RNA. The second domain along the polypeptide chain is the shell (S) domain, which has a similar structure in TBSV, SBMV, and STNV but is not necessarily homologous in amino acid sequence for these plant viruses. The surface domain is about 35 A thick in the radial direction of the virus. TBSV has an additional protruding (P) domain, which is folded similarly to the S domain. There is a hinge between the S and P domains that permits changes of conformation in the C subunit compared to the A and B subunits of TBSV. The greatest difference in subunit conformations for SBMV occurs at the carboxyl termini in the "vestigial hinge" of the C subunit relative to the A and B subunits (Rossmann et al., 1983a). The packing of the three quasi-equivalent A, B, and

Table 6 Three-dimensional structural data on simple spherical viruses

Virus	Resolution, Å	References	
		Structure	Amino acid sequence
TBSV	3.0	7	11
Swollen TBSV	8.0	14	11
TCV	3.5(?)	10	
SBMV	2.8	1,15	9
Metal-free SBMV	5.0	2	9
SBMV $T = 1$ assembly	6.0	5	9
STNV	2.4	12	8,18
AMV	4.5	6	3,17
Polyoma	22.5	13	4,16

References: (1) Abad-Zapatero et al., 1980; (2) Abdel-Meguid et al., 1981; (3) Collot et al., 1976; (4) Deininger et al., 1979; (5) Erickson and Rossmann, 1982; (6) Fukuyama et al., 1983; (7) Harrison et al., 1978; (8) Henriksson et al., 1981; (9) Hermodson et al., 1982; (10) Hogle and Harrison, private communication; (11) Hopper et al., 1984. (12) Liljas et al., 1982; (13) Rayment et al., 1982; (14) Robinson and Harrison, 1982; (15) Rossmann et al., 1983a; (16) Soeda et al., 1980; (17) Van Beynum et al., 1977; (18) Ysebaert et al., 1980.

C subunits in the viral coat is identical in TBSV and SBMV (Figure 3). The quaternary organization of the STNV and AMV coat proteins is somewhat different from TBSV and SBMV (Figure 6). The topology of the S domain in TBSV, SBMV, and STNV is an 8-strand antiparallel β-barrel consisting of two back-to-back four-stranded sheets. The topology of these domains is that of a "jelly-roll," in which a pair of consecutive antiparallel strands are wound into a helix (Figure 5). The nomenclature for the secondary structural elements of the S domain is shown in Figure 5.

Subunit Interactions in SBMV

Inspection of Figure 3 shows that the following types of subunit contacts exist (Rossmann et al., 1983a):

a. Contact between subunits related by the quasi-threefold axes, AB, BC, and CA (Figure 7a). These regions were

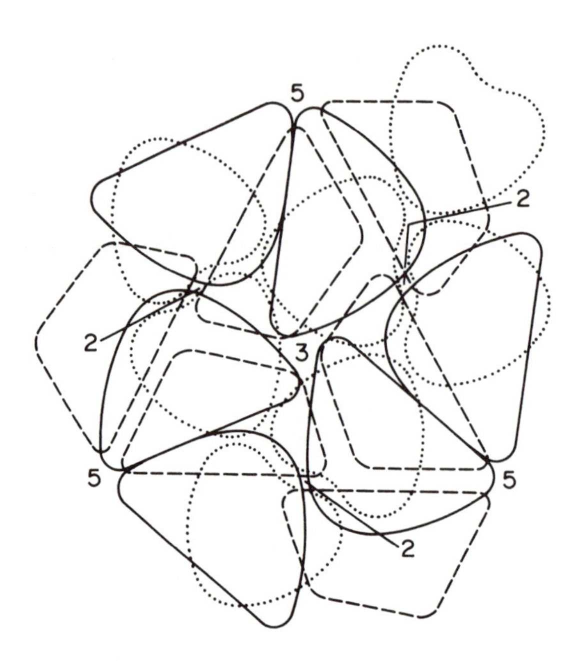

FIGURE 6 The organization of TBSV and SBMV surface domains (dashed) in the viral shell are identical and situated at the same radius. However, STNV (continuous) and AMV (dotted) subunits have a slightly different arrangement within the icosahedral framework of symmetry elements. (From Rossmann et al., 1983b.)

found to be very similar, though the BC contacts have some small differences compared to the AB and CA contacts.

b. Contacts between subunits related by the icosahedral fivefold axes AA_5 (Figure 7b).

c. Contacts between subunits related by the quasi-sixfold axes, B_5C and CB_2 (Figure 7c). These contacts bear little resemblance to each other. The B_5C and AA_5 interactions are very similar. The CB_2 interface contains the additionally ordered βA arm of the C subunit.

d. Contacts between subunits related by the icosahedral threefold axes, CC_3 (Figure 7d). The CC_3 interface comprises the "β-annulus" structure (Harrison et al., 1978). There is no contact between subunits B_5 and B_2.

e. Contacts between subunits related by the icosahedral twofold axes, CC_2 (Figure 7e and 8b). There are few contacts between S domains. In TBSV the additional P domains provide the strong dimer interaction (Golden and Harrison, 1982).

f. Contacts between subunits related by the quasi-twofold axes, AB_5 (Figure 8a). The principal AB_5 and CC_2 interactions are mediated by a hydrophobic pocket in one subunit (created in part by Trp107) into which is inserted Trp99 of the other subunit. Nevertheless, there are substantial differences between the AB_5 and CC_2 contacts.

The quasi-threefold-related interactions deform the contact region by only 6°, consistent with the concepts of Caspar and Klug. However, the twofold and quasi-sixfold contacts are related by rotations of the interacting subunits about a hinge region (Rossmann et al., 1983a).

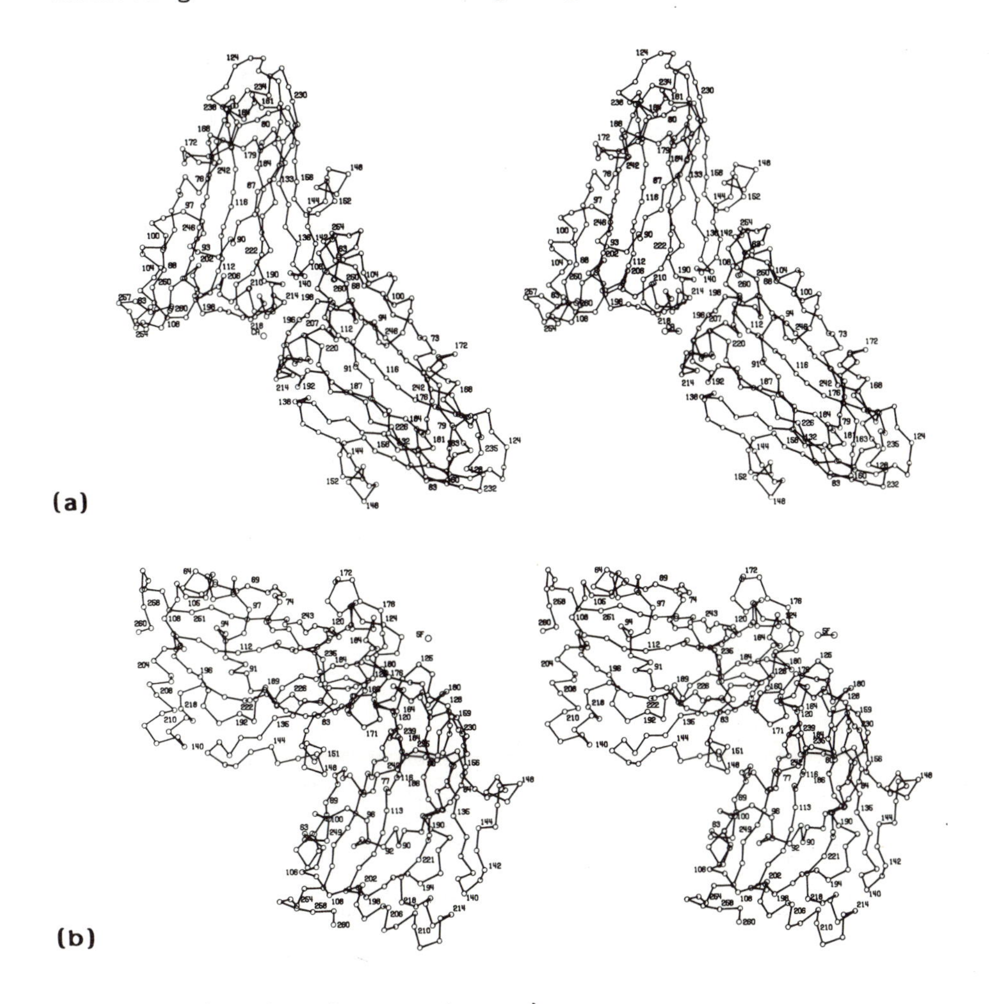

FIGURE 7 (continued on next page)

FIGURE 7 (continued)

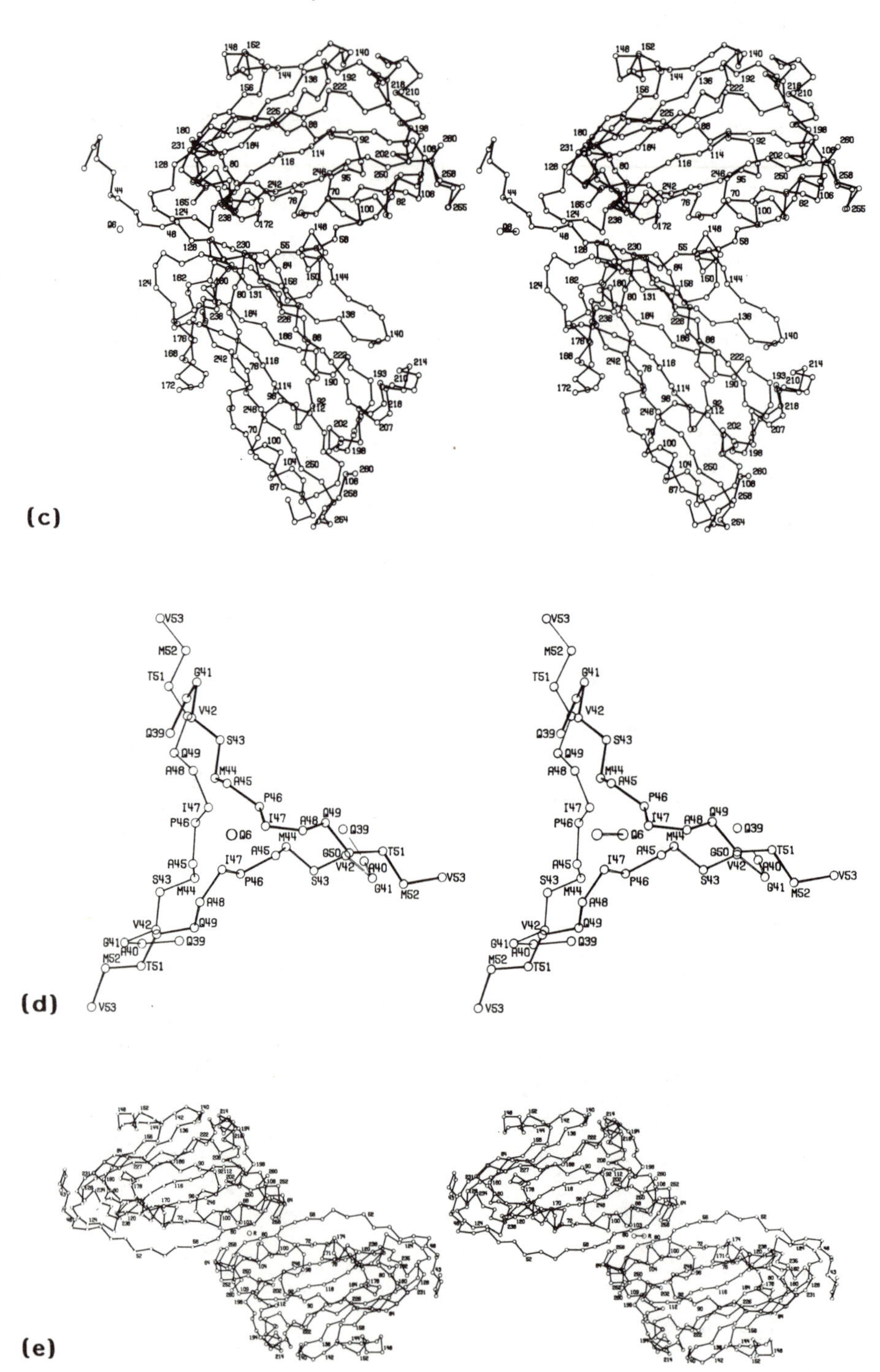

FIGURE 7 Cα backbone diagrams of pairs of SBMV subunits viewed from the exterior of the capsid in approximately the same orientation as used in Figure 3, showing the different types of contacts. (a) Subunits A (above) and B (below) showing the quasi-threefold-generated contacts as well as the bound calcium atoms. (b) Subunits A (above) and A_5 (below) showing the icosahedral fivefold-generated contacts. These contacts are almost identical to those between subunits B_5 and C around the quasi-sixfold axes. (c) Subunits C (above) and B_2 (below), showing the quasi-sixfold contacts where the additionally ordered βA arm of subunit C is interspersed. (d) The β-annulus generated by the three C subunits related by an icosahedral threefold axis. (e) Subunits C (above) and C_2 (below) related by the icosahedral twofold axes. These contacts are similar, but by no means identical, to the quasi-twofold contacts between subunits A and B_5. (From Rossmann et al., 1983a.)

It is then apparent that the "quasi"-symmetry is accommodated primarily as a result of the rolling of two subunits with respect to each other. This rolling permits two distinct types of dimer contacts (Figure 4 and 8) (the AB_5 subunits are each rotated by 18° relative to the CC_2 subunits) and two different types of contacts between the quasi-sixfold-related subunits (the B_5C subunits are each rotated by 19° relative to the CB_2 subunits).

In light of the preference for SBMV dimer formation in solution (Verduin and Erickson, unpublished results; Sehgal and Hsu, 1977), it would seem reasonable to assume that SBMV assembles in a series of cooperative dimer-assimilation steps. Fluorescence-quenching studies (Verduin, unpublished results) suggest that the contacts between the subunits of dimers in solution resemble those of the intact particle. Whether or not both AB_5 and CC_2 dimer types are present in solution is unknown, but quite generally they can be assumed to be in equilibrium. Each assembly step selects an AB_5 or CC_2 dimer according to the "autosteric" requirements of binding to the already partially assembled structure.

Fukuyama et al. (1983) have speculated that there are also two types of dimers in AMV (Figure 9). They proposed that the flexible arm of AMV (Kan et al., 1982; Andree et al., 1981) is partially ordered in one-third of the dimers in the hexagonal sections of the bacillus-like particles.

The amino-terminal arm of the coat protein of SBMV and many other plant viruses contains a large number of lysines and arginines, no aromatic residues, and many prolines and glutamines. Such a sequence is likely to be disordered in aqueous solution but is suitable for binding to nucleic acids. The β-annulus and βA arm also contain this unusual composition. Thus, the βA arm is likely to be disordered in solution, becoming ordered only in the C subunit during assembly. A hydrophobic cavity, open on one side to RNA, is formed between assembled CC_2 dimers and is suitable for inducing icosahedral symmetry onto the βA section of the arm. Some evidence for the ordering of the arm of the C subunits, during assembly can be obtained from circular dichroism of SBMV and the related turnip

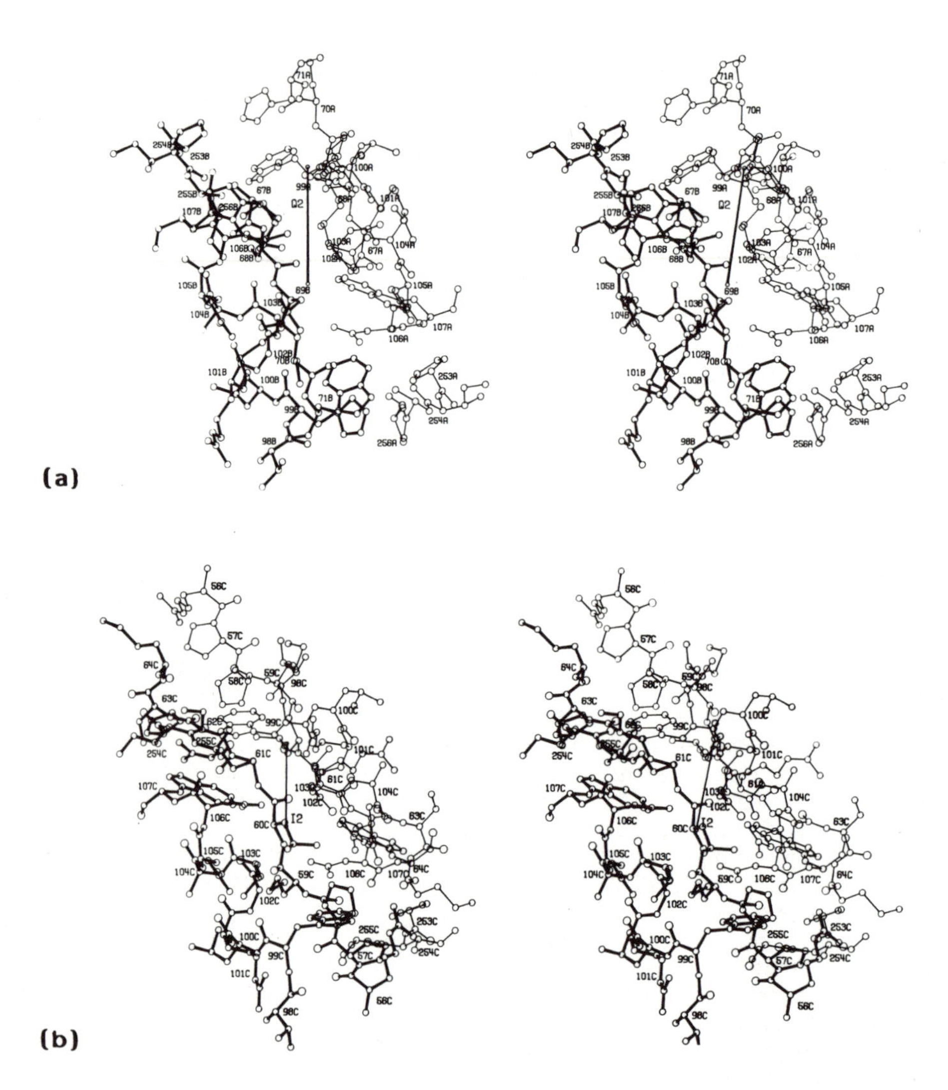

FIGURE 8 The hinge for the two types of twofold contacts (AB_5 or CC_2) is Trp99 inserted into a hydrophobic pocket. This residue is seen interacting with Trp107 in the neighboring subunit in (a) the quasi-twofold interaction (AB_5) and (b) the icosahedral twofold interaction (CC_2). The orientation and position of the two tryptophan residues can be recognized easily, showing that there is good π electron interaction at least in the CC_2 interactions. (From Rossmann et al., 1983a.)

rosette virus (Denloye et al., 1978; Odumosu et al., 1981). Since the CC_2 dimer requires the ordering of the βA arm for some of the subunit contacts, it can be tentatively assumed that the AB_5 dimer is present in solution. Therefore, it has been proposed that the formation of the β-annulus and ordering of the βA arm occurs during assembly, and this could control the subsequent additions of dimers (Savithri and Erickson, 1983) (Figure 10). Such an assembly mechanism is not

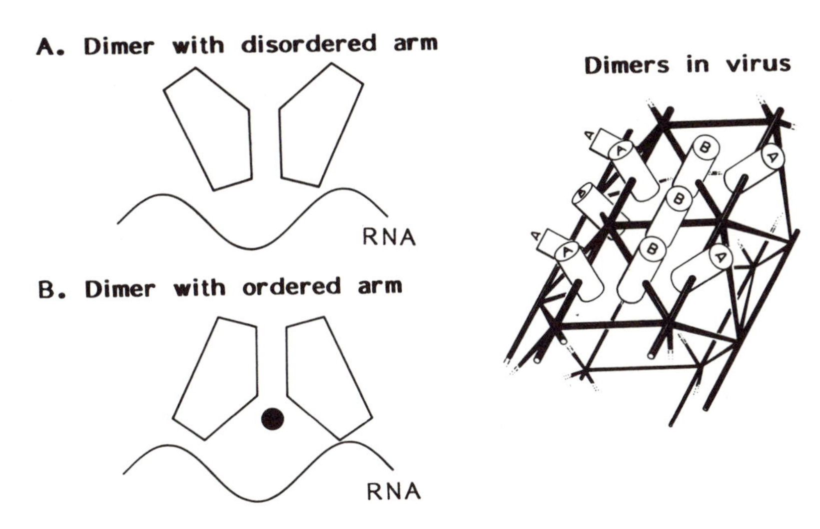

FIGURE 9 The two types of dimer environments that occur in the cylindrical portions of AMV. The A-type dimers have less space in the interior. The B-type dimers have greater space and, therefore, may have ordered amino-terminal arms in the subunit interface. The icosahedral virus ends would contain only A-type dimers. The large hole at the sixfold axes excludes the possibility of forming a β-annulus-type structure around these axes. (From Fukuyama et al., 1983.)

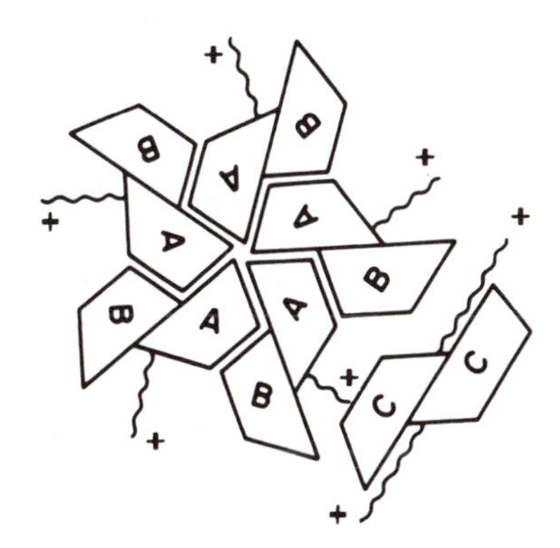

FIGURE 10 A 10-mer cap composed of five AB_5 dimers might assemble initially. All the amino-terminal arms are disordered. Addition of calcium on the quasi-threefold axes at alkaline pH might throw a switch, charging His132, and thereby alter the properties of the peripheral surfaces of the cap. Additional dimers must then "tense" in order to bind, leaving a hydrophobic channel between subunits C and B. The βA arm will then be induced to fold into this cavity and cause the formation of the β annulus. (From Rossmann et al., 1983a.)

consistent with the proposal of Harrison et al. (1978), in which all the C subunits form an initial skeletal icosahedral shell, which then acts as a scaffold for the subsequent assembly of the A and B subunits.

The availability of the complete amino acid sequences for both SBMV and STNV permits comparison of the interactions between subunits (Rossmann et al., 1983b). An assessment of the various forces acting between subunits in SBMV and STNV is shown in Table 7. The number of contacts between subunits is comparable in both viruses. The fivefold contacts have the greatest frequency. However, STNV has substantially more twofold and fewer threefold contacts than SBMV. The number of STNV threefold contacts is boosted by an additional amino-terminal α helix, buried in the RNA. Calcium ions

Table 7 Frequency of interatomic distance between subunits

Type of contact[a]	Axial type[b]	Frequency dist. <4.5 Å[c]	Hydro-phobic[d]	Hydrogen bonds	*e*	*f*	Ca^{2+}
SBMV							
AB	Q3	256	19	9	2		1-1/3
BC	Q3	428	16	7	3		1-1/3
CA	Q3	302	17	10	3		1-1/3
AA_5	I5	450	23	15	2	1	
B_5C	Q6	259	22	9	1		
CB_2	Q6	276	20	11		1	
CC_2	I2	173	8	6			
AB_5	Q2	127	13	5			
STNV							
AA_3	I3	195	9	9			2[g]
AA_5	I5	274	16	9	2		1[h]
AA_2	I2	238	7	7	1		

[a]Capital letters denote subunits; subscripts denote symmetry operations applied to subunits within the standard icosahedral asymmetric unit (see Figure 3). AA_3 contacts in STNV, not shown in Figure 3, approximately correspond spatially to the AB, BC, CA contacts of SBMV, all of which are identical in a *T* = 1 structure. Similarly, AA_2 STNV contacts correspond to AB_5 SBMV contacts.
[b]Q3 signifies subunit contacts related by a quasi-threefold axis; I5 denotes subunit contacts related by an icosahedral fivefold axis, etc.
[c]The number of interatomic distances between subunits <4.5 Å.
[d]The number of residues, not involved in polar interactions, within 4.5 Å of any other residue in a neighboring subunit.
[e]Interactions between oppositely charged residues.
[f]Interactions between similarly charged residues.
[g]One of these is on the icosahedral threefold axis.
[h]On icosahedral fivefold axis.

maintain the integrity of these contacts at high pH in both viruses. Additional subunit-binding forces are provided by salt linkages (Tables 8 and 9). The different charge interactions between the B_2C and CB_5 quasi-sixfold-related subunits in SBMV may provide a control mechanism in which the assembly process switches an AB_5 to a CC_2 dimer.

Of exceptional interest are the binding sites of the metal ions that control the swelling and disassembly of the virus (see

Table 8 Interaction of charged residues and Ca^{2+} between subunits*

Subunit	Residue	Atom[a]	Subunit[b]	Residue	Atom[a]	Interaction
STNV						
A	140 Glu	$O_{\varepsilon 1}$	A_5	143 Lys	N_ζ	-+
A	75 Arg	$N_{\varepsilon 2}$	A_5	178 Asp	$O_{\delta 1}$	+-
A	96 Arg	$N_{\varepsilon 1}$	A_2	92 Asp	$O_{\delta 2}$	+-
A	55 Asp	$O_{\varepsilon 1}$	On I3		Ca ion	-Ca
A	25 Glu	$O_{\varepsilon 1}$	Near I3		Ca ion	-Ca
A	194 Asp	$O_{\delta 1}$	Near I3		Ca ion	-Ca
A	138 Thr	O	On I5		Ca ion	-Ca
SBMV						
A,B_5	181 Arg	$N_{\varepsilon 2}$	A_5,C	163 Glu	$O_{\varepsilon 1}$	+-
A,B_5	181 Arg	$N_{\varepsilon 2}$	A_5,C	181 Arg	$N_{\varepsilon 1}$	++
A,B_5	132 His	$N_{\varepsilon 2}$	A_5,C	241 Arg	$N_{\varepsilon 2}$	++
A,B_5	229 Glu	$O_{\varepsilon 2}$	A_5,C	241 Arg	$N_{\varepsilon 2}$	-+
A,B,C	191 Glu	$O_{\varepsilon 1}$	B,C,A	196 Arg	$N_{\varepsilon 1}$	-+
A,B,C	138 Asp	$O_{\delta 1}$	B,C,A	200 Lys	N_ζ	-+
A,B,C	191 Glu	$O_{\varepsilon 1}$	B,C,A	249 Arg	$N_{\varepsilon 2}$	-+
B_2	229 Glu	$O_{\varepsilon 1}$	C	77 Glu	$O_{\varepsilon 1}$	--
A,B,C	194 Glu	$O_{\varepsilon 1}$	On Q3		Ca ion	-Ca
				108 Ser	O_γ	
A,B,C	138 Asp	O_δ	B,C,A	259 Ala	O	-Ca
	141 Asp	O_δ		199 Ala	O	

*Only those charge interactions with an interaction of less than 4.5 Å across the subunit boundary have been included here.

[a]See IUPAC-IUB Commission on Biochemical Nomenclature (1970) for notation of atoms in amino acids.

[b]I and Q refer to icosahedral or quasi-symmetry axes. A, B, and C are subunits in the standard icosahedral asymmetric unit. Subscripts denote two-, three-, or fivefold icosahedral symmetry operations.

Table 9 Intrasubunit charge clusters in the cowpea strain of SBMV

	Residue number	Subunit	Residue	Charge cluster
A. Quasi-threefold interactions (AB, BC, and CA contacts)				
1st Ca binding site (at Q3)	194	A	Glu	-
	194	B	Glu	-
	194	C	Glu	-
2nd Ca binding site	138	A	Asp	-
	141	A	Asp	-
	199	B	Carbonyl	
	259	B	Carbonyl	
1st charge cluster	196	A	Arg	+
	249	A	Arg	+
	252	A	Glu	-
	188	C	Asp	-
	191	C	Glu	-
2nd charge cluster	200	A	Lys	+
	204	A	Asp	-
	138	C	Asp	-
3rd charge cluster	65	A	Asp	-
	154	C	Lys	+
B. Icosahedral fivefold interactions (AA_5 and B_5C contacts)				
1st charge cluster	181	C	Arg	+
	181	B_5	Arg	+
	163	B_5	Glu	-
2nd charge cluster	132	C	His	+
	229	C	Glu	-
	241	B_5	Arg	+
C. Quasi-sixfold interactions (CB_2 contacts)				
1st charge cluster	132	B_2	His	+
	229	B_2	Glu	-
	77	C	Glu	-

Refer to Figure 3 for spatial description of subunit interactions.

section, *Dissassembly and Assembly*) (Hsu et al., 1976; Rayment et al., 1979). In SBMV one calcium ion is bound between the A, B, and C subunits on each quasi-threefold axis, coordinating with Glu194 from each of the three subunits (Abdel-Megued et al., 1981) (Figure 3). This site accounts for 60 Ca^{2+} ions that can be removed at low pH by dialysis against EDTA. A second cation site is between adjacent quasi-threefold-related subunits, in the AB-BC-CA subunit-contact regions (Figure 3). The site is surrounded by Asp-138 and Asp-141 of the C (or A or B) subunits of the A (or B or C) subunits, respectively (Table 9). The second site accounts for another 180 Ca^{2+} ions per capsid and corresponds to an equivalent Ca^{2+} binding site in TBSV (Harrison, 1980). The total amount of calcium per virion is then in rough agreement with the measurements of Abdel-Meguid et al. (1981), Hsu et al. (1976), and Hull (1978). The positions of the magnesium ions are not yet known; these may be bound, in part, to the disordered RNA.

The swelling in TBSV results from a disengagement of the quasi-threefold subunit contacts, AB, BC, and CA. Each of these contains a pair of adjacent Ca^{2+} binding site formed by a cluster of carboxylate side chains. The electrostatic repulsion resulting from the removal of the Ca^{2+} at high pH, when the carboxylates are ionized, probably provides the driving force for the structural rearrangements that occur during swelling. Robinson and Harrison (1982) showed that when TBSV swells, the fivefold (AA_5) and quasi-sixfold (B_5C and CB_2) contacts are maintained, while all the other contacts between S domains are broken. Thus, the process of disassembly is likely to differ from that of assembly.

RNA-Protein Interactions

It is not clear whether the RNA molecules fold into a unique tertiary structure. There is no clear evidence for any specific structure in the RNA region of crystallized viruses (Table 6). Thus, either there is no unique RNA structure or the external surface of the icosahedral coat imposes 60, equally probable orientations onto the virus during crystallization. Nevertheless, the interior of the coat is invariably extremely basic (Hermodson et al., 1982; Liljas et al., 1982) and, therefore, must aid in organizing the RNA during assembly. Rossmann et al. (1983c) have suggested a mode of A-form RNA binding to the inner surface of the SBMV coat based on the organization of basic residues. However, this can only be a preferred binding mode; otherwise, it would be more clearly visible in the electron density map.

Low-angle neutron scattering has shown the random, basic domain to be deeply buried in the RNA of TBSV (Chauvin et al., 1978) and of SBMV (Kruse et al., 1982). NMR experiments for SBMV (McCain et al., 1982a,b) have shown that as the virus swells, the basic domains become mobile along with the RNA. In contrast, the amino-terminal arms in AMV capsids are mobile in the native virion (Kan et al.,1982). The total positive charge on the 180 N-terminal arms of SBMV, plus that on the interior surface of the virus, nearly completely cancels the charge on the nucleic acid. In those viruses in which there is no basic amino-terminal arm the concentration of polyamines such as spermidine (Cohen and McCormick, 1979) is usually higher compared to viruses that have a basic arm. Therefore, the arm may be considered "histone-like", helping the RNA to fold during the assembly of the virus.

A portion of the basic arm in STNV is ordered and forms an α helix (the βR arm in Figure 5c). The α helices from sets of three neighboring subunits are clustered around threefold axes in the interior of the virus. These helices are hydrophobic on the side nearest the threefold symmetry axes, permitting the helices to bind together. However, they expose a strongly basic face to the RNA. Abad-Zapatero et al. (1980) and Argos (1981) have suggested that the random domains might fold into α helices nestling in the minor groove of the double-helical RNA. Argos (1981) has observed that the occurrence of basic residues in R domains is often periodic, which would concentrate basic residues onto one side of a presumptive α helix, as found in STNV.

Coat-protein dimers bind to specific sites on the RNA components of AMV (Houwing and Jaspars, 1982). Coat protein still binds AMV-RNA even when the flexible arm is removed with trypsin (Zuidema et al., 1983), though it loses its specificity for a particular segment of RNA. Thus, the presumably basic interior surface of AMV coat protein binds RNA in a manner that may be like that predicted by Rossmann et al. (1983c) for SBMV, whereas the flexible amino-terminal R domain, which possibly is folded into an α helix, has the function of recognizing a specific RNA sequence. The apparent function of such binding is to initiate RNA replication and may also be to form an initiation complex for capsid assembly (Houwing and Jaspars, 1980). For instance, TMV RNA (Zimmern, 1977) and papaya mosaic virus RNA (AbouHaidar and Bancroft, 1978) have specific coat-protein binding sequences that act as initiator sites for assembly of these rod-shaped particles.

Polyoma virus

Polyoma virus is the only animal virus whose structure is known in any detail, albeit only to 22.5 A resolution. It is a member of the papova group and is closely related to the much-studied SV40 virus. Virions contain seven different proteins, four of which are cellular histones that organize the viral DNA into nucleosomes. The other three (VP1, VP2 and VP3) are coded by the viral genome. VP1 is the major coat protein and has a molecular weight of about 42,000.

Analysis of electron micrographs have suggested that the capsid has a $T = 7$ (dextro) surface lattice (Figure 11) (Klug, 1965; Finch, 1974), which would require 12 pentamer and 60 hexamer units. However, single-crystal X-ray diffraction analysis (Rayment et al., 1982) has shown that the predicted hexamers are in fact pentamers (Figure 12). These results were not immediately accepted by the scientific community, both because they contradicted the concept of quasi-equivalence (Figure 13) and because the crystallographic analysis was unconventional. Unfractionated preparations of virus normally contain hollow tubular particles. There are two types of tubes approximately 500 A long (close to the normal virus particle) and with a diameter of 300 A, these were thought to be assembled of hexamers (Finch and Klug, 1965) and pentamers (Kiselev and Klug, 1969), respectively. Thus, the tubes appeared to result from the separation

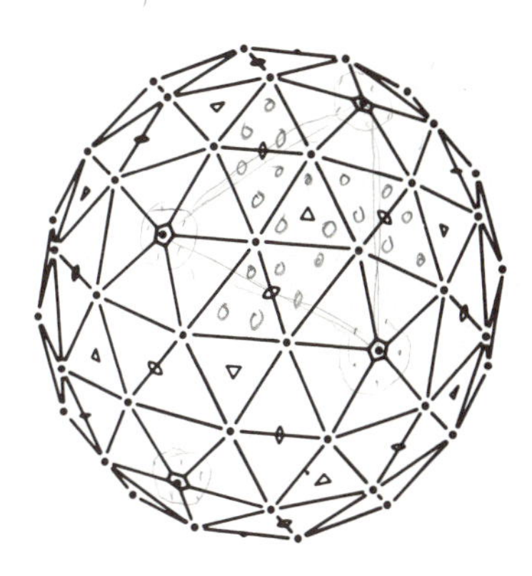

FIGURE 11 $T = 7$ (dextro) icosahedral surface lattice with five-, three-, and twofold axes marked. The drawing shows one side of the polyhedral surface consisting of 60 six-coordinated and 12 five-coordinated lattice points at the same radius. The location of the six-coordinated point is that determined for the hexavalent morphological unit in the polyoma capsid. (From Rayment et al., 1982.)

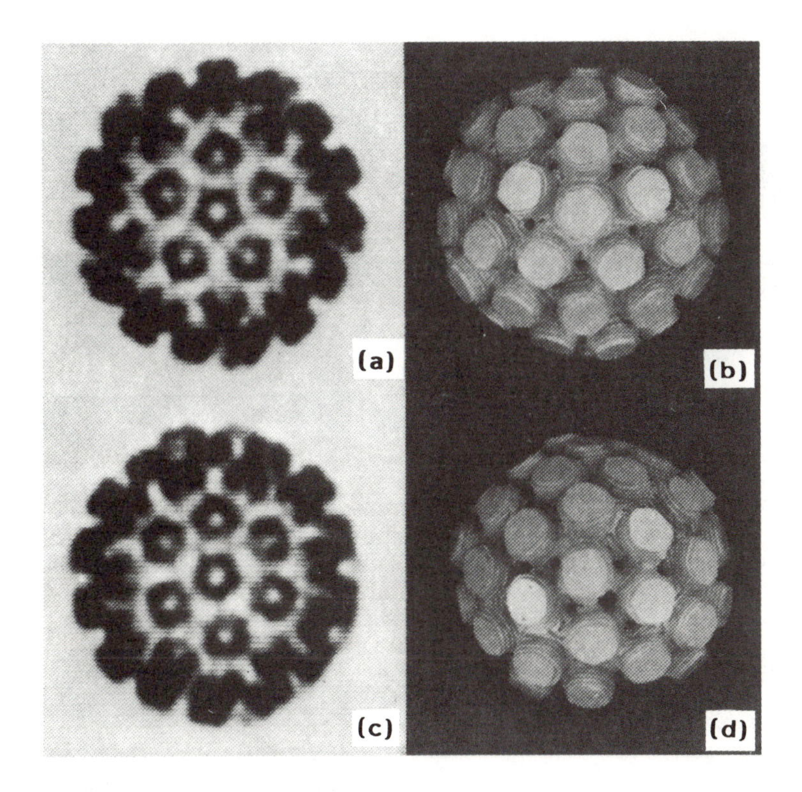

FIGURE 12 Views of half the capsid (495 A diameter) down the fivefold axis (a,b) and down the axis of the hexavalent unit (c,d). (a) and (c) are computer graphics projections of the electron density map at 22.5 A resolution, (b) and (d) are photographs of a model of stacked sections of the map at 30 A resolution with the external contour level set at 0.2 of the maximum density. The half model was built symmetric about the fivefold axis (b); thus, when it is tilted to view down the hexavalent unit (d), the bottom portion of the image is incomplete. (From Rayment et al., 1982.)

of naturally occurring pentamers and hexamers. However, recently Baker et al. (1983) have shown that the larger diameter tubes are also composed of pentamers, and that the pentamers do not have quasi-equivalent environments in either type of tube, thus confirming the conclusions of the x-ray work.

The analysis of polyoma virus is in good agreement with the observations for plant viruses that show that a single type of protein subunit can have different environments. Polyoma virus subunits would be expected to assemble into pentamers, whereas $T = 3$ plant-virus coat proteins aggregate into dimers in solution. Bacteriophage Qβ has been disassembled into pentamers and hexamers stabilized by disulfide bonds in a $T = 3$ surface lattice (Takamatsu and Iso, 1982). Thus, different viruses may assemble from different coat-protein oligomers.

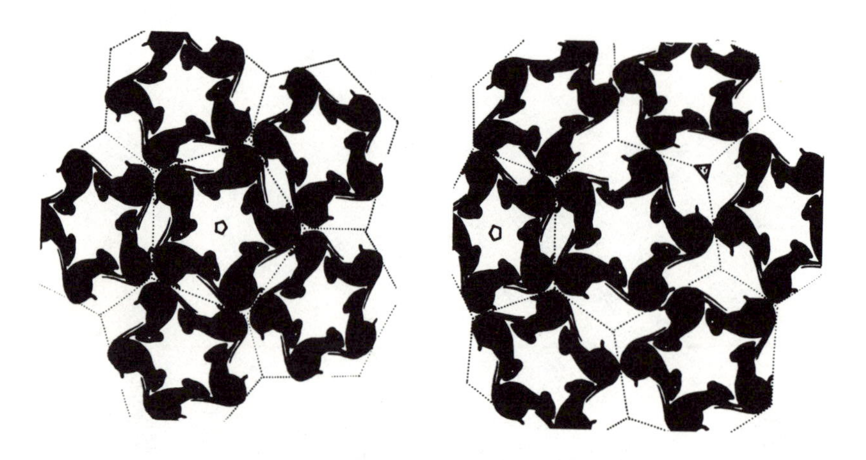

FIGURE 13 The environment of each major protein subunit (represented as a mouse) in the polyoma virus coat. Left panel: all subunits are equivalent about an icosahedral fivefold axis. Right panel: each subunit has a nonequivalent environment in the onefold position of the pentamer. (From Rayment et al., 1982.)

DISASSEMBLY AND ASSEMBLY

Stabilizing and Regulatory Forces

Plant viruses have been grouped (Kaper, 1975) according to the relative contribution of protein-protein and protein-RNA interactions toward virion stability. Cucumoviruses (viruses in the cucumber mosaic virus group) typify virions dominated by strong protein-nucleic acid interactions (Table 10, category I1). They can be readily dissociated in high salt concentrations and are frequently sensitive to nuclease attack. Their coat proteins often have very basic amino-terminal random domains that vary in length from 20–70 residues and are closely associated with the RNA interior (Tremain and Ronald, 1978; Agrawal and Tremaine, 1972; Bol et al., 1974). The other extreme is

Table 10 Assembly and disassembly strategies for spherical viruses

I. Disassembly

1. Strong protein-nucleic acid interactions

   a. Viruses that swell, have basic arm, require cation for stability.
      Examples: SBMV, TBSV, TCV, STNV, CCMV, BMV.
   b. Viruses that are swollen, have basic arm.
      Examples: AMV, CMV.

2. Strong protein-protein interactions

   a. Viruses that extrude nucleic acid, do not have basic amino-terminal arm, do not require cation for stability from empty capsids *in vivo*.
      Examples: TYMV, BdMV (RNA); f2, R17, MS2, Qβ (RNA); picornaviruses (RNA); cauliflower mosaic virus (DNA).

II. Assembly

$$\underset{1}{\text{RNA}} + \underset{2}{\text{dimer}} \longrightarrow \underset{3}{\text{[RNA-dimer] complex}} \longrightarrow \underset{4}{\text{Cap}} \longrightarrow \underset{5}{\text{Virus}}$$

Examples of 2: CCMV, AMV, SBMV
Examples of 3: AMV, TYMV, CCMV
Examples of 4: SBMV, BMV

represented by the tymoviruses (viruses in the TYMV group), which are dominated by strong protein-protein interactions (Table 10, category I2). They assemble empty capsids *in vivo*, expel their RNA long before conditions are reached for shell disassembly *in vitro* and are very resistant to a variety of structural perturbants. Furthermore, their coat proteins do not contain a basic amino-terminal arm and, thus, are limited in their interaction with viral RNA. Many viruses, including SBMV, TBSV, and STNV, fall between these to extremes.

Besides stabilizing forces, regulatory interactions must exist that permit disassembly and assembly. One sort of regulatory interaction, originally proposed for TMV (Caspar, 1963) and cowpea chlorotic mottle virus (CCMV) (Bancroft, 1979), involves the juxtaposing of carboxylic acid side chains from adjacent subunits to form a carboxyl "cage" or cluster. Such carboxyl clusters are seen in the structures of TMV, SBMV, and TBSV. Due to the elevated pK_a values of carboxyl groups in such an environment, the state of protonation of the cluster will be sensitive to pH changes near neutrality, as well as to the concentration of Ca^{2+} ions that will be attracted to negative ionized clusters. The charge of the carboxyl clusters at subunit contacts may regulate the assembly and disassembly of the virions, since they lie in the regions of contact between subunits (Figure 3).

Viruses That Swell

The utilization of carboxylic clusters in regulating protein-protein interactions is manifested in many plant viruses by swelling (Table 10, category I1a). This phenomenon was first observed for CCMV as a 10-15 percent increase in hydrodynamic radius of the virion (Bancroft, 1970). Swelling reactions are favored at elevated pH levels on removal of bound Ca^{2+} and Mg^{2+} ions by chelating agents. Swollen virions are stabilized mainly by electrostatic protein-RNA linkages, as indicated by their salt lability (Kaper, 1975). They can also exhibit altered antigenic properties (Tremaine et al., 1980). In the swollen state both the basic amino-terminal arm of the coat protein and the RNA are often rendered susceptible to enzymatic cleavage (Bancroft, 1970; Tremaine and Ronald, 1978; Sehgal et al., 1979; Chidlow and Tremaine, 1971; Agrawal and Tremaine, 1972). An increase in mobility of the RNA and basic amino-terminal arms is also often observed upon swelling.

Certain viruses, such as AMV and cucumber mosaic virus (CMV), which neither swell nor bind divalent cations, (Oostergetel et al., 1981) nevertheless exhibit salt and enzyme lability (Kaper, 1975; Hull, 1969) (Table 10, category I1b). Such viruses might be regarded as naturally swollen with an inability to shrink. Possible analogs of the swelling reaction of plant viruses have been observed in animal viruses. In particular, the icosahedral $T = 4$ core of Sindbis virus is composed of about 240 identical protein subunits and viral RNA. It unfolds into a filamentous structure in response to EDTA treatment at low ionic strength (Söderlund et al., 1975).

Viruses That Do Not Swell

Tymoviruses (Kaper, 1975; Virudachalam et al., 1983a,b), comoviruses, RNA phages, and animal picornaviruses do not swell but expel their RNA in response to pH changes (Table 10, category I2). Expulsion is regulated presumably by charge interactions between the RNA phosphate backbone and carboxylate groups on the interior of the protein shell. This is illustrated here by examples of plant, bacterial, and animal viruses.

The weak protein-RNA links in tymoviruses are broken near neutral pH, after which a variety of treatments causes the virions to release their RNA. However, RNA expulsion from belladonna mottle virus (BdMV) can be prevented if low concentrations of spermidine, Mg^{2+} or Ca^{2+} ions, or high concentrations of K^{+} ions, are present (K. Sitaraman, K. L. Heuss, R. Virudachalam, J. L. Markley, and P. Argos, unpublished work). The coat protein of RNA-free shells formed *in vitro* as well as those found *in vivo* exhibit increased motion of a segment containing acidic residues, suggesting possible involvement of carboxyl groups in mediating the protein-RNA interaction (Virudachalam et al., 1983b; Vriend et al., 1982).

Picornaviruses can release their RNA, along with VP4, under a variety of conditions concomitant with an antigenic conversion (Rueckert, 1976; Putnak and Phillips, 1981). ϕX174, an isometric single-stranded DNA phage, undergoes a conformational change upon binding to cell wall fragments, resulting in ejection of the DNA (Incardona, 1978). R17 phage partially extrudes its RNA upon heating, which results in slowly sedimenting particles that retain their infectivity (Steitz, 1968a,b; Verbraeken and Fiers, 1972). Complete expulsion of the RNA from R17 can be affected at high pH and ionic strength (Samuelson and Kaesberg, 1970). Reversible extrusion of DNA has also been observed with cauliflower mosaic virus (Al Ani et al., 1979).

Assembly Mechanisms

The first stage in virus assembly could, in some cases, be the binding of a limited number of oligomers to a selected nucleotide sequence (Table 10, II). Successful assembly of icosahedral viral coat proteins with heterologous RNAs, DNA, and even polyvinyl sulfate suggests that the normal viral RNA provides merely a minimum-length polyanionic backbone to aid in condensation of the subunits (CCMV and BMV: Bancroft, 1970; f2: Hohn, 1969; CMV: Kaper and Geelen, 1971; AMV: Driedonks et al., 1978b; sindbis core: Wengler et al., 1982; SBMV: Tremaine and Ronald, 1977). However, in competition experiments using several species of nucleic acids with the coat proteins of RNA phages (Sugiyama et al., 1967) and of BMV, each protein shows a strong preference for encapsulating its own RNA (Cuillel et al., 1979). Furthermore, the coat proteins of AMV, TMV, and PMV bind to specific sequences on their own RNA, and RNA extracted from CCMV (Verduin, private communication) and TYMV (Keeling and Matthews, 1982) contains several tightly bound coat-protein subunits. Therefore, it seems likely that, at least in some cases, assembly *in vivo* is initiated by the specific recognition of viral nucleic acid by coat protein.

The second stage of assembly (Table 10, II) will be the association of a number of dimers (or other oligomers) to form a "cap" (Figure 10). Where a protein subunit can form different particles, it would seem reasonable that its environment in each particle would be similar. At most, one of three types of contacts can be preserved in a cap suitable for $T = 1$ or $T = 3$ formation. The predominance of dimers in solution suggests that the AB_5 dimer is common to both assembly types. However, the substantial intersubunit contacts generated by the fivefold axes would lead to caps with common AA_5 contacts in $T = 1$ and $T = 3$ particles. Nevertheless, the control of assembly of these particles by pH and Ca^{2+} would suggest preservation of the trimer contacts (Erickson et al., 1985).

In the third stage (Table 10, II), addition of subsequent dimers would create threefold contacts (AB, BC, CA), in which the regulatory carboxyl clusters are situated. At low charge, the incoming dimers could add to the growing structure in the same AB_5 configuration, leading to $T = 1$ assembly. If the carboxyls are ionized, the resulting electrostatic repulsion at the threefold contact would lead to a bonding strain that would be unequally distributed about the three subunits. Thus, the incoming dimer would have to absorb the strain, resulting in the addition of a CC_2 dimer to the growing $T = 3$ cap. Ensuing folding of the βA arm and β annulus could then help direct further the assembly of the growing $T = 3$ shell. Electron micrographs of partially assembled shells of SBMV are consistent with the hypothesis that both $T = 1$ and $T = 3$ shells grow from "caps" (Figure 14).

Observations on SBMV assembly support the structural model of stepwise assembly through intermediate dimers and a growing cap. Several strains have been successfully reassembled *in vitro* (Tremaine and Ronald, 1977; Hsu et al., 1977) The products of low-ionic-strength assembly of SBMV protein (Savithri and Erickson, 1983) with low- and high-MW fractions of heterogeneous SBMV-RNA (Rutgers et al., 1980; Weber and Sehgal, 1982; Savithri and Erickson, 1983) depend critically on pH and the presence of Ca^{2+} and Mg^{2+} ions (Table 11; Figure 15). The pH and divalent cation effects on assembly may be regarded as a combined charge effect on the coat protein in which the charge state of ionizable carboxylate side chains, which are found in the quasi-threefold subunit interfaces of the SBMV structure, regulate the mode of assembly (Figure 16). The low charge configuration, in which $T = 1$ particles are favored, is produced by protonation of carboxylates at pH 5 or by neutralization of partially ionized carboxyls with Ca^{2+} ions at pH 7. The high charge state, which favors $T = 3$ formation, is produced by partial ionization of carboxylates at pH 7 or by Ca^{2+}-ion binding to fully ionized carboxyls at pH 9. The fully ionized state (pH 9) precludes assembly.

SBMV coat protein does not assemble either into $T = 1$ or $T = 3$

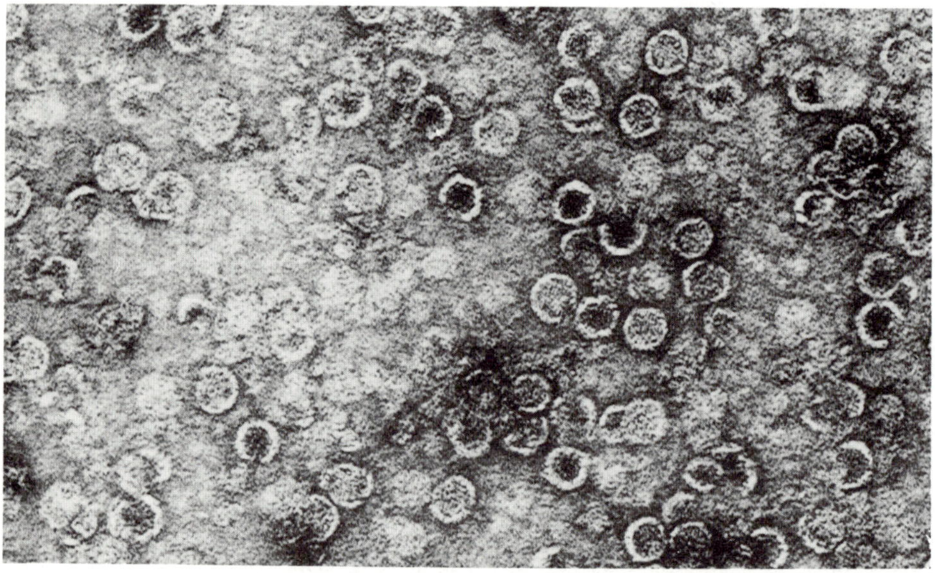

FIGURE 14 Electron micrograph of incompletely assembled SBMV $T = 3$ nucleoprotein particles. Shells at various stages of growth are visible.

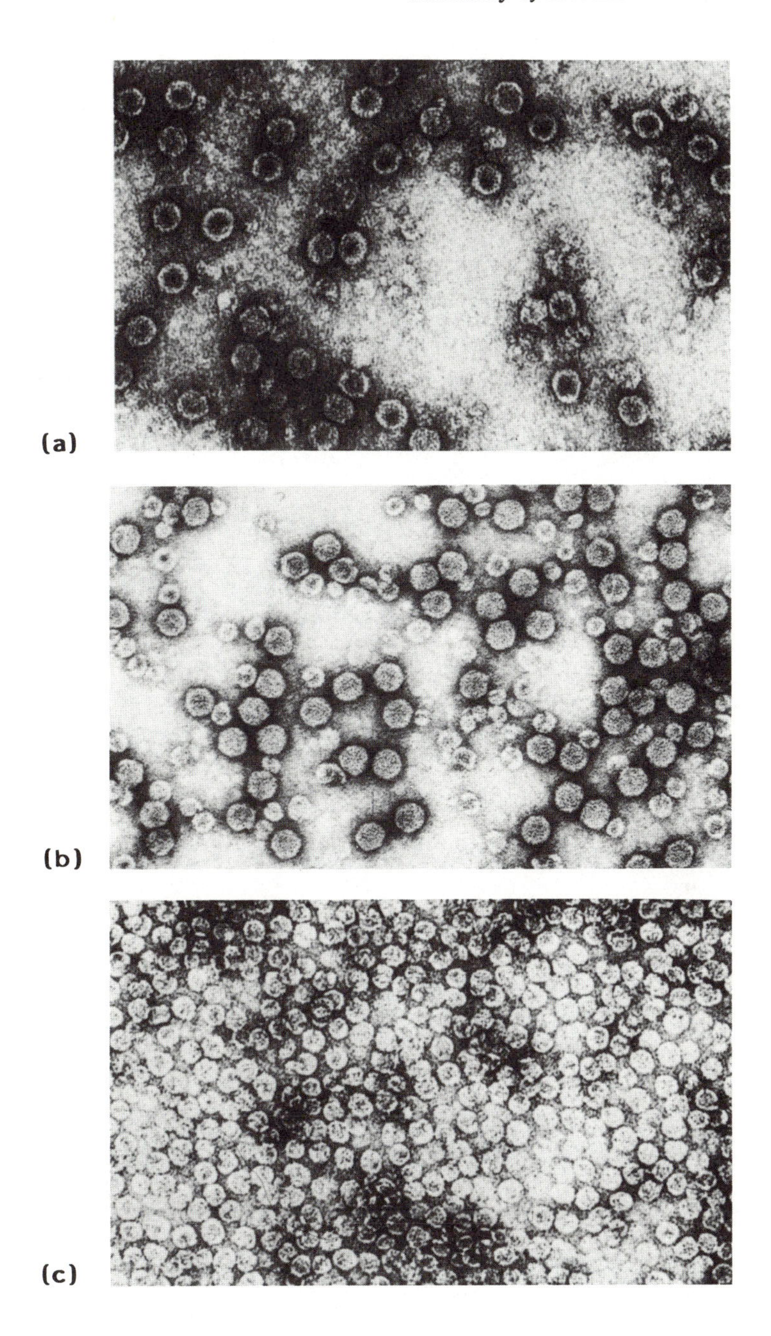

FIGURE 15 Electron micrographs of SBMV assembly products using heterogeneous SBMV RNA in the presence of calcium and magnesium ions. (a) $T = 3$ particles at pH 9.0. (b) $T = 3$ and $T = 1$ particles at pH 7.0. (c) $T = 1$ particles at pH 5.0. The $T = 3$ particle diameter is about 300 A. (From Savrithri and Erickson, 1983.)

Table 11 SBMV assembly products

Protein P22		Protein P28 (native)			
pH	No RNA	No RNA	Unfractionated SBMV RNA	Low-MW RNA[a]	High-MW RNA[b]
1. Divalent cations (Ca^{2+} and Mg^{2+}) present					
5	$T = 1$	—*	$T = 1$	$T = 1$	—
7	$T = 1$	—	$T = 1$, $T = 3$	$T = 1$	$T = 3$
9	$T = 1$	—	$T = 3$	$T = 3$	$T = 3$
2. Divalent cations (Ca^{2+} and Mg^{2+}) absent					
5	$T = 1$	—[c]	$T = 1$	$T = 1$	—
7	$T = 1$	—	$T = 3$	—	$T = 3$
9	—	—	—	—	—

The table summarizes the assembly studies of Savrithi and Erickson (1983). P28 and P22 refer to SBMV subunits of molecular weight ca. 28,000 and 22,000, respectively. P28 is the native protein, and P22 has been cleaved at Arg-61. The assembly products were produced by dialysis.
[a]Unfractionated yeast tRNA; low-MW fractions (3×10^5 daltons) of SBMV RNBA.
[b]High-MW fractions (> 1 million) of SBMV RNA.
[c]Dick Verduin and John Erickson, unpublished results.
*— indicates no assembly.

empty capsids in the absence of RNA (Table 11). In contrast, RNA is not required for the assembly of capsids by the coat proteins of CCMV, BMV (Bancroft, 1970), AMV (Driedonks et al., 1977), or group-I RNA bacteriophages (Knolle and Hohn, 1975). Evidently the regulatory controls on capsid assembly for SBMV are more restrictive than for these other viruses. Assembly of SBMV coat protein into $T = 1$ or $T = 3$ shells is controlled primarily by the regulatory interactions between subunits, not by the RNA (Table 11), though RNA is essential for assembly at all times. Thus, when $T = 3$ particle formation is favored, it will occur with subgenomic RNAs and even tRNAs (J. W. Erickson, unpublished work). On the other hand, viral RNA cannot induce the formation of $T = 3$ particles when conditions favor $T = 1$ assembly.

An alternative assembly model is the "scaffolding" proposal of Harrison (1980) in which the 60 subunits of TBSV form an initial open $T = 1$ nucleoprotein lattice linked by the P domain interactions ("clamps") and β annuli. The A and B subunits would subsequently be inserted into the holes of this structure. Still another suggestion has been that the empty capsids observed with tymoviruses, comoviruses, and picornaviruses *in vivo* are preformed for subsequent

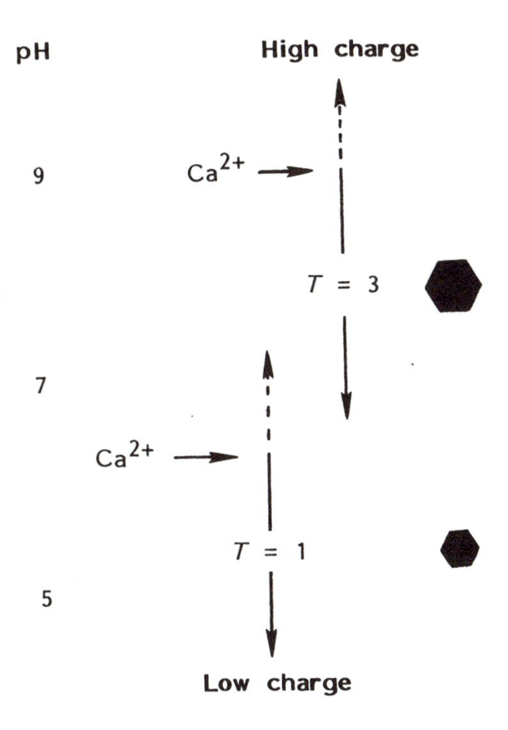

FIGURE 16 Illustration of the combined charge effects of pH and Ca^{2+} ions on SBMV assembly. The preferred mode of particle formation, $T = 1$ or $T = 3$, is indicated by the vertical arrows. Dashed arrows represent a requirement for Ca^{2+} ions. The end points of the arrows are approximate, but the overlap at pH 7 is accurate. (From Savrithi and Erickson, 1983.)

packaging of the nucleic acid (Jonard et al., 1972; Jocobson and Baltimore, 1968). However, convincing evidence for this last model is lacking, since none of these viruses have been reassembled *in vitro*.

Capsid Polymorphism

The coat proteins of icosahedral viruses often can assemble into lattices other than those in which they occur naturally. Spheres of different sizes, as well as sheets and tubes, can all be formed (Bancroft, 1970; Chapter 5, this volume). $T = 1$ particles can be formed from AMV, CCMV, BMC, and TCV coat-protein subunits with or without the presence of the basic arm or nucleic acid (Bol et al., 1974; Driedonks et al., 1976; Chidlow and Tremaine, 1971; Agrawal and Tremaine, 1972; Leberman and Finch, 1970). The RNA phage coat proteins may assemble double shells consisting of a $T = 1$ inner coat and a $T = 3$ outer coat stabilized by a complementary charge distribution on the inner and outer surfaces of the shells (Matthews and Cole, 1972).

The coat protein for SBMV can be cleaved at Arg-61 with trypsin to remove the amino-terminal arm. If the assembly is performed above neutral pH in the presence of Ca^{2+}, the shortened coat protein (MW = 22,000) aggregates only into $T = 1$ particles (Figure 17). The pH and divalent cation requirements (Figure 18) for $T = 1$ assembly suggest that the regulatory interactions for assembly of the $T = 3$ virus are maintained for the $T = 1$ capsid. Thus, the quasi-threefold interactions of the native $T = 3$ virus will presumably become icosahedral threefold interactions in the $T = 1$ shells (Erickson et al.,

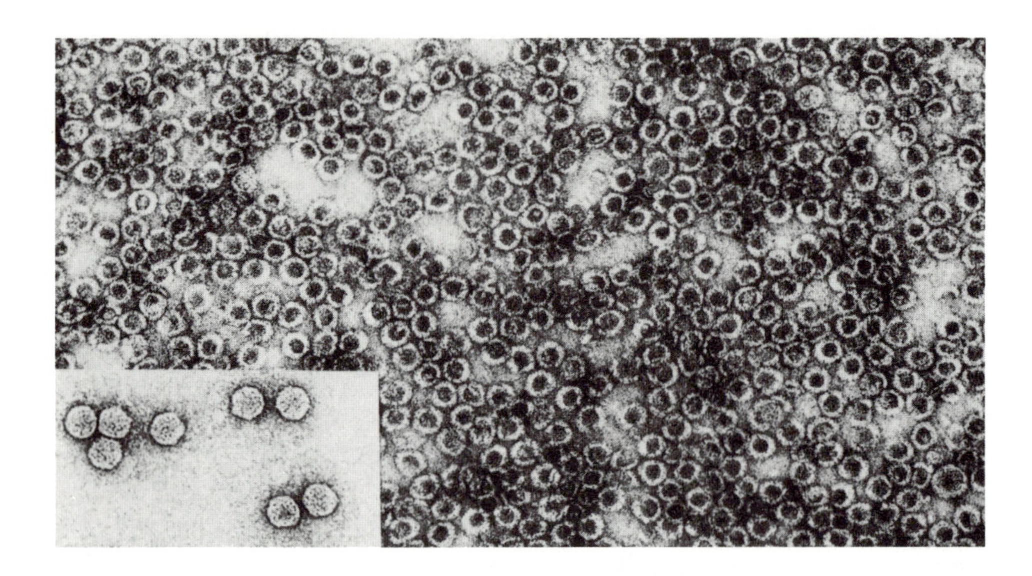

FIGURE 17 Electron micrograph of $T = 1$ particles formed by SBMV coat protein after removal of the N-terminal 61-residue arm. Native SBMV (300 A diameter) is shown in the inset at the same magnification. (From Erickson and Rossmann, 1982.)

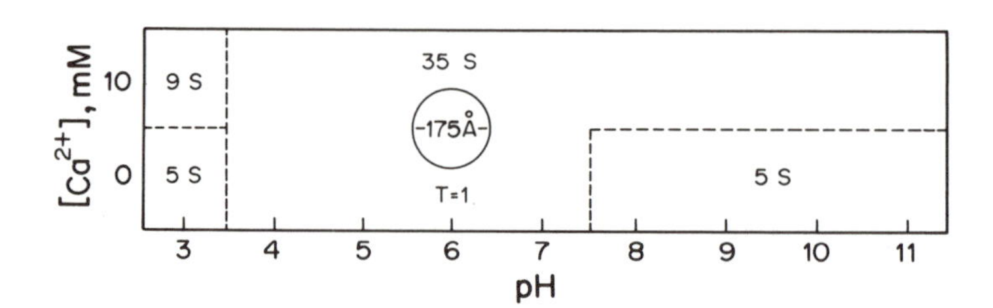

FIGURE 18 Effects of pH and Ca^{2+} on the assembly of SBMV proteolytically shortened coat protein (P22) lacking 61 N-terminal amino acids. The diagram summarizes the low ionic strength assembly products observed in the analytical ultracentrifuge or in the electron microscope. (From Erickson and Rossmann, 1982.)

1985). Coat proteins with similar folds can pack differently, as shown by comparing STNV with SBMV and TBSV (Rossmann et al., 1983b). The difference in packing reflects the different amino acid sequences and the differing surface lattices.

CONCLUSIONS

The first virus structures to be solved to atomic resolution have shown remarkable conservation of structural principles: (1) The shell domain is clearly designed to aggregate into a variety of icosahedrally symmetric structures to package nucleic acids of varying size; (2) the β-barrel fold seen for the subunits of SBMV, TBSV, and STNV has been evolutionarily selected for building shells of deltahedra, the simplest being the T = 1 icosahedra; and (3) construction of shells with higher T numbers requires additional information. It is to be anticipated that many, if not most, spherical plant viruses will derive their structures from the same ancestral fold, even in the absence of amino acid homology. Therefore, the diversity of assembly and stability properties results from the variation in the amino acids decorating the folded polypeptide chain and the occurrence of a basic amino-terminal arm, while the unifying concepts are derived from the unique shape of the S domains.

The assembly processes also seem to have common features: (1) It may need to be initiated by the specific binding of dimers, trimers, or pentamers to the nucleic acid; and (2) it proceeds in a path-dependent process, controlled by environmentally sensitive switches; and (3) dimers are the most likely oligomeric intermediates for the assembly of T = 3 shells. Viral disassembly proceeds via capsid expansion or nucleic acid expulsion, dependent on the degree of RNA charge neutralization provided by the coat protein.

ACKNOWLEDGEMENTS

We are most grateful for the detailed and careful discussion and comments by the editor, Sherwood Casjens. We also thank Sharon Wilder for help in the preparation of this manuscript. The work was supported by grants from the National Institutes of Health, the National Science Foundation, and the United States Department of Agriculture.

REFERENCES

Abad-Zapatero, C., S.S. Abdel-Meguid, J.E. Johnson, A.G.W. Leslie, I. Rayment, M.G. Rossmann, D. Suck, and T. Tsukihara. 1980. *Nature*, 286, 33.

Abdel-Meguid, S.S., T. Yamane, K. Fukuyama, and M.G. Rossmann. 1981. *Virology*, 114, 81.

Abdel-Meguid, S.S., K. Fukuyama, and M.G. Rossmann. 1982. *Acta Crystall.*, B38, 2004.

AbouHaidar, M. and J.B. Bancroft. 1978. *Virology*, 90, 54.

Adams, M.J., G.C. Ford, R. Koekoek, P.J. Lentz Jr., A. McPherson Jr., M.G. Rossmann, I.E. Smiley, R.W. Schevitz, and A.J. Wonacott. 1970. *Nature*, 227, 1098.

Agrawal, H.O. and J.H. Tremaine. 1972. *Virology*, 47, 8.

Al Ani, R., P. Pfeiffer, G. Lebeurier, and L. Hirth. 1979. *Virology*, 93, 175.

Anderson, C.M., F.H. Zucker, and T.A. Steitz. 1979. *Science*, 204, 375.

Andree, P.J., J.H. Kan, and J.E. Mellema. 1981. *FEBS Lett.*, 130, 265.

Argos, P. 1981. *Virology*, 110, 55.

Baker, T.S., D.L.D. Caspar, and W.T. Murakami. 1983. *Nature*, 303, 446.

Baldwin, J. and C. Chothia. 1979. *J. Mol. Biol.*, 129, 175.

Bancroft, J.B. 1970. *Adv. Virus Res.*, 16, 99.

Banyard, S.H., D.K. Stammers, and P.M. Harrison. 1978. *Nature*, 271, 282.

Bloomer, A.C., J.N. Champness, G. Bricogne, R. Staden, and A. Klug. 1978. *Nature*, 276, 362.

Boedtker, H. and R.F. Gesteland. 1975. In *RNA Phages*. N.D. Zinder, ed. p. 1. Cold Spring Harbor Laboratory.

Bol, J.F., B. Kraal, and F.T. Brederode. 1974. *Virology*, 58, 101.

Bränden, C.I., H. Jörnvall, H. Eklund, and B. Furugren. 1975. In *The Enzymes*. P.D. Boyer, ed. Vol. XI, p. 103. Academic Press.

Buehner, M., G.C. Ford, D. Moras, K.W. Olsen, and M.G. Rossmann. 1974. *J. Mol. Biol.*, 90, 25.

Casjens, S. 1979. *J. Mol. Biol.*, 131, 1.

Caspar, D.L.D. 1956. *Nature*, 177, 475.

Caspar, D.L.D. 1963. *Adv. Prot. Chem.*, 18, 37.

Caspar, D.L.D. 1980. *Biophys. J.*, 32, 103.

Caspar, D.L.D. and A. Klug. 1962. *Cold Spring Harb. Symp. Quant. Biol.*, 27, 1.

Champness, J.N., A.C. Bloomer, G. Bricogne, P.J.G. Butler, and A. Klug. 1976. *Nature*, 259, 20.

Chauvin, C., J. Witz, and B. Jacrot. 1978. *J. Mol. Biol.*, 124, 641.

Chidlow, J. and J.H. Tremaine. 1971. *Virology*, 43, 267.

Chothia, C. and J. Janin. 1981. *Proc. Nat. Acad. Sci. USA*, 78, 4146.

Chothia, C. and J. Janin. 1982. *Biochem.*, 21, 3955.

Cohen, F.E., M.J.E. Sternberg, and W.R. Taylor. 1980. *Nature*, 285, 378.

Cohen, S.S. and F.P. McCormick. 1979. *Adv. Virus Res.*, 24, 331.

Collot, D., R. Peter, B. Das, B. Wolff, and H. Duranton. 1976. *Virology*, 74, 236.

Cotton, F.A. 1971. *Chemical Applications of Group Theory*. Wiley.

Cremers, A.F.M., G.T. Oostergetel, M.J. Schilstra, and J.E. Mellema. 1981. *J. Mol. Biol.*, 145, 545.

Crick, F.H.C. and J.D. Watson. 1956. *Nature*, 177, 473.

Crick, F.H.C. and J.D. Watson. 1957. In *Ciba Foundation Symp. on the Nature of Viruses*. G.E.W. Wolstenholme and E.C.P. Millar, eds. p. 5. Little Brown.

Cuillel, M., M. Herzog, and L. Hirth. 1979. *Virology*, 95, 146.

Cuillel, M., M. Zulauf, and B. Jacrot. 1983a. *J. Mol. Biol.*, 164, 589.

Cuillel, M., C. Berthet-Colominas, B. Krop, A. Tardieu, P. Vachette, and B. Jacrot. 1983b. *J. Mol. Biol.*, 164, 645.

Cusack, S., G.T. Oostergetel, P.C.J. Krijgsman, and J.E. Mellema. 1983. *J. Mol. Biol.*, 171, 139.

Deininger, P., A. Esty, P. LaPorte, and T. Friedmann. 1979. *Cell*, 18, 771.

Denloye, A.O., R.B. Homer, and R. Hull. 1978. *J. Gen. Virol.*, 41, 77.

DeRosier, D.J., R.M. Oliver, and L.J. Reed. 1971. *Proc. Nat. Acad. Sci. USA*, 68, 1135.

Dickerson, R.E. and I. Geis. 1983. *Hemoglobin: Structure, Function, Evolution, and Pathology*. Benjamin Cummings.

Driedonks, R.A., P.C.J. Krijgsman, and J.E. Mellema. 1976. *Phil. Trans. Roy. Soc. Lond.*, B276, 131.

Driedonks, R.A., P.C. J. Krijgsman, and J.E. Mellema. 1977. *J. Mol. Biol.*, 113, 123.

Driedonks, R.A., P.C.J. Krijgsman, and J.E. Mellema. 1978a. *J. Mol. Biol.*, 124, 713.

Driedonks, R.A., P.C.J. Krijgsman, and J.E. Mellema. 1978b *Eur. J. Biochem.*, 82, 405.

Eiserling, F.A. 1979. In *Comprehensive Virology*. H. Fraenkel-Conrat and R.R. Wagner, eds. Vol. 13, p. 543. Plenum.

Eklund, H., C. I. Bränden, and H. Jörnvall. 1976. *J. Mol. Biol.*, 102, 61.

Erickson, J.W. and M.G. Rossmann. 1982. *Virology*, 116, 128.

Erickson, J.W., A.M. Silva, M.R.N. Murthy, I. Fita, and M.G. Rossmann. 1985. *Nature*, submitted.

Fehlhammer, H., W. Bode, and R. Huber. 1977. *J. Mol. Biol.*, 111, 415.

Fermi, G. and M.F. Perutz. 1981. In *Atlas of Protein Structure*. D.C. #. D.C. Phillips and F.M. Richards, eds. Vol. 2. Clarendon Press.

Finch, J.T. 1974. *J. Gen Virol.*, 24, 359.

Finch, J.T. and A. Klug. 1965. *J. Mol. Biol.*, 13, 1.

Fukuyama, K., S.S. Abdel-Meguid, J.E. Johnson, and M.G. Rossmann. 1983. *J. Mol. Biol.*, 167, 873.

Fuller, M.T. and J.A. King. 1980. *Biophys. J.*, 32, 381.

Geisselsoder, J., J.M. Sedivy, R.B. Walsh, and R. Goldstein. 1982. *J. Ultrastruct. Res.*, 79, 165.

Gibbs, A.J., H.L.Nixon, and R.D. Woods. 1963. *Virology*, 19, 441.

Ginsberg, H.S. 1979. In *Comprehensive Virology*. H. Fraenkel-Conrat and R.R. Wagner, eds. Vol. 13, p. 409. Plenum.

Golden, J.S. and S.C. Harrison. 1982. *Biochem.*, 21, 3862.

Harrison, S.C. 1980. *Biophys. J.*, 32, 139.

Harrison, S.C., A.J. Olson, C.E. Schutt, F.K. Winkler, and G. Bricogne. 1978. *Nature*, 276, 368.

Hayashi, M. 1978. In *The Single-Stranded DNA Phages*. D.T. Denhardt, D. Dressler, and D.S. Ray, eds. p. 531. Cold Spring Harbor Lab.

Heidner, E.G., T.G. Frey, J. Held, L.J. Weissman, R.E. Fenna, M. Lei, M. Harel, H. Kabsch, R.M. Sweet, and D. Eisenberg. 1978. *J. Mol. Biol.*, 122, 163.

Henriksson, D., R.J. Tanis, R.E. Tashian, and P.O. Nyman. 1981.

J. Mol. Biol., 152, 171.

Hermodson, M.A., C. Abad-Zapatero, S.S. Abdel-Meguid, S. Pundak, M.G. Rossmann, and J.H. Tremaine. 1982. *Virology*, 119, 133.

Heuss, K.L., M.R.N. Murthy, and P. Argos. 1981. *J. Mol. Biol.*, 153, 1161.

Hill, E., D. Tsernoglou, L. Webb, and L.J. Banaszak. 1972. *J. Mol. Biol.*, 72, 577.

Hodgkin, D.C. 1974. *Proc. Roy. Soc. Lond.*, B186, 191.

Hohn, T. 1969. *J. Mol. Biol.*, 43, 191.

Hopper, P., S.C. Harrison, and R.T. Sauer. 1984. *J. Mol. Biol.*, 177, 701

Houwing, C.J. and E.M.J. Jaspars. 1980. *Biochem.*, 19, 5255.

Houwing, C.J. and E.M.J. Jaspars. 1982. *Biochem.*, 21, 3408.

Hsu, C.H., O.P. Sehgal, and E.E. Pickett. 1976. *Virology*, 69, 587.

Hsu, C.H., J.A. White, and O.P. Sehgal. 1977. *Virology*, 81, 471.

Huber, R. 1979. *Trends Biochem. Sci.*, 4, 271.

Huber, R. and W.S. Bennett Jr. 1983. *Biopolymers*, 22, 261.

Huber, R., J. Deisenhofer, P.M. Colman, M. Matsushima, and W. Palm. 1976. *Nature*, 264, 415.

Hull, R. 1969. *Adv. Virus. Res.*, 15, 365.

Hull, R. 1978. *Virology*, 89, 418.

Hull, R., G.J. Hills, and R. Markham. 1969. *Virology*, 37, 416.

Hung, P.P. 1976. In *Comprehensive Virology*. H. Fraenkel-Conrat and R.R. Wagner, eds. Vol.6, p. 65. Plenum.

Huxley, H.E. and G. Zubay. 1960. *J. Mol. Biol.*,2, 189.

Incardona, N.L. 1978. In *The Single-Stranded DNA Phages*. D.T. Denhardt, D. Dressler, and D.S. Ray, eds. p. 549. Cold Spring Harbor Laboratory.

IUPAC-IUB Commission on Biological Nomenclature. 1970. *J. Biol. Chem.*, 245, 6489.

Jacobson, M.F. and D. Baltimore. 1968. *J. Mol. Biol.*, 33, 369.

Johnson, J.E. and C. Hollingshead. 1981. *J. Ultrastruct. Res.*, 74, 223.

Jonard, G., J. Witz, and L. Hirth. 1972. *J. Mol. Biol.*, 67, 165.

Kääriäinen, L. and H. Söderlund. 1978. *Curr. Topics Microbiol. Immunol.*, 82, 15.

Kan, J.H., P.J. Andree, L.C. Kouijzer, and J.E. Mellema. 1982. *Eur. J. Biochem.*, 126, 29.

Kaper, J.M. 1975. *The Chemical Basis of Virus Structure, Dissociation, and Reassembly.* North-Holland.

Kaper, J.M. and J.L.M.C. Geelen. 1971. *J. Mol. Biol.*, 56, 277.

Kasvinsky, P.J., N.B. Madsen, J. Sygusch, and R.J. Fletterick. 1978. *J. Biol. Chem.*, 253, 3343.

Keeling, J. and R.E.F. Matthews. 1982. *Virology*, 119, 214.

Kiselev, N.A. and A. Klug. 1969. *J. Mol. Biol.*, 40, 155.

Klug, A. 1965. *J. Mol. Biol.*, 11, 424.

Klug, A. and J.T. Finch. 1965. *J. Mol. Biol.*, 11, 403.

Knolle, P. and T. Hohn. 1975. In *RNA Phages*. N.D. Zinder, ed. p. 147. Cold Spring Harbor Lab.

Krüse, J., P.A. Timmins, and J. Witz. 1982. *Virology*, 119, 42.

Leberman, R. and J.T. Finch. 1970. *J. Mol. Biol.*, 50, 209.

Lesk, A.M. and C. Chothia. 1982. *J. Mol. Biol.*, 160, 325.

Liljas, L. T. Unge, T.A. Jones, K. Fridborg, S. Lövgren, U. Skoglund, and B. Strandberg. 1982. *J. Mol. Biol.*, 159, 93.

Matthews, B.W. and M.G. Rossmann. 1985. *Methods Enzymol.*, in press.

Matthews, B.W., R.E. Fenna, M.C. Bolognesi, M.F. Schmid, and J.M. Olson. 1979. *J. Mol. Biol.*, 131, 259.

Matthews, K.S. and R.D. Cole. 1972. *J. Mol. Biol.*, 65, 1.

Matthews, R.E.F. 1979. *Intervirol.*, 12, 129.

Mavridis, I.M., M.H. Hatada, A. Tulinksy, and L. Lebioda. 1982. *J. Mol. Biol.*, 162, 419.

McCain, D.C., R. Virudachalam, J.L. Markley, S.S. Abdel-Meguid, and M.G. Rossmann. 1982a. *Virology*, 117, 501.

McCain, D.C. R. Virudachalam, R.E. Santini, S.S. Abdel-Meguid, and J.L. Markley. 1982b. *Biochem.*, 21, 5390.

Mellema, J.E. and H.J.N. van den Berg. 1974. *J. Supramolec. Struct.*, 2, 17.

Metcalf, P. 1982. *J. Ultrastruct. Res.*, 78, 292.

Monaco, H.L., J.L. Crawford, and W.N. Lipscomb. 1978. *Proc. Nat. Acad. Sc. USA*, 75, 5276.

Munowitz, M.G., C.M. Dobson, R.G. Griffin, and S.C. Harrison. 1980. *J. Mol. Biol.*, 141, 327.

Nixon, H.L. and A.J. Gibbs. 1960. *J. Mol. Biol.*, 2, 197.

Odumosu, A.O., R.B. Homer, and R. Hull. 1981. *J. Gen. Virol.*, 53, 193.

Oostergetel, G.T., P.C.J. Krijgsman, J.E. Mellema, S. Cusack, and A. Miller. 1981. *Virology*, 109, 206.

Pfeiffer, P. and L. Hirth, 1974, *Virology*, 61, 160.

Phillips, D.C., P.S. Rivers, M.J.E. Sternberg, J.M. Thornton, and I.A. Wilson. 1977. *Biochem. Soc. Trans.*, 5, 642.

Putnak, J.R. and B.A. Phillips. 1981. *Microbiol. Rev.*, 45, 287.

Rayment, I., P. Argos, and J.E. Johnson. 1977. *J. Ultrastruct. Res.*, 61, 240.

Rayment, I., J.E. Johnson, and M.G. Rossmann. 1979. *J. Biol. Chem.*, 254, 5243.

Rayment, I., T.S. Baker, D.L.D. Caspar, and W.T. Murakami. 1982. *Nature*, 295, 110.

Robinson, I.K. and S.C. Harrison. 1982. *Nature*, 297, 563.

Roizman, B. and D. Furlong. 1974. In *Comprehensive Virology*. H. Fraenkel-Conrat and R.R. Wagner, eds. Vol. 3, p. 229. Plenum.

Rossmann, M.G. 1984. *Virology*, 4, 134.

Rossmann, M.G. and P. Argos. 1981. *Ann. Rev. Biochem.*, 50, 497.

Rossmann, M.G., M.J. Adams, M. Buehner, G.C. Ford, M.L. Hackert,

P.J. Lentz Jr., A. McPherson Jr., R.W. Schevitz, and I.E. Smiley. 1972. *Cold Spring Harb. Symp. Quant. Biol.*, 36, 179.

Rossmann, M.G., D. Moras, and K.W. Olsen. 1974. *Nature*, 250, 194.

Rossmann, M.G., C. Abad-Zapatero, M.A. Hermodson, and J.W. Erickson. 1983a. *J. Mol. Biol.*, 166, 37.

Rossmann, M.G., C. Abad-Zapatero, M.R.N. Murthy, L. Liljas, T.A. Jones, and B. Strandberg. 1983b. *J. Mol. Biol.*, 165, 711.

Rossmann, M.G., R. Chandrasekaran, C. Abad-Zapatero, J.W. Erickson, and S. Arnott. 1983c. *J. Mol. Biol.*, 166, 73.

Rueckert, R. 1976. In *Comprehensive Virology*. H. Fraenkel-Conrat and R.R. Wagner, eds. Vol. 6, p. 131. Plenum.

Rutgers, T.T. Salerno-Rife, and P. Kaesberg. 1980. *Virology*, 104, 506.

Samuelson, G. and P. Kaesberg. 1970. *J. Mol. Biol.*, 47, 87.

Savithri, H.S. and J.W. Erickson. 1983. *Virology*, 126, 328.

Sehgal, O.P. and C.H. Hsu. 1977. *Virology*, 77, 1.

Sehgal, O.P., C.H. Hsu, J.A. White, and M. Van. 1979. *Phytopathol. Z.*, 95, 167.

Söderlund, H., L. Kääriäinen, and C.H. von Bonsdorff. 1975. *Med. Biol.*, 53, 412.

Soeda, E., J.R. Arrand, N. Smolar, J.E. Walsh, and B.E. Griffin. 1980. *Nature*, 283, 445.

Steitz, J.A. 1968a. *J. Mol. Biol.*, 33, 923.

Steitz, J.A. 1968b. *J. Mol. Biol.*, 33, 947.

Steitz, T.A., W.F. Anderson, R.J. Fletterick, and C.M. Anderson. 1977. *J. Biol Chem.*, 252, 4494.

Steven, A.C., P. Serwer, M.E. Bisher, and B.L. Trus. 1983. *Virology*, 124, 109.

Stubbs, G., S. Warren, and K. Holmes. 1977. *Nature*, 267, 216.

Sugiyama, T., R.R. Hebert, and K.A. Hartman. 1967. *J. Mol. Biol.*, 25, 455.

Takamatsu, H. and K. Iso. 1982. *Nature*, 298, 819.

Tremaine, J.H. and W.P. Ronald. 1977. *Canad. J. Botany*, 55, 2274.

Tremaine, J.H. and W.P. Ronald. 1978. *Virology*, 91, 164.

Tremaine, J.H., W.P. Ronald, and E.M. Kelly. 1980. *Can. J. Microbiol.*, 26, 1450.

Van Beynum, G.M.A., J.M. De Graaf, A. Castel, B. Kraal, and L. Bosch. 1977. *Eur. J. Biochem.*, 72, 63.

van Schaick, E.J.M., W.G. Schutter, W.P.J. Gaykema, A.M.H. Schepman, and W.G.J. Hol. 1982. *J. Mol. Biol.*, 158, 457.

van Vloten-Doting, L., R.I.B. Francki, R.W. Fulton, J.M. Kaper, and L.C. Lane. 1981. *Intervirol.*, 15, 198.

Varghese, J.N., W.G. Laver, and P.M. Colman. 1983. *Nature*, 303, 35.

Verbraeken, E. and W. Fiers. 1972. *Virology*, 50, 690.

Virudachalam, R., K. Sitaraman, K.L. Heuss, J.L. Markley, and P. Argos. 1983a. *Virology*, 130, 351.

Virudachalam, R., K. Sitaraman, K.L. Heuss, P. Argos, and J.L. Markley. 1983b. *Virology*, 130, 360.

Vriend, G., B.J.M. Verduin, M.A. Hemminga, and T.J. Schaafsma. 1982. *FEBS Lett.*, 145, 49.

Weber, K.A. and O.P. Sehgal. 1982. *Phytopathol.*, 72, 909.

Wengler, G., U. Boege, G. Wengler, H. Bischoff, and K. Wahn. 1982. *Virology*, 118, 401.

Williams, R.C. and K.E. Richards. 1974. *J. Mol. Biol.*, 88, 547.

Wilson, I.A., J.J. Skehel, and D.C. Wiley. 1981. *Nature*, 289, 366.

Ysebaert, M., J. van Emmelo, and W. Fiers. 1980. *J. Mol. Biol.*, 143, 273.

Zimmern, D. 1977. *Cell*, 11, 463.

Zuidema, D., M.F.A. Bierhuizen, and E.M.J. Jaspars. 1983. *Virology*, 129, 255.

3
Nucleic Acid Packaging by Viruses

Sherwood Casjens
University of Utah Medical Center

I. INTRODUCTION

Early studies on the nature of viruses showed that they are composed mainly of protein and nucleic acid. As structural details of various virus particles were delineated, there was simultaneous interest in the mechanisms by which these particles are assembled. Such interest stemmed from both the inherent interest of the subject and the potential of viruses as models for the nature of nucleic acid-protein interaction and the assembly of supramolecular structures. Fraenkel-Conrat and Williams (1955) first showed that the separated protein and nucleic acid components of a simple virus could be caused to reassemble into true virions spontaneously. More recent experiments with the double-stranded DNA (dsDNA) bacteriophages have suggested that in these cases nucleic acid packaging does not proceed spontaneously, but requires input of energy (reviewed by Earnshaw and Casjens, 1980). As nucleic acid functions were elaborated in the 1960s a paradox emerged: how could virus particles be assembled within infected cells, and yet be able to disassemble into protein and nucleic acid components to expose the nucleic acid and initiate another round of infection? This problem may have been solved by different groups of viruses in different ways. For example viruses may spontaneously package nucleic acid in infected cells and then require that changes in the structure or intracellular environment be made to allow uncoating, or they may supply energy during the packaging process to place the packaged nucleic acid in a higher metastable energy state, so that the particle will spontaneously "uncoat" in response to specific signals not present during assembly.

In addition to these energetic possibilities, different general biochemical pathways of packaging are also possible. Viruses whose structures are understood contain nucleic acid within structures whose protein subunits are arranged with either helical or icosahedral symmetry (see Chapters 1 and 2 this volume). Each of these types could, in theory, package nucleic acid by any of three overall mechanisms: (I) co-condensation of nucleic acid and protein, (II) condensation of the nucleic acid followed by addition of the surrounding protein subunits or (III) assembly of the protein subunits into a preformed container followed by nucleic acid entry into the structure. Where it is understood, the assembly of helical viruses occurs by the first mechanism (Durham et al., 1971). Some icosahedral viruses appear to use the first mechanism, while others use the third, and possibly the second, mechanism (Luftig et al., 1971; Knolle and Hohn, 1975; Kaiser et al., 1975). Perhaps these differences should not be surprising in view of the fact that various viruses have very different lifestyles. Furthermore, viruses are known with different types, forms, and numbers of nucleic acids: for example, single-stranded or double-stranded; DNA or RNA; one or several chromosomes; circular or linear chromosomes (with or without terminal proteins); and either haploid or diploid, with at least two basic protein arrangements.

A problem that *all* viruses must solve is how to package only viral nucleic acid. The solution to this is possibly universal: all virus "packaging apparatuses" appear to recognize the nucleic acid to be packaged by virtue of a specific base sequence in that nucleic acid.

Such packaging-recognition signals will be referred to here as *pac* sites. Another problem which all viruses face is how to avoid attempts to package nucleic acid molecules that are in the process of serving as templates for DNA, RNA, or protein synthesis. Little is known about this aspect of the control of packaging, but it may contribute to the complexity of the packaging apparatus in some viruses.

In the following discussion, I will examine the similarities and differences between viruses in terms of what is known about mechanisms of nucleic acid packaging and structure of the intravirion nucleic acid. Since this topic has not been the primary focus of a previous review, I will for completeness occasionally mention systems that are under study but have not yet contributed greatly to our overall knowledge of nucleic acid encapsidation. In the interest of conciseness, historical significance, direct applicability to the point being developed, and of giving access to literature only review articles and/or recent or original references are cited in the text.

II. NUCLEIC ACID PACKAGING BY HELICAL VIRUSES

The helical viruses are of two types: rigid and flexible. These are discussed in the following two subsections.

IIA. Rigid Helical Viruses

IIA.1. Tobamoviruses

Tobacco mosaic virus (TMV), vulgare strain, is the prototypic member of this group. Its assembly is a "simple" nucleic acid packaging reaction in which one ssRNA molecule and one species of protein co-assemble spontaneously *in vitro* to form virus particles apparently identical to those formed *in vivo* (Fraenkel-Conrat and Williams, 1955). The TMV virion has been analyzed by x-ray diffraction to 4 A resolution (Stubbs et al., 1977). The coat protein subunits form a one-start, right-handed helix with 16-1/3 subunits per turn, and the RNA lies in a groove between successive helical turns of protein subunits (Figure 1). Each protein subunit binds three nucleotides. The RNA binding site in this structure has been shown by Stubbs and Stauffacher (1981) to consist of a hydrophobic region in which methyl and methylene groups of the LR alpha helix of one subunit lie flat against the bases, and three arginine side chains of the adjacent subunit probably form salt bridges with the phosphate groups. In addition, the ribose 2'-OH on the two outside nucleotides is probably hydrogen-bonded to a protein side chain. Sufficient resolution to see more detailed contacts was not obtained. Stubbs and Stauffacher (1981) also concluded that the sugar rings are puckered in the 3'-endo configuration (that of A-form RNA), and the two most-5' bases bound to each coat subunit are in the *anti* configuration, while the third base is in the unusual *syn* configuration, at least when purines occupy this position. Nuclear magnetic resonance studies have confirmed that there are three distinct phosphate environments in the virion (Cross et al., 1983). TMV RNA has been sequenced and contains 6395 bases (Goelet et al., 1982); its 5' end is capped (Zimmern, 1975). It seems likely that the coat protein causes the third

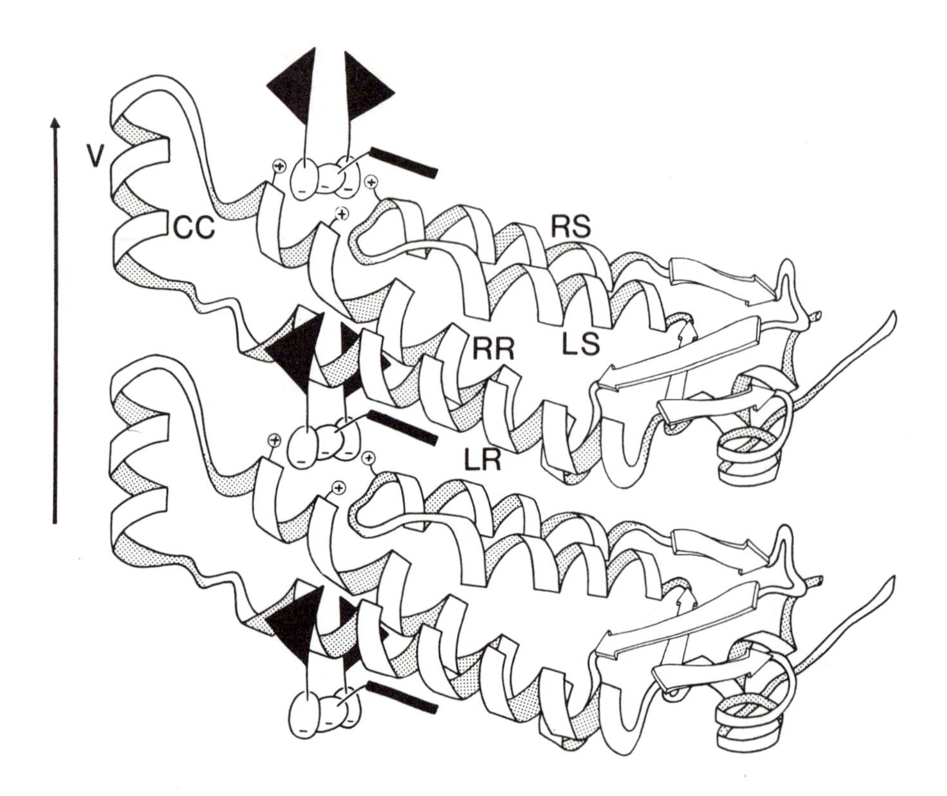

FIGURE 1 Structure of tobacco mosaic virus. Two coat–protein subunits in two successive turns of the virion helix are shown diagrammatically, viewed from the side (according to Caspar and Stubbs, 1983; Hirth and Richards, 1981). The polypeptide folding is depicted as in Richardson (1981), where the flat arrows indicate beta structure and the coiled ribbons indicate alpha helix (the figure is not meant to be accurate in detail, but show the general features of the protein). The alpha helix names (RS, LS, RR, LR, and V) and the carboxyl cage (CC) thought to control assembly mode are indicated on the upper subunit. The left third of the RR helix and the V helix are disordered and thus thought to be more flexible in the disk structure, perhaps allowing RNA to enter the internal binding site. Three successive turns of RNA are diagrammed as they pass through the RNA binding sites of the two protein subunits (according to Stauffacher and Stubbs, 1981). The sugar-phosphate portions are indicated with ovals and the bases by triangles. The 5' end of the RNA is closest to the viewer, and the most-3' base (behind the LR helix) is shown in the *syn* conformation. The center of the virion corresponds approximately to the position of the vertical arrow on the left, which indicates the direction of the long (screw) axis of the virion.

base to adopt the *syn* form, since repeating features were not found in the sequence (Goelet et al., 1982).

In solution the coat-protein monomer is in equilibrium with small oligomers of coat protein and larger structures, including discrete "34-mers" called disks (consisting of two stacked, flat 17-subunit rings) at slightly basic pH, and in helical rods of variable length at slightly acid pH (Durham and Klug, 1971; Durham, 1972; Bloomer et al., 1978). In these helical rods the coat protein subunits occupy positions nearly identical to those in the virus particle, showing that the information necessary for construction of a rigid rod the diameter (but not the length) of TMV lies in the primary structure of the coat protein (Franklin, 1955; Mandelkow et al., 1981). Protonation of anomalously titrating carboxyl groups, with pK values near 7, controls the mode (disk or helix) of polymerization (Caspar, 1963; Butler et al., 1972; see Figure 1).

The mechanism by which the RNA chromosome is packaged is surprisingly complex. It is clear that the coat protein molecules do not add at random positions on the RNA, but assemble directionally, with the individual nucleotides assuming virion orientation as coat protein molecules add to the growing structure. The RNA and coat protein can thus be said to co-assemble. Assembly initiates at a specific *pac* site or "assembly origin" about 1000 nucleotides from the 3' end (Zimmern and Butler, 1977), with subsequent bidirectional coat protein addition from this site (Lebeurier et al., 1977; Otsuki et al., 1977). The rate of assembly is "nucleation controlled" (Schuster et al., 1979). Controlling nucleation is an efficient way of controlling any assembly process (Crane, 1950; Caspar, 1980; King 1980; Chapter 4, this volume). TMV and other tobamoviruses produce several subgenomic mRNA's in infected cells, and in each case only the ones that contain the *pac* site are packaged (Beachy and Zaitlin, 1977; Meshi et al., 1981; Fukuda et al., 1981). This supports the notion that recognition occurs in this manner *in vivo*.

As Butler and Lomonossoff (1980) have pointed out, recognition of a trinucleotide by a coat protein monomer is insufficient to give the observed specificity of packaging. Thus, if coat protein must perform the recognition by itself, recognition of a larger sequence by an oligomer of coat protein is necessary. This can explain in part why TMV assembly initiates by the complex mechanism discussed below. Zimmern (1977) noticed that the initiation region possesses substantial homology with a region of the coat-protein gene, and Kaplan et al. (1982) isolated a mutant that appears to utilize the homologous site as well as the normal *pac* site. Analysis of the structure and behavior of this mutant should help in understanding what is required of a recognition signal.

There is general agreement among workers in the field that a large coat-protein oligomer, perhaps the disk or a short protein helix, is the first protein species to interact with the RNA during encapsidation (Butler and Klug, 1971; Richards and Williams, 1976; Butler and Durham, 1977; Casjens and King, 1975; Raghavendra et al., 1985). The manner by which this initial interaction takes place is obscure. Coat-protein monomers or disks appear to bind various RNAs only very weakly, while long helical protein rods bind trinucleoside diphosphates quite well (Erickson and Bancroft, 1982; Steckert and Schuster, 1982). This binding is quite sequence-dependent, and while simple rules for binding strength are not obvious, trimers containing no G could not be shown to bind, and trimers of the form XAG, XGA, GAX, and XUG bind relatively well. Ledneva et al. (1980) found that U and G residues inserted in poly(A) caused a more stable structure after packaging than poly(A) alone. The sequence in the region of the *pac* site has a potential stem-loop that contains two AAG's in the loop (Zimmern, 1977; Jonard et al., 1977). This is the most tightly bound trinucleotide in the experiments of Steckert and Schuster (1982). It is not known whether *pac*-site recognition during initiation of assembly occurs because of a concentration of favorable trimers, or via a specific recognition by oligomers that could have a different sequence selectivity than subunits in the large helical assemblies. From their work with such helical protein assemblies Steckert and Schuster (1982) suggested that this secondary structure and pair of AAG's may define a "binding frame" for the protein on the RNA. Such a frame has not yet been demonstrated experimentally. It has also been shown that during assembly the RNA molecule does not thread completely through the central hole, but both 3' and 5' ends extend from one end of the growing structure (Lebeurier et al., 1977; Otsuki et al., 1977; Butler et al., 1977). Zimmern (1977) has speculated that the stem-loop might protrude into the central hole of the disk during the initial protein-RNA contact. More recent results have suggested that a short protein helix might be the nucleating protein species (Correia et al., 1981; K. Raghavendra et al., 1985). Subsequent assembly from either initiator would require separation of the base pairs of the stem. The actual mechanism of initial RNA binding and possible switching from the disk to a helical or "lockwasher" structure is unknown (Butler, 1971; Caspar, 1980). However, Butler et al. (1977) pointed out that in the disk, the RNA-binding region of the coat protein (region at small radius) is more open and flexible than in the virion. Thus, RNA could reach into the central hole and contact the RNA-binding sites without disassembly of the subunits of the disk. Such initial RNA contact could cause the switch to a helical structure or stabilization of pre-existing short helical structures. The short RNA-protein helix formed during initiation presumably provides a nucleation point for more rapid cooperative addition of coat-protein molecules to the nucleated structure, thereby giving a potential physical basis to the nucleation-controlled kinetics of assembly (Schuster et al., 1980). King (1980) and Caspar (1980) have called this type of addition of subunits to a nucleated structure "self-regulated" and "autosteric," respectively, and have discussed its ramifications.

The extent of the "nucleation" reaction and the state of the coat protein added during the elongation reaction are not agreed upon by workers in this field. The addition of coat protein towards the 5' end of the RNA from the nucleated structure occurs rapidly, with completion being reached in 5-10 min (Lomonossoff and Butler, 1979; Fukuda et al., 1978). Butler and colleagues (Butler and Lomonossoff, 1978; 1980) have argued that 5' elongation occurs by the addition of more disks, while Richards and Williams (1972) and Ohno et al. (1977) have argued that smaller oligomers are more active in this reaction. Shire et al. (1981) and Schuster et al. (1980) have concluded that elongation may occur from either disks or smaller oligomers (at somewhat different rates). These uncertainties, which arise from the fact that disks, oligomers, and monomers are in equilibrium under assembly conditions, have yet to be fully resolved. Addition of coat protein in the 3' direction is also not yet understood, though recent experiments by Dr. Y. Okada and coworkers (pers. commun.) have shown that the rate of subunit addition in the 3' direction is at least 5 to 10 times slower than that in the 5' direction. Fukuda et al. (1981), Fukuda and Okada (1982), and Meshi et al. (1983) have studied the assembly of watermelon strain cucumber green mottle mosaic virus (a tobamovirus whose coat protein amino acid sequence is 36% homologous to TMV) and found that the equilibrium between disks and oligomers of coat protein lies very far to the side of disks under nucleation conditions, which makes the analysis of elongation-active species somewhat less ambiguous. They concluded that in this case the disk is the active coat-protein species in elongation in the 3' direction, but smaller oligomers or monomers are active in elongation in the 5' direction.

Elongation terminates when the end of the RNA is reached. Under assembly conditions the RNA contact is required for continued protein assembly in the helical mode (see Chapter 5, this volume). There is no evidence that there is any special protein structure at either end of the virion, however Collmer et al. (1983) and Collmer and Zaitlin (1983) have recently found that there is about one molecule of coat protein per virion which may have a covalent peptide adduct.

IIA.2. Other rigid helical viruses

The assembly of one other group of rigid helical viruses has been studied to a limited extent. Tobacco rattle virus (TRV), the prototypic tobravirus, has a protein structural organization similar to TMV; however, there are 25-1/3 coat-protein subunits per turn, and apparently four bases interact with each subunit (reviewed by Hirth, 1976; Harrison and Robinson, 1978). It is a multicomponent virus or covirus: two ssRNA chromosomes are packaged into separate particles, and both are needed for successful infection). TRV can be assembled *in vitro*, and a disk-like structure seems to be the most active coat protein form in the assembly reaction (Abou Haidar et al., 1973; Morris and Semancik, 1973). Mayo and De Marcillac (1977) suggested that these disks contain 3 or 4 layers of subunits. Abou Haidar and Hirth (1977) found that assembly initiates very near the 5' end of a small packaged RNA. Bisaro and Siegel (1982) found that *in vivo* only subgenomic RNA's with intact 5' ends are packaged. Thus, TRV assembly appears to be analogous in an overall sense to that of TMV, with variation in the location of the *pac* site and possibly disk structure.

IIB. Flexible Helical Viruses

IIB.1. Flexible plant viruses

A large number of the ssRNA plant viruses have coat proteins that assemble into flexible helices. They consist of naked flexible rods that vary in length from about 480 to 1250 nm and have diameters between 11 and 12 nm. In part, on the basis of particle length they have been divided into four major classes, potexviruses, potyviruses, carlaviruses and closteroviruses. All appear to contain one type of coat protein of molecular weight 20-30,000 which is arranged helically with 8-11 subunits per turn and a pitch of 3.3 to 3.8 nm (Hirth, 1976; Goodman et al., 1976; Bar-Joseph and Hull, 1974; Varma et al., 1968; Richardson et al., 1981). Calculations have in general indicated 4 or 5 bases per coat-protein subunit in the virion (Tollin et al., 1979). All contain a message (+) strand chromosome (Matthews, 1981); one member of the potexviruses is capped at the 5' end (Abou Haider and Bancroft, 1978a), while one potyvirus has a 6000-dalton protein covalently linked to its 5' terminus and is 3'-polyadenylated (Hari, 1981). These differences could be significant for RNA encapsidation, but only papaya mosaic virus (PMV, a potexvirus) has been studied in this respect.

Novikov et al. (1972) demonstrated that PMV could be reconstituted, and Bancroft and AbouHaidar and coworkers have studied its assembly in some detail (AbouHaidar and Erickson, 1983). *In vitro* PMV protein will assemble into normal-appearing virions only with PMV or closely related RNAs, indicating specificity of assembly (Erickson et al., 1978; Abouhaidar and Bancroft, 1979). However, under the same conditions PMV coat protein appears to bind nonspecifically to other RNAs and even to DNA, to form thin, kinked structures (Erickson and Bancroft, 1981). Under other conditions it appears to encapsidate other RNAs properly (Abouhaidar and Bancroft, 1980). The specific encapsidation reaction initiates within 200 bp of the 5' end of the RNA (Abouhaidar and Bancroft, 1978b, 1980). Interestingly, the nucleotide pentamer N(C/A)AAA is present in eight adjacent repeats at the 5' end of the RNA (M. Abouhaidar and S. Lok, pers. commun.). Like TMV, PMV coat protein has several groups that are protonated at anomalous pH values near neutrality (Durham and Bancroft, 1979). The coat protein of a close PMV relative, clover yellow mosaic virus, like that of TMV, assembles into helical structures at acid pH and stacked-disk structures at basic pH (Bancroft et al., 1979). PMV coat protein monomer is in equilibrium with a large, possibly double-layer, disk structure (Erickson and Bancroft, 1978; Abouhaidar and Erickson, 1983), but it is not known if this oligomer is required for assembly. Bancroft et al. (1979) argued that monomers or small oligomers are active in elongation. Thus, insofar as they have been studied, the flexible plant viruses have a number of similarities with TMV and may utilize a similar overall packaging strategy.

IIB.2. Helical nucleocapsids

Many animal and some plant viruses package their nucleic acid within protein structures that are subsequently surrounded by a lipid "envelope" (see Chapter 7, this volume). In such viruses the internal structure, within which the nucleic acid is packaged, is referred to as the *nucleocapsid*. In this section I will examine the structure and assembly of helical nucleocapsids but will not consider in detail the assembly of the surrounding envelope. Viruses in the orthomyxo-, paramyxo-, rhabdo-, arena- and bunyavirus groups have apparently helical nucleocapsids, and anti-message sense ssRNA chromosomes with 5'-triphosphates. The coronaviruses contain a message sense ssRNA that is capped and polyadenylated. In general, the nucleocapsid structures and modes of assembly are not understood in detail for these viruses, though interesting features of packaging are known. In all cases, there is a single protein, usually called N protein, that coats the ssRNA. The reader is advised to consult the following reviews for details of the life cycles of these viruses: orthomyxoviruses, Compans and Choppin (1975); paramyxoviruses, Choppin and Compans (1975); rhabdoviruses, Wagner (1975); coronaviruses, Robb and Bond (1979); arenaviruses, Rawls and Leung (1979); bunyaviruses, Bishop and Shope (1979).

The only detailed structural study on these nucleocapsids has been that of Finch and Gibbs (1970). Their optical diffraction analysis of electron micrographs suggested that the paramyxovirus N protein coats RNA in a helical fashion, with 11—13 subunits per turn. The helix is probably left-handed (Nonomuna and Kohama, 1974). These structural studies and estimates of the RNA size allow the calculation of 6-9 bases per N-protein subunit in the paramyxovirus nucleocapsid structure. Similar calculations for rhabdoviruses (Cartwright et al., 1972) and bunyaviruses (Obijeski et al., 1976) give about 6 bases per subunit. The orientation of the RNA in the nucleocapsids is also unknown, however, bound rhabdovirus N protein does not appear to interfere with G-methylation by dimethyl sulfate (Keene et al., 1981). This observation implies that the protein may not form specific contacts with the bases, and these authors suggested that N protein may bind to the phosphate backbone. In several cases, salt conditions have dramatic effects on the nucleocapsid structure. For example nucleocapsids of an orthomyxovirus are loosened greatly by high NaCl (Heggeness et al., 1982), while those of a paramyxovirus are loosened by low salt (Heggeness et al., 1981), and those of a rhabdovirus are loosened by chelating agents (Naeve et al., 1980). Such observations suggest that the N proteins may have several conformational states or RNA-binding modes. This point may be related to the observation that in all cases studied, N protein appears to coat RNA *in vivo*, even while the RNA is functioning as a template for RNA synthesis. The amino acid sequence of the N protein for an orthomyxovirus (Winter and Fields, 1981), a rhabdovirus (Gallione et al., 1981), a bunyavirus (Bishop et al., 1982) and a coronavirus (Armstrong et al., 1983) are known. All are enriched for basic amino acids, but no sequence homology has been reported.

Compans et al. (1972) calculated that there are about 20 nucleotides per N-protein subunit in the orthomyxovirus nucleocapsid, and related calculations suggest that the amount of RNA per unit length is about 5 to 20 times that of other helical virus structures. In addition, these nucleocapsids are ribonuclease sensitive (Pons et al., 1969). These observations have led to the idea that these structures may be basically different from the other helical nucleocapsids, perhaps with the RNA wrapped around the outside of the structure (Jennings et al., 1983).

In the virion, most nucleocapsids are thought to have additional structure; for example, rhabdovirus nucleocapsids are coiled regularly in the virion (Cartwright et al., 1972; Newcomb et al., 1982), orthomyxovirus nucleocapsids are supercoiled in a right-handed double helix (Heggeness et al., 1982), and bunyavirus and arenavirus nucleocapsids are circular and possibly supercoiled (Pettersson and von Bonsdorff, 1975; Palmer et al., 1977; Hewlett et al., 1977; Patterson et al., 1983b). Paramyxovirus nucleocapsids may under some circumstances aggregate end-to-end (Hosaka et al., 1966); however, Lynch and Kolakofsky (1978) have shown that neither end of the RNA protrudes from nucleocapsids.

Recently *in vitro* assembly of the rhabdovirus vesicular stomatitis virus (VSV) nucleocapsids in crude extracts has shown probable RNA-specific nucleocapsid assembly (Hill et al., 1981; Ghosh and Ghosh, 1982; Patton et al., 1983). In addition to the N-protein helix, several other proteins, thought to be part of the RNA-dependent RNA polymerase, are present in small amounts (1 to 10% of the number of N-protein molecules). *In vivo* studies have not been able to distinguish clearly between models in which N protein assembles on the RNA first, followed by addition of the RNA polymerase subunits, and models in which the different proteins co-assemble on the RNA (Hsu et al., 1979; Rubio et al., 1980). It is clear, however, that these minor proteins are not required for integrity of the N-protein RNA structure, and that *in vitro* they can add after N protein (Mellon and Emerson, 1978). In general the locations of the minor proteins are not known, though Keene et al. (1981) have shown that one of the VSV minor proteins protects a region 16 to 30 nucleotides from the 3' end from G-methylation by dimethyl sulfate.

Many studies with viruses containing chromosomal deletions have shown that *in vivo* the length of the RNA controls the length of the nucleocapsid, as with TMV coat protein (for example, see Kiley and Wagner, 1972). The study of deletion viruses and the structure of various subgenomic RNAs made in infected cells can help locate the *pac* sequence. Large deletions have been isolated for rhabdoviruses (Lazzarini et al., 1981), orthomyxoviruses (Nayak, 1980) and paramyxoviruses (Amesse et al., 1982). Since these deletions are packaged into nucleocapsids, the regions deleted must not contain the putative *pac* sites. There is a stipulation in that the manner by which such deletions are selected may also select for various point mutations in the genome (Hagen and Huang, 1981). Similarly, subgenomic RNAs are usually not packaged, and hence must not contain *pac* sites. Again there is a stipulation: since most subgenomic RNAs are capped and polyadenylated, these or other modifications could block packaging. For several of these virus groups such analyses allow the tentative deduction that the putative packaging recognition site must be near one of the ends of the ssRNA molecule, probably the 5' end (Blumberg and Kolakofsky, 1981; Blumberg et al., 1981; Rao and Huang, 1982; Leibowitz et al., 1981; Amesse et al., 1982; Jennings et al., 1983). Furthermore, Hill et al. (1979) have shown that the general direction of VSV-nucleocapsid assembly *in vivo* is 5' to 3'. Recently, Blumberg et al. (1983) have obtained some specificity of packaging of RNAs *in vitro* by purified VSV N protein, suggesting that the specificity of nucleocapsid assembly lies in the N protein. Their experiments also show that (1) the N protein can assemble into nucleocapsid-like structures without RNA (it is not yet known if these are helical or stacked-disk structures), (2) nucleocapsid assembly is apparently nucleation-controlled, with elongation being more rapid than initiation, (3) a cap structure on the 5' end of the RNA does not block packaging, and (4) initiation occurs within the 14 5'-terminal nucleotides of the RNA. However, bunyaviruses appear not to have a recognition signal this close to the 5' end (Patterson et al., 1983a). As Blumberg et al. (1983) point out, it is tantalizing to notice that several of these features are similar to what has been seen with TMV; however, such analogies should be treated with caution until more detailed structural and biochemical studies have been completed.

Even if N protein has the ability to recognize and assemble onto virus specific RNA, at least two packaging problems remain:

1. *How do these viruses select a specific RNA strand for envelopment to form virions, even though both strands are packaged by N protein in the infected cell?* For example, VSV virions contain virtually no message-strand chromosomes, while at least 20% of the full-length viral RNAs in the infected cell are messenger sense strands, which are found in nucleocapsids (Simonsen et al., 1979). In addition, two short "leader" RNAs are made that correspond in sequence to the 50 5'-terminal nucleotides of both strands, which are coated with N protein, but are not found in virions (Blumberg and Kolakofsky, 1981). To date, no structural differences have been found between the nucleocapsids containing the different strands (Naeve et al., 1980), and no model to explain the selectivity of the final budding process has been suggested.

2. *Do the viruses with more than one chromosome (bunya-, orthomyxo-, and arenaviruses) partition one copy of each chromosome in each virion, and, if so, how is this accomplished?* Four models for packaging a segmented genome seem reasonable: (i) A parental genome acts as a template for the assembly of progeny virions, perhaps through replication of joined segments (Pons, 1970). (ii) Specific virion protein-nucleic acid interactions dictate specific assortment during assembly. (iii) Specific nucleic acid-nucleic acid interactions dictate specific assortment during assembly. (iv) Incorporation of segments is random, with only a fraction of the virions receiving all of the chromosomes. No support for models i, ii or iii is available, since no aggregates of nucleocapsids have been reported from virions or infected cells. In the orthomyxoviruses all virion proteins appear to have other roles, so if model ii were correct, the proteins active in assortment would have to be multi-functional. If model iii were correct, one might expect the

RNAs to pair in specific ways. In the case of an arenavirus, the first 19 nucleotides at the 3' end of the two chromosomes are identical (Auperin et al., 1982), and for a bunyavirus the first 11 nucleotides at the 3' and 5' ends are the same for the three chromosomes (Obijeski et al., 1980); thus, specific end-to-end base pairing is not possible. In fact, in no case has it been directly shown that there is precise chromosome assortment. Hirst (1973) has suggested that orthomyxoviruses may often be missing chromosomes, and arenaviruses and bunyaviruses do not always package equal amounts of their different chromosomes (Romanowski and Bishop, 1983; Pettersson and Kääriäinen, 1973). Thus, none of the models can be ruled out at present, and the problem remains unanswered.

IIB.3. Filamentous bacteriophages

The filamentous phage are quite different from the helical ssRNA-containing viruses. In particular, they contain a circular ssDNA chromosome, but the particles are not circular, so there must be two antiparallel DNA strands within the helical virion. The complete base sequence of several of these phages is known, and the opposing strands cannot simply be base-paired to form a double helical molecule (Hill and Petersen, 1982). The morphogenesis of these phages is covered in detail in Chapter 7 of this volume. The virions are about 6 nm wide and 800 to 2000 nm long, and the particles are built with helical arrays of a single coat protein (Day and Wiseman, 1978). There appear to be at least two types of coat-protein arrangements. Type I can be described as a five-start helix in which all helices start at the same level. Type II can be described as a six-start helix, in which the six helices start at different levels (Marvin et al., 1974; Makowski and Caspar, 1981; Nave et al., 1981; Peterson et al., 1982). It is important to note that in both virion types the coat proteins have similar three-dimensional shapes and generally similar interactions with neighboring subunits (Marvin, 1978). However, the protein-DNA contacts may not be the same in all filamentous phages. For example, some appear to contain two nucleotides per coat-protein subunit, while others have one (Day and Wiseman, 1978; Wiseman and Day, 1977; Newman et al., 1982). The number of nucleotides per subunit does not correlate with the helix type. At each end of the virion of the one filamentous phage studied in this regard, f1, are 5-10 copies of specific additional structural proteins (Grant et al., 1981; Simons et al., 1981).

It has been argued that the DNA is rigidly held in place by the coat protein (DiVerdi and Opella, 1981), and that the bases are stacked upon one another in some phages, whereas in others the bases stack on coat-protein tyrosines. This difference may correlate with the number of bases per subunit (Day and Wiseman, 1978). Cases have been made for the two antiparallel DNA strands being oriented with the bases directed inward for phages fd and Xf (Casadevall and Day, 1982) and outward for Pf1 (Makowski et al., 1981; Day et al.,

1979). The amino acid sequence of several type I coat proteins are known, and each contains a negatively charged N-terminal region, a hydrophobic central region, and a positively charged C-terminal region (Nakashima et al., 1981). Thus, the C-terminal region is thought to interact with the DNA in the interior of the virion (Wachtel et al., 1974; Marvin and Wachtel, 1975; Nakashima and Konigsberg, 1980). Marzec and Day (1980) and Banner et al. (1981) have suggested that in fd the DNA may not follow the regular helical rise of the coat-protein subunits. In addition, since the DNA is circular, in the body of the virion there must be at least one strand of either polarity. Since the coat proteins all have the same polarity, it seems likely that there are two types of DNA-protein interactions in the particle.

The fd and f1 DNA molecules are oriented so that a particular sequence always occupies a specific end of the particle, the end containing the gene-*IX* protein (Shen et al., 1979; Webster et al., 1981). Deletion analyses have shown that this sequence includes the longest (60 bp) stem-loop structure in the genome, and is required for successful virus assembly (Shen et al., 1979; Dotto et al., 1981; G. Dotto and N. Zinder, pers. commun.). If this stem-loop is inverted with respect to the origin of replication, it fails to function, which suggests some connection between replication and packaging (G. Dotto and N. Zinder, pers. commun.).

The detailed mechanism by which these viruses package DNA is quite mysterious. Although particle length is determined by DNA length (Enea and Zinder, 1975), the assembly mechanism appears to be substantially different from that of the other helical viruses. Intracellular, newly replicated progeny ssDNA is coated with the phage-encoded gene-*V* protein, and held in a rod shaped particle by this protein (Alberts et al., 1972; Pratt et al., 1974; Gray et al., 1982). Since no free DNA intermediate has been found, it is thought that the coat protein somehow replaces the gene-*V* protein during the packaging reaction (Webster and Cashman, 1978). The structure of the protein-DNA complex is not well understood, but it is clear that there are 4-5 nucleotides per protein subunit. Alternative models have been formulated in which the DNA is wrapped around the outside of a rod of gene-*V* protein (McPherson et al., 1980) and in which the DNA is inside the protein (Torbet et al., 1981; Kneale et al., 1982). Both of these models have two antiparallel, double helical chains of V protein molecules, each of which has identical DNA-protein interactions. This structure can be reconstituted *in vitro*, with the gene-*V* protein assembling onto the DNA cooperatively (Alma et al., 1983). It is not known if the DNA has a specific orientation in this complex, as it does in the virion.

The coat protein is synthesized as a longer polypeptide that is inserted into the inner membrane and then proteolytically processed to its mature size (Zimmermann et al., 1982). It is thought to lie across the membrane with the N-terminus in the periplasmic space and the C-terminus in the cytoplasm. The assembly of the virions appears to occur by replacement of the V protein by coat protein as the DNA traverses the inner membrane. Since, the C-terminal end of the coat protein is thought to interact with the DNA in the virion, it may interact with each segment of DNA before it passes into the membrane.

Lopez and Webster (1983) have shown that the portion of the DNA molecule that contains the putative *pac* site passes through the membrane early, but it is not known if the gene-*IX* protein determines the specificity of packaging. The mechanism by which the extrusion of virions occurs must be complex, because coat-protein subunits of one polarity are replacing V-protein subunits of two polarities. There may also be a conformational change in the coat protein during assembly (Dunker, 1981), and several other phage gene products are required (Chapter 7, this volume).

III. ICOSAHEDRAL VIRUSES

The protein architecture of icosahedral virions has been dealt with in detail elsewhere (Caspar and Klug, 1962; Caspar, 1965; Chapters 1 and 2, this volume). Icosahedral viruses are known that contain all types of nucleic acid chromosomes (ss and ds RNA and DNA). Thus, icosahedral structures appear more versatile than helical ones, which contain only ss nucleic acids. Packaging a nucleic acid molecule within a closed shell presents somewhat different topological problems from packing in a helical structure. Two obvious differences are (1) the closed structure may place an upper limit on the size of the chromosome which can be packaged, while the helical structures do not, and (2) it is more difficult to imagine how a nucleic acid can be arranged with icosahedral symmetry than with helical symmetry. Another feature of packaging by closed structures is that a mechanism seems to exist that prevents the very concentrated cellular components (Kennell and Riezman, 1977; Fulton, 1982) from being included in the interior space of the virion that is occupied only by the nucleic acid. With helical viruses this problem is avoided by polymerizing the coat subunits from one point outward, so displacement of cellular components occurs as each coat-protein subunit adds to the growing structure. With closed structures "outward assembling" coat-protein subunits must meet to enclose an interior volume. This problem appears to be overcome by the ssRNA viruses by an assembly mechanism in which either the RNA is quite condensed before coat protein surrounds it, or in which the RNA occupies the core of the particle in a directional manner from a nucleation point. The dsDNA phages build a solid protein structure and subsequently remove the core to create space for the nucleic acid (see Earnshaw and Casjens, 1980).

IIIA. Icosahedral Viruses with ssRNA Genomes

Icoashedral viruses with single-stranded RNA as genetic material are found among the plant and animal viruses and the bacteriophages.

IIIA.1. ssRNA plant viruses

This supergroup includes a number of families of viruses, which are shown in Table 1. All virions contain message sense ssRNA chromosomes, which in most cases are capped or covalently linked to a virus-encoded protein (Vpg) at the 5' end. Some have a poly(A) tract at the 3' end, and others have a tRNA-like sequence of unknown function (reviewed by Joshi et al., 1983). A single coat protein is

Table 1 Icosahedral ssRNA Plant Virus Characteristics

Virus group	*T*	Species of coat protein	Number of chromosomes	Structure of end 5'	3'
Bromoviruses	3	1	3	Cap	tRNA-like
Comoviruses	1	2	2	Vpg	poly(A)
Cucumoviruses	3	1	3	Cap	tRNA-like
Ilarviruses*	1,3,4(?)	1	3	Cap	Hairpins
Nepoviruses	4(?)	1	2	Vpg	tRNA-like
TNV•	3	1	1	ppA	?
STNV•	1	1	1	ppA	tRNA-like
Tombusviruses▼	3	1	1	Vpg	tRNA-like
Tymoviruses	3	1	1	Cap	tRNA-like

These viruses all carry + strands in the virion. Designations in table are according to Mathews (1981; 1982); other groups exist that are less well characterized structurally.

*Alfalfa mosaic virus has been included in this group.
▼Southern bean mosaic virus has been included in this group.
•Tobacco necrosis virus and satellite tobacco necrosis virus.

present in the particles in all but the comoviruses, which have two. Many of these viruses are coviruses: they package several chromosomes individually in separate particles.

The structure of three viruses of this type have been determined by x-ray diffraction to high resolution, tomato bushy stunt virus (TBSV, Harrison et al., 1978), southern bean mosaic virus (SBMV, Abad-Zapatero et al., 1980), and satellite tobacco necrosis virus (STNV, Liljas et al., 1982)(see Chapter 2, this volume). In all cases studied, one domain of the coat protein forms a tightly packed shell that is not penetrated greatly by the RNA (Kruse et al., 1982; Jarcot et al., 1977). In several cases an N-terminal arm of the coat protein, which is rich in basic amino acids, extends into the interior of the virion from the outer shell (Chauvin et al., 1978; Harrison et al., 1978; Abad-Zapatero et al., 1980; Liljas et al., 1982). The amino acid sequences and spectral analyses imply that the N-terminal regions may form helices or tight turns with basic residues on one side (Abad-Zapatero et al., 1980). Argos (1981) suggested that these structures could bind in one groove of double-stranded regions of the RNA, in a manner analogous to the binding of repressors to dsDNA (Matthews et al., 1982). A consequence of this idea is that double-stranded regions might be required at particular locations along the RNA molecule. The N-terminal arm is, however, not simply present to interact with the RNA, since its removal significantly affects the way in which the protein alone assembles (Fukuyama et al., 1981; Erickson and Rossmann, 1982). In addition, the interior surface of the coat-protein shell domain is basic and may interact with the RNA. Rossmann et al. (1983) presented a detailed, speculative model for RNA binding in SBMV. The combined positive charges of the N-terminal arm and interior surface are nearly sufficient to neutralize the negative charges of the RNA (Chapter 2, this volume). In those cases in which there is no N-terminal arm, polyamines may help neutralize the RNA charges (Cohen and McCormick, 1978). Although the RNA is held tightly in place within the virion (Munowitz et al., 1980; McCain et al., 1982), it does not contribute to the icosahedrally symmetric x-ray scattering, so it cannot be arranged within the virion with icosahedral symmetry (or it has a different icosahedrally symmetric arrangement in each particle, a seemingly unlikely design).

The nucleotide sequence has been completed for a number of plant virus RNA molecules, and when analyzed, substantial potential secondary structure appears likely (Ysebaert et al., 1980; Gordon and Symons, 1983; Haseloff and Symons, 1982). In addition, in STNV and one tymovirus the 3' and 5' ends are complementary, so the RNA could at times circularize (Strazielle et al., 1965; Briand et al., 1978; Ysebaert et al., 1980). Physical measurements on isolated RNA molecules have generally supported the existence of considerable base pairing in the free RNAs (Obumosu et al., 1981). Using a relative of SBMV, Denloye et al. (1978) concluded that some base pairs must be broken upon encapsidation! Thus, these RNA molecules are likely to be quite compact in solution, and may not need to be greatly condensed upon encapsidation. An interesting example of secondary structure is the unique circular viroid-like chromosome of velvet tobacco mottle virus, which is packaged and has more than 85% potential base pairing in its RNA (Haseloff and Symons, 1982). If this RNA has the normal viroid rod shape (see Diener, 1979), as the sequence suggests, it must bend and/or break base pairs to be encapsidated.

The assembly mechanism of these viruses has remained elusive, in spite of the large amount of structural knowledge. Perhaps the most perplexing aspect of the *in vitro* assembly of these viruses is that the coat proteins often do not preferentially package homologous RNA molecules. Some specificity of *in vitro* packaging has been found for a bromovirus and a tymovirus coat protein (Cuillel et al., 1979; Briand et al., 1975), but more often investigators have found that various RNAs (or even other polyanions) can be packaged *in vitro* (Bancroft et al., 1969; Hull, 1970; Adolph and Butler, 1977). This finding contrasts with the situation for to viruses produced in infected cells, for these contain viral RNA almost exclusively. Either proper conditions have not been found *in vitro* in spite of extensive searches, or additional factors in the cell increase the specificity of packaging. Such additional factors could, in principle, include "compartmentalizing agents," which bypass the need for specific recognition of RNA molecules by the coat protein. It is interesting to note that a number of the coviruses, for example alfalfa mosaic virus (AMV), package their coat-protein message with an efficiency comparable to that of the chromosomes (Van Vloten-Doting and Jaspars, 1977), so any recognition mechanism cannot simply be a selection against mRNA.

The relationship of tobacco necrosis virus (TNV) to its satellite (STNV) is instructive with respect to *in vivo* specificity of RNA packaging. TNV codes for the synthesis of a coat protein that assembles with its single ssRNA chromosome into what are probably icosahedral $T = 3$ virions (reviewed by Atabekov, 1977). When cells are co-infected with TNV and STNV, two kinds of virions are produced: one containing the TNV coat protein and chromosome, and a smaller $T = 1$ virion containing the STNV coat protein and chromosome. The STNV genome codes for a new coat protein which specifically packages its chromosome, but it is dependent upon the TNV-coded replicase for RNA replication. The later observation may account for the fact that TNV and STNV appear to have at least some 5'-terminal homology (Lesnaw and Reichmann, 1970), but are not encapsidated by the same coat proteins. Thus, specificity of packaging by a compartmentalization mechanism becomes more unwieldy, since there would have to be specific compartments for TNV and STNV RNA molecules.

In only one case has specificity of RNA binding by coat protein been demonstrated directly. The alfalfa mosaic virus (AMV) coat protein has been shown by Jaspars and colleagues to bind several sites on AMV RNA. AMV is a covirus; its three chromosomes and coat protein message (which corresponds to the 3'-portion of chromosome 3) are packaged in four separate particles. Sequence analysis has shown that there are about 150 nucleotides of homology at the 3' end of the AMV packaged RNAs and much shorter homology at the 5' ends (Koper-Zwarthoff et al., 1980; Cornelissen et al., 1983). Houwing and Jaspars (1982) and Zuidema et al. (1983a) studied coat-protein binding on the coat-protein message and chromosome 1. It binds to the 3' ends of these RNAs with high affinity and to a few additional internal sites with a somewhat lower affinity. There is a common combination of sequence and secondary structure within each of the binding sites (Koper-Zwarthoff and Bol, 1980; Zuidema et al., 1983a, 1984). If the N-terminal 25 amino acids are proteolytically removed, the coat protein no longer exhibits this specific binding (Zuidema et al., 1983b) and may assemble *in vitro* into only empty protein shells (Bol et al., 1974). There are no direct experiments supporting the hypothesis that this binding represents a nucleation complex for assembly (Koper-Zwarthoff and Bol, 1980), and there is reason to be wary of this conclusion. The three isolated RNA chromosomes of this virus (and its relatives the ilarviruses) require the presence of a small amount of coat protein or coat protein message to be infectious. Coat protein is not made from chromosome 3, but only from coat message. Therefore, coat protein can only be present in the newly infected cell if coat protein itself or its message enters the cell with the three chromosomes. These observations have given rise to the notion that an oligomer of coat protein must be bound to the RNA for the viral replicase to function. It remains possible that these studies are relevant to the nucleation of assembly, but this has not been proven. It has been shown that the tRNA-like sequences present at the 3' terminus of several of these RNA chromosomes can sometimes be aminoacylated by specific tRNA synthetases *in vitro*. Although it is not known if these RNAs are aminoacylated *in vivo*, the hypothesis that the presence of a 3'-amino acid on the RNA is required for replication and blocks packaging (since virion RNAs do not have attached amino acids) has been put forward, but not tested (Hall, 1979).

The coat proteins of these ssRNA plant viruses possess much of the information needed to assemble into icosahedral shells of the correct size. Empty structures with correct dimensions can sometimes be assembled by coat proteins alone (Adolph and Butler, 1974; Driedonks et al., 1976 and 1978b; Cuilell et al., 1983). In the two cases in which it has been analyzed, the coat protein from disrupted virions is mostly dimeric (AMV, Driedonks et al., 1977; BMV, Adolph and Butler, 1974) and is in equilibrium with monomers and higher oligomers (Cuillel et al., 1983). No potential special nucleating protein structure has been identified in any of these cases.

Although it has been suggested that preassembled coat-protein shells might be precursors of virions (Jonard et al., 1972), more recent studies suggest this is not the case. Current ideas are consistent with models in which the ssRNA and coat proteins co-condense and with models in which the RNA is condensed first by virtue of its own secondary structure. The main arguments in favor of these hypotheses are as follows: (1) under conditions of *in vitro* assembly no empty protein shells are present (Adolph and Butler, 1974; Cuilell et al., 1983), and (2) in several cases RNA appears to nucleate the assembly reaction or affect the size of the final coat protein shell. In the case of AMV which has elongated particles, the length of the RNA controls the length of the virion both *in vitro* and *in vivo* (Bol and Kruseman, 1969; Lebeurier et al., 1971; Driedonks et al., 1978a), and in SBMV the RNA length affects the diameter (T-number) of the particle assembled (Savithri and Erickson, 1983). If one assumes that these viruses utilize basically similar mechanisms of packaging, the following overall picture emerges (which should be viewed with considerable skepticism). In this working hypothesis the coat protein specifically recognizes some feature of the 3' end of the RNA. Additional coat protein molecules then add to the nucleated structure, in which the RNA is already quite compact, until the virion is completed. Since assembly is nucleation controlled, each adding protein subunit should interact with RNA after or simultaneously with its first interaction with the previously assembled coat proteins.

IIIA.2. Picornaviruses

In spite of the facts that the poliovirus complete nucleotide sequence is known (Kitamura et al., 1981; Racaniello and Baltimore, 1981) and its virion crystallizes easily (Hogle, 1982), little is known about packaging of picornavirus RNA. The message sense ssRNA chromosome is about 7400 bases long, and has a poly(A) tract at its 3' end and a covalently bound protein at its 5' end (reviewed by Wimmer, 1982). The capsid is a $T = 1$ structure in which the basic structural unit contains four polypeptides, called VP1, VP2, VP3, and VP4. These four polypeptides are proteolytically cleaved from a large precursor. Cleavage of VP1 and VP3 from the precursor polypeptide occurs before assembly of the protein shell, and cleavage to VP2 and VP4 occurs after RNA encapsidation into "provirions", so proteolytic cleavage appears not to be temporally connected to packaging. Various workers have argued for co-assembly of the RNA with coat proteins or for preassembly of coat protein shells followed by insertion of the RNA (summarized by Putnak and Philips, 1981).

It has been suggested that the RNA, free in solution, has more base-pairing (up to 60%) and base-stacking than RNA in the virion, a finding reminiscent of the ssRNA plant viruses (Bachrach, 1964;

Frisby et al., 1977). Analysis of the nucleotide sequence supports the finding of considerable secondary structure in the RNA (Racaniello and Baltimore, 1981). Although core-like structures have been seen in preparations of disrupted picornavirus particles (Dubra et al., 1982; Boublik and Dzerniek, 1976), it is not known if any specific virion protein is associated with them. Cross-linking experiments have given inconclusive answers to this question (Wetz and Habermehl, 1982). Deletions of poliovirus RNA have been isolated that lack up to 20% of the genome and are packaged (Lundquist et al., 1979). Combined, they cover the coat protein region of the chromosome (about 5-50% from the 5' end), implying that sequences there are not required for the RNA to be packaged.

Poliovirus provides a framework for the discussion of an intriguing protein, called Vpg, that is covalently linked to the 5' end of the chromosome. This protein is thought to function as a specific primer during the initiation of RNA synthesis. Analogous proteins exist for number of other ssRNA and some dsDNA viruses (Wimmer, 1982). In no case is the location of the Vpg known in the virion; for example, it does not appear in the structure of SMBV determined by x-ray diffraction. It is attractive to imagine that, in addition to their priming function, Vgp molecules might serve as recognition signals for packaging. Replication obligatorily attaches a Vpg to each progeny RNA strand. Newly made strands are thereby specifically marked as viral nucleic acid molecules by the presence of these proteins. This possibility has in no case been fully ruled out, but with polio it appears unlikely, at least in its simplest form, since full-length minus strands, which never appear in virions, also have an attached Vpg (Wimmer, 1982).

IIIA.3. Togaviruses

Togaviruses have enveloped virions that contain icosahedral nucleocapsids (Harrison et al., 1974; Enzmann and Weiland, 1979). They contain a single message sense ssRNA chromosome about 12,000 nucleotides long, which is capped and has poly(A) at the 3' end (Strauss et al., 1984). The cores are made from a single protein, called C protein that, like the ssRNA plant-virus coat proteins, has a highly basic N-terminal amino acid sequence (Garoff et al., 1980; Rice and Strauss, 1981). C protein will reassemble into core-like structures *in vitro*, apparently without specificity for the type of nucleic acid used (Wengler et al., 1982), for example, tRNA or ssDNA nucleate assembly as well as viral RNA, and some particles containing ssDNA appeared significantly larger in negatively stained preparations, even though the chain length was less than that of viral RNA. Particles containing abnormal RNAs or polyanions are less stable than those containing normal virus chromosomes (Wengler et al., 1984). A number of deletion strains of togaviruses have been isolated. The combined deletions cover all of the genome except the extreme 5' and 3' termini, and individual isolates have up to 80% of the genome deleted (reviewed by Stoller, 1980). The study of these deletions suggests that the sequences required for replication and packaging are near one of the ends. The fact that an intracellular subgenomic mRNA containing only the 3' end of the chromosome is not packaged (Kennedy, 1980) may indicate that the 5' region is recognized during packaging. Measurements of the density of deletion virions and of cores has shown that these particles are more variable in density than wildtype particles. Thus, a variable number of deletion chromosomes and/or host RNAs can be packaged into cores when full length RNA is not available (Wengler et al., 1982). We cannot draw any overall conclusions concerning the mechanism of togavirus nucleocapsid assembly.

IIIA.4. Nodaviruses

This newly recognized family of icosahedral ssRNA insect viruses has not not been studied extensively with respect to RNA packaging, but is of interest because there appear to be two small chromosomes packaged in each virion (Newman and Brown, 1977). Since only one type of coat protein has been identified in these virions, they are a particularly simple case for the study of how a small number of proteins (one) can recognize and encapsidate a larger number of virus chromosomes (two), a problem that is unsolved in the more-studied but more complex orthomyxoviruses and rotaviruses.

IIIA.5. ssRNA bacteriophages (Leviviruses)

Single-stranded RNA phage virions contain a single RNA molecule, a $T = 3$ coat protein shell, and one molecule of an additional virus-encoded polypeptide called A protein (Steitz, 1968; Crowther et al., 1975; Dunker and Paranchych, 1975; Takamatsu and Iso, 1981). The isolated coat-protein subunits of these viruses, like those of the icosahedral ssRNA plant viruses, can reconstitute shells similar to those in virions under the appropriate conditions. In other conditions assembly of virus-like particles requires RNA or other polyanion to "nucleate" assembly (reviewed by Knolle and Hohn, 1975). The particles assembled in this way are not infectious. If A protein is included in the reaction, the yield of infectious particles increases 100-fold, but it is still far below the number of RNA molecules assembled into virus-like structures (Roberts and Steitz, 1967). More efficient *in vitro* assembly has not been achieved, but additional experiments in which the order of addition of the protein components was varied showed that A protein had to be added before coat protein and RNA were mixed (Knolle and Hohn, 1975). The reason that better packaging has not been achieved is not clear, but there is an indication that efficient packaging *in vivo* may require a host protein (Schonlaker-Schwarz and Engelberg-Kulka, 1981).

The nucleotide sequence of one of these viruses has been analyzed for potential secondary structure, and a model was devised in which most of the nucleotides are base paired in 30-40 hairpins (the "flower" model). Fiers (1979) has reviewed both this model and the abundant evidence that the RNA is up to 85% base-paired (or stacked in a helical conformation) in solution and in the virion. Of particular interest is the observation that in addition to secondary structure, the RNA in solution appears to adopt a fairly compact tertiary structure; thus, it is condensed only twofold during packaging (Fiers,

1979). Like the plant icosahedral RNA viruses, some empty coat-protein shells are made during infection (Paranchych et al., 1970), but most evidence supports a "co-condensation" packaging mechanism (Knolle and Hohn, 1975).

Ling et al. (1970) showed specific packaging of homotypic RNA molecules for two "species" of RNA phages both *in vitro* and *in vivo*. Unfortunately, they did not separate coat and A protein and hence were not able to tell which protein gave rise to the observed specificity. Both proteins are candidates for the packaging recognition protein. Krahn et al. (1972) found that during the normal infection process A protein enters the cell with infecting RNA molecules and hence could be bound to the RNA. Indirect experiments led Shiba (in Fiers, 1979) to conclude that A protein will bind phage RNA in a species-specific manner. The absence of A protein during infection leads to the formation of particles from which the 5' end of the RNA is thought to protrude (Heisenberg, 1966; Lodish, 1968; Fiers, 1979). Added A protein cannot efficiently convert these particles into virions (Knolle and Hohn, 1975), even though at least a portion of the A-protein molecule is probably on the exterior of the virion (Curtiss and Krueger, 1974). Attempts to analyze *in vivo* assembly have shown that A protein is present in the first intermediates (thought to contain a small number of coat-protein subunits) that could be identified (Bonner, 1974). These observations suggested to Knolle and Hohn (1975) that A protein might be responsible for specificity of packaging, but the fact that particles assembled *in vivo* without A protein appear to contain mainly viral RNA argues against this idea. Fiers and coworkers found that A protein interacts with several different sites on the RNA and suggested that the role of A protein might be to alter the tertiary structure of the RNA for packaging (Fiers, 1979).

The coat protein has been shown to bind specifically to a short region near the beginning of the replicase gene, where it functions as a translational repressor for that gene (reviewed by Steitz, 1979; Krug et al., 1982). This translational repression complex between coat protein and RNA could also be the initiation complex for assembly. Hohn (1969) and Spahr et al. (1969) found that such complexes were not favored over free RNA during *in vitro* assembly following addition of more coat protein, so this question remains to be completely resolved. Thus, initiation of proper RNA packaging by these phages may involve recognition of the RNA by coat protein, specific alteration of the RNA tertiary structure by A protein binding, and even perhaps additional host components.

IIIB. Icosahedral Viruses with ssDNA Genomes

Three types of icosahedral ssDNA viruses have been identified. These are: (1) the microviruses, relatives of bacteriophage øX174, which have circular chromosomes; (2) the parvoviruses, which have linear chromosomes; and (3) the geminiviruses, which have two circular chromosomes and virions in which two icosahedral shells appear to be joined (reviewed by Goodman, 1981; Hatta and Franki, 1979). DNA packaging of the first two groups will be discussed below. Packaging of the geminiviruses, even though the genome of one member has been completely sequenced (Stanley and Gray, 1983), has not been studied.

IIB.1. Icosahedral ssDNA bacteriophages (microviruses)

These small $T = 1$ viruses with circular ssDNA chromosomes 5000-6000 nucleotides long (two have been completely sequenced; Sanger et al., 1977; Godson et al., 1978) are among the most thoroughly understood viruses. The protein shell of the virions is built from 60 molecules of F protein, and each of the twelve vertices has a "spike" composed of 5 molecules of G protein and 1 molecule of H protein. About 60 molecules of the J protein are thought to be in the interior of the particle (Burgess, 1969; Aoyama et al., 1983a). Hayashi, Dressler and coworkers have studied the assembly of one member of this group, ϕX174, extensively and have found that it assembles a coat-protein shell and then inserts the DNA into the shell. These protein precursor particles are called *proheads* or *procapsids*. Such structures were originally identified as precursors to the heads of the tailed phages (see below) and are now a common theme in phage assembly. The procapsids contain F, G, and H proteins and no J protein. In addition, they contain a large number (I crudely estimate >150 from published gels) of D-protein molecules and a smaller number (about 60?) of B-protein molecules (Fugisawa and Hayashi, 1977b; Mukai et al., 1979). Virion antigenic determinants are not available on the exterior of the procapsids suggesting that B and D proteins may cover the outside surface of these structures (Fugisawa and Hayashi, 1977a). During the DNA packaging reaction the B and D proteins are lost from the structure. It is not known if either of these proteins is directly involved in inserting the DNA into the procapsid, but all molecules of D protein do not need to be removed before DNA can be inserted, since particles can be isolated that contain DNA and retain some D protein (Weisbeek and Sinsheimer, 1974). The role of J protein in DNA packaging is also uncertain. It is extremely basic (Freymeyer et al., 1977), it is not found in procapsids, it may be bound to replicating DNA (Fugisawa and Hayashi, 1976), and it is required for DNA packaging (Aoyama et al., 1983a). Thus, it has been speculated that J protein is bound to DNA within the particle.

The nature of the physical mechanism that inserts the chromosome into the procapsid remains unknown. Two generalized models for such a reaction can be envisioned. (1) The DNA enters and stays inside the shell because this is its lowest energy state inside the cell. (2) Something forces the DNA in and traps it there in a metastable high-energy state. Neither of these hypotheses is currently favored or disfavored. To date, successful *in vitro* packaging of ϕX174 DNA has only been accomplished with concomitant DNA synthesis. In addition, there is considerable evidence that asymmetric accumulation of viral strands requires the presence of proheads (Aoyama and Hayashi, 1982; Aoyama et al., 1983a; Koths and Dressler, 1980). Furthermore, since packaging can occur while a strand is being synthesized (Koths and Dressler, 1980), and DNA replication could supply the energy required to insert the chromosome, hypothesis 2 is attractive. Koths and Dressler (1980) and Aoyama et al. (1981) have proposed the following general model for this DNA encapsidation reaction (Figure 2). Viral strands are known to be made from a closed circular dsDNA replicative intermediate. Synthesis of such a strand is initiated by the nicking of the viral strand by the virus-encoded A protein, which becomes

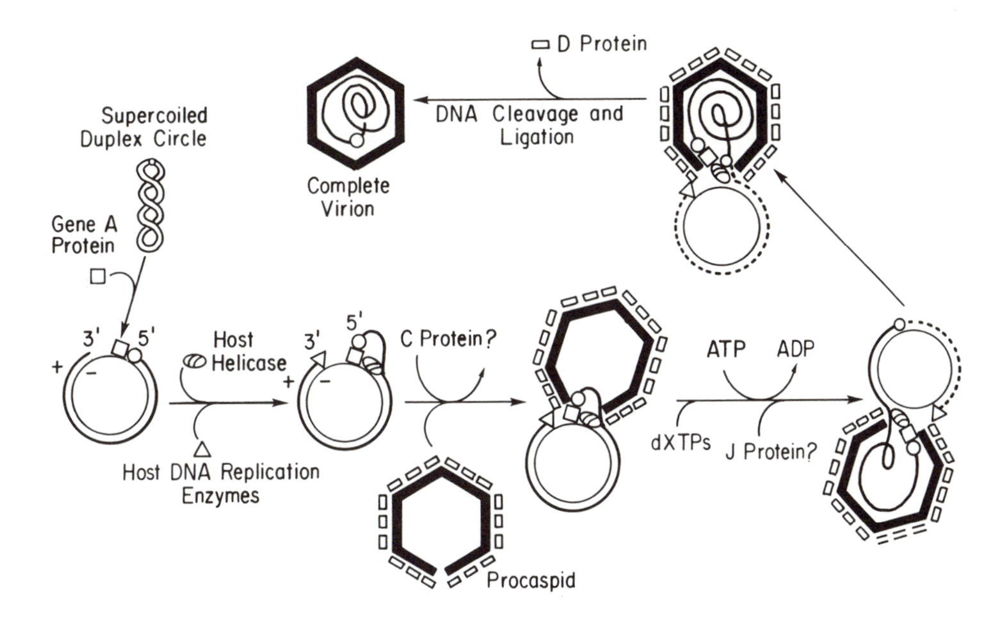

FIGURE 2 Packaging of bacteriophage øX174 DNA. Pathways of phage DNA metabolism in cells infected with øX174 are shown. The viral (+) DNA strand is a bold line and the complementary strand is a fine line. Newly synthesized strands are shown as dashed lines. The solid circle on the viral strand represents the A protein cleavage site. The figure is derived from Fujisawa and Hayashi (1977), Koths and Dressler (1980), and Aoyama et al. (1981).

covalently bound to the 5' end of that strand. The 3' end serves as a primer for synthesis of a new viral strand while the pre-existing viral strand is displaced. The A protein appears to stay with the replication fork, so a loop of the parental viral strand is displaced. The model suggests that the procapsid binds to the A protein, and the viral strand is inserted into the procapsid as it is displaced. When a round of synthesis is completed, the A protein ligates the two ends of the displaced strand and is released. The J protein enters the virion sometime during the packaging reaction. It is small, only 37 amino acids, and its possible presence in replicating DNA is puzzling. If its role is to bind to and condense DNA within the virion would it not also do this to DNA outside the virion? (The question of how uncondensed DNA can enter the structure with supposedly perfect icosahedral symmetry is unaddressed as well.) The last component required for packaging is C protein, which is not present in virions or procapsids. Aoyama et al. (1983b) have suggested that it may help the procapsid bind to the A protein-replicating DNA complex. Since packaging occurs with only these purified proteins, no additional components are needed.

Van der Ende et al. (1982) found that *in vivo* the packaging apparatus will recognize and initiate packaging on a plasmid containing a 300-bp fragment of phage DNA. This fragment also includes the A-protein binding site and nicking site. These authors suggested that the A protein–DNA complex is recognized by procapsids. Two additional features of this packaging reaction were explored using this system. First, if proper termination of the "loop displacement replication" is blocked, no DNA is stably packaged. Although it was not shown directly that initiation occurred normally in these experiments, they do point out that termination of replication is clearly important to proper packaging according to the current model. Second, the size of the DNA to be packaged was varied. As might have been expected, if the DNA is too long (greater than 106% of the length of the genome), it was not successfully packaged (see also Russel and Muller, 1984). Presumably, this is due to the limits on space within the protein shell, which is a discrete structure and cannot expand indefinitely. On the other hand, molecules as short as 12% genome length were packaged efficiently. Whether particles with such short chromosomes contain only one or multiple copies of such chromosomes was not reported. Surprisingly, with ϕX174, particles containing molecules less than 70-80% of the length of the genome are not functional, while if similar fragments are packaged by a relative (G4), molecules as small as 60% of genome length are functional. This could represent a failure to uncoat in the next round of infection in the ϕX174 case. Such a failure might be interesting in view of the energetic aspects of packaging and unpackaging.

IIIB. 2. Parvoviruses

Parvovirus particles contain a linear ssDNA chromosome with hairpins of 100–200 base pairs at both ends (summarized in terms of the complete sequence by Astell et al., 1983), a coat-protein shell composed of probably 60 or 180 molecules of the major coat protein, and 6-9 molecules of a minor protein, which may be clustered in the virion (reviewed in Tattersall, 1978; Paradiso, 1983). Replicating DNA contains a protein covalently attached to the 5' end of the strand not present in virions. A strand-displacement type of DNA synthesis has been suggested for these viruses (Revie et al., 1979; Astell et al., 1982). Empty capsids are made during infection with kinetics compatible with their being virus precursors (Richards et al., 1978), replicating molecules have been seen with capsids attached (Gunther and Revet, 1978), and mutations thought to alter one of the capsid proteins block progeny single-stranded DNA synthesis (Rhode, 1976). Although this packaging reaction has not yet been analyzed in detail, these similarities to ϕX174 have led workers in the field to speculate that parvovirus packaging is analogous to that of ϕX174 (Tattersall, 1978; Gunther and Revet, 1978).

A number of packageable deletions of mouse minute virus DNA have been studied by Faust and Ward (1979). They found two types of deletions: (1) internal deletions, the longest of which had only 5% of the genome remaining, and (2) substitution of the 3' end by the complement of the 5' end. Taken together, they suggest that any *pac* site must be within a few hundred bases of the 5' end of the normal chromosome. Virions containing short chromosomes had very light densities in CsCl, suggesting that each contained only one or a few chromosomes. One interpretation of this is that packaging can initiate only once per capsid.

IIC. Icosahedral dsRNA Viruses

Four groups of icosahedral dsRNA viruses have been characterized. These are: (1) reoviruses and their relatives, which usually have 10 chromosomes and a double-layer coat-protein shell (Silverstein et al., 1976; Metcalf, 1982); (2) the birnaviruses, which have two chromosomes and a $T = 9$ shell (Dobbs et al., 1977); (3) bacteriophage ϕ6, which has three chromosomes, a probably icosahedral nucleocapsid and a lipid envelope (Van Etten et al., 1974; Mindich, 1978); and (4) fungal viruses with multiple chromosomes, which have not been studied in detail (Saksena and Lemke, 1978). Although the assembly of these viruses is not well understood, we know that all except the fungal viruses (Bozarth et al., 1981) appear to carry all of their chromosomes in one virion. However, in no case has there been a *direct* demonstration of selective packaging of one of each of the chromosomes into every virion. The probability of a virion receiving one each of its n chromosomes if it randomly packages n chromosomes per virion is $n!/n^n$ (not $1/n^n$ as suggested by Silverstein et al., 1976). Thus, reovirus, which has 10 RNA molecules would receive a correct complement of RNA in 1 of 2756 particles if the chromosomes were packaged randomly. This is in fact somewhat less than observed ratios of plaque-forming units to particle number; in addition, since other animal viruses with one chromosome have similarly low ratios, the infectivity of reovirus particles containing all of the chromosomes could also be low. This assumption, plus physical measurements that show that about 10 chromosomes per virion are in fact packaged (Harvey et al., 1981), suggests that these viruses do show some level of "segment selection" during packaging. How this selection occurs is unknown. The ends of all of the segments of ϕ6 , and those of several reoviruses have been sequenced (McCrae, 1981; Kuchino et al., 1982; Antczak et al., 1982; Gaillard et al., 1982; Iba et al., 1982). For each virus, all of the genome segments have identical 5'-terminal sequences and different identical 3'-terminal sequences 4-10 bp long, so the extreme ends of the RNAs cannot be recognition sites for segment selection in packaging.

With the reoviruses there are fewer virion-protein species than there are RNA molecules; for example, a possible virus precursor contains only 4 or 5 polypeptide types and all 10 RNAs (Clark et al., 1980). Either several proteins can recognize more than one RNA, nonvirion proteins participate in packaging, or a specific complex of RNAs is recognized. The RNAs of the reoviruses and ϕ6 are made as follows: Message sense strands are copied from the parental dsRNA. These function as messengers and/or assemble into structures that are progeny virion precursors, where they are used as templates for the synthesis of the antimessage strands (reviewed by Silverstein et al., 1976; Emori et al., 1982). The nature of the initially assembled structure is not clear, but since the assembling RNA is single-stranded, it could be available for intersegment base-pairing.

Antczak et al. (1982) determined the sequence of 60-90 bp on the ends of all reovirus segments and searched among these sequences for complementarities that could cause specific aggregation of the segments. They concluded that if such sequences exist, most must be internal to the sequenced regions.

IIID. Viruses That Use Reverse Transcriptase for Replication

IIID.1. Hepadnaviruses and cauliflower mosaic virus

These viruses are mentioned together, because they have incomplete strands in their circular virion DNA chromosome and *appear* to replicate by reverse transcription (Summers and Mason, 1982; Pfeiffer and Hohn, 1983; Hull and Covey, 1983a). The hepadnaviruses (hepatitis B virus and relatives) are enveloped with nucleocapsids that are probably icosahedral, and cauliflower mosaic virus is a naked icosahedral virus. Both contain one major polypeptide and perhaps small numbers of additional proteins (Robinson, 1979; Al Ani et al., 1979). The chromosome of hepatitis B virus is small and contains (1) a single antimessage DNA strand about 3000 nucleotides long with a protein covalently linked to the 5' end and (2) an incomplete message-sense strand that is base paired to the antimessage strand in such a way as to hold it in a circular form (Figure 3). Summers and Mason (1982) have presented evidence for a unique model in which a full-length message-sense strand of RNA is packaged by the core proteins, and the partially double-stranded DNA chromosome is copied from it in the core. The RNA is then lost, leaving only the DNA in the virion. The mature cauliflower mosaic virus chromosome is similar to that of the hepadnaviruses, except that the antimessage strand does not have a protein on its terminus, and two overlapping, partial message-sense DNA strands are paired to it (Figure 3) (Franck et al., 1980). Although its life cycle is not yet fully understood, Pfeiffer and Hohn (1983) and Hull and Covey (1983a) have presented indirect evidence that reverse transcription is involved and have speculated that packaging occurs after the DNA is made by reverse transcription. Unlike the hepadnaviruses, no reverse transcriptase activity has been found in cauliflower mosaic virions. However, replicating RNA-DNA could be encapsidated, with the enzyme being lost before virus release (Hull and Covey, 1983b). Very few details of the packaging process are known with either virus. Gronenborn et al. (1981) did find that cauliflower mosaic virus insertion mutations longer than 250 bp did not give rise to progeny, and suggested that this was due to the inability of longer chromosomes to be packaged. Howarth et al. (1981) found that deletions up to 500 bp can be packaged.

IIID.2. Retroviruses

Retroviruses replicate by reverse transcription, but the message-strand RNA is packaged and remains in the virion (reviewed by Varmus, 1982). Retroviruses are enveloped, and RNA is packaged into nucleocapsids that then bud out of the cell. The structure of these nucleocapsids is not fully understood, but it is thought that they have icosahedral shells (Nermut et al., 1972; Bader et al., 1970;

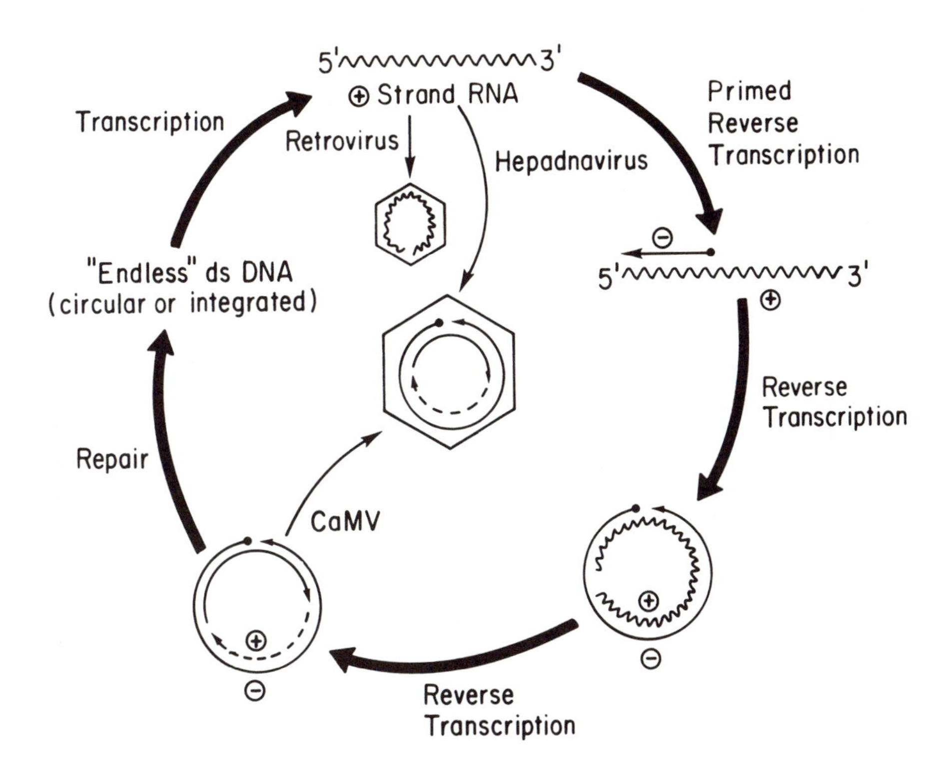

FIGURE 3 Nucleic acid packaging by viruses that use reverse transcriptase for their replication. The outside circle shows reactions undergone by unencapsidated nucleic acid; inside the circle are the packaging reactions and reactions that occur within immature virions. RNA is represented by a wavy lines and DNA by solid lines. Arrowheads indicate the direction of synthesis. The solid circle represents the primer for initiating reverse transcriptase, a tRNA for retroviruses and a protein for the other viruses. "+" and "-" indicate message and anti-message sense strands, respectively. The figure is derived from Pfeiffer and Hohn (1983), Hull and Covey (1983), and Summers and Mason (1982).

Yoshinaka and Luftig, 1977; Durbin and Manning, 1982). Uniquely, these viruses are diploid; there are two identical RNA chromosomes in each virion. In addition, specific tRNAs are found in the virion, and these function as primers for replication in the next round of infection (Varmus, 1982). The two chromosomes are apparently packaged "independently", but form hydrogen-bonded structures with each other and with the packaged tRNAs after budding (described, with models for the final chromosome pairing, in Coffin, 1982). The proteins present in the mature nucleocapsid are cleaved from a single large polypeptide, the *gag* gene product (summarized by Dickson et al., 1982; Hunter et al., 1983). The available evidence is consistent with a model in which the uncleaved protein binds to the RNA chromosome to initiate packaging, and cleavage into the mature nucleocapsid proteins follows (Nasco et al., 1983). The mammalian retrovirus Gag-protein precursor is cleaved into four peptides and the avian precursor into five. This cleavage appears not to be necessary for successful RNA packaging (Demsey et al., 1980). In each case, one of the cleavage products is the nucleocapsid shell protein, and one is a small, basic protein (see Dickson et al., 1982; Hunter et al., 1983). It has been found to bind specifically homotypic RNA *in vitro* (Sen and Todaro, 1977; Leis et al., 1978; Sen et al., 1978; Meric et al., 1984; see, however, Sykora and Moelling, 1981). Nucleocapsids also contain a small amount of reverse transcriptase, which is required for the packaging of the specific tRNAs but not the viral chromosomes (Levin and Seidman, 1979; Sawyer and Hanafusa, 1979).

The specificity of RNA binding by a Gag-protein cleavage product suggests that it might be responsible for the specificity of chromosome packaging in these viruses. Supporting this idea is a mutation, thought to modify the Gag protein, which lowers the specificity of packaging *in vivo* (Harms et al., 1977). Only a fraction of this cleavage product is found bound to the RNA, so it may occupy two types of positions in the virion. In related findings, Sen et al. (1977) and Erikson et al. (1977) noted that these proteins are phosphorylated, and the degree of phosphorylation seems to affect the extent, but not the specificity of RNA binding. Cleavage of the Gag protein and perhaps phosphorylation of a portion of the "specifically bound" protein may lead to structural rearrangements of the protein and RNA components (this may occur after budding of the virions; Yoshinaka and Luftig, 1977 and 1982; Jamjoon et al., 1977). In one retrovirus the specifically binding cleavage product binds to 12 sites scattered across the genome (Darlix and Spahr, 1982). These sites are near one another in a "flower" model for the RNA deduced from the nucleotide sequence by Darlix et al. (1980). In contrast, recently it has been observed that these viruses have a single site, required in *cis* for packaging, that lies near the beginning of the most-5' gene, the *gag* gene (Shank and Linial, 1980; Watanabe and Temin, 1982; Mann et al., 1983), and possibly one near the 3' end of the chromosome (Sorge et al., 1983). The first type of site is spliced out of the messenger RNAs that are not packaged. It is not known if this site(s) represents a critical site for protein-RNA interaction during packaging or a critical site for RNA-RNA interactions that might occur during partial condensation of the RNA before packaging.

IIIE. Icosahedral dsDNA Viruses

This supergroup includes some of the most complex viruses known; for example, bacteriophage T4 has a linear chromosome about 166,000 bp long, and the virions contain over 40 different protein species. On the other hand the papovaviruses have circular chromosomes less than 6000 bp long, and their virions contain only 3 virus-encoded proteins. With this wide a variety of viruses it seems unlikely that they all package DNA by similar mechanisms, but one feature of packaging by these viruses is constant. In spite of the fact that sequence can affect the overall structure of the double helix, by affecting its flexibility or even dictating bends (Marini et al., 1982; Dickerson, 1983; Hogan et al., 1983; Wu and Crothers, 1984), no dsDNA virus for which information is available requires specific base sequences scattered throughout the genome in order to be packaged. This was surmised from early *in vivo* genetic experiments with phages lambda, P22, and P1 that showed that essentially any bacterial host DNA sequence could be condensed within these virions (Kayajanian and Campbell, 1966; Kayajanian, 1968; Susskind and Botstein, 1978; Sternberg and Hoess, 1983). More recent genetic engineering techniques have allowed virtually any DNA from any species to be packaged by these phage. Similar, though less extensive, experiments allow the same conclusion to be drawn for adenoviruses and papovaviruses.

IIIE.1. Tailed dsDNA bacteriophages or myoviruses

A detailed historical review of the search for the mechanism of DNA packaging by the tailed phages has been published (Earnshaw and Casjens, 1980), so this discussion will only summarize this very large body of work. Included in this group of phages are lambda, T1, T4, T5, T7, P1, P2, P22, Mu, ϕ29, SP01, SPP1 and their relatives. These viruses have proven to be ideal subjects for the study of the nature of assembly pathways and control processes active in these pathways (chapters 4, 5 and 8 this volume; Wood and King, 1979; King, 1980). In spite of this intense study, and the availability of extensive genetics to complement biochemical studies, the physicochemical mechanism of DNA condensation during packaging by these phages is not known. However, it is clear that all of these phages assemble icosahedral coat-protein shells (*proheads*) first, and then insert the DNA into these structures. Streisinger et al. (1967) formulated this idea, when it appeared from their careful genetic analyses that the coat-protein shell was measuring the length of the phage T4 chromosome to be cut and packaged from concatemeric DNA. This mechanism has become known as "headful packaging". Proheads are apparently always assembled with internal protein cores or scaffolds (at this stage, called immature proheads), which are removed in the formation of "mature" proheads before the act of DNA packaging. The details of prohead assembly have been recently reviewed (Chapter 5 this volume; Wood and King, 1979; King et al., 1980; Black and Showe, 1983; Georgopoulos et al., 1983).

Although empty coat-protein shells have been found in nearly all cells infected with icosahedral viruses, *absolute proof* that these are used as preformed containers for nucleic acid requires that it be shown that there is no breakdown and reassembly of these structures. Such a proof is virtually impossible *in vivo* (Luftig et al., 1971). However, since with phage lambda DNA packaging proceeds in extracts of infected cells, Hohn et al. (1974) and Kaiser et al. (1975) were able to show unambiguously that for lambda phage the precursor particles are filled directly with DNA. Such a proof is technically difficult and has not been repeated for other members of the group, but all available evidence supports the notion that all members package

DNA by basically similar mechanisms (Earnshaw and Casjens, 1980). This premise will be used in the following discussion. Two current problems in this area are: (1) what is the DNA substrate for packaging, and how is it recognized and processed by the packaging apparatus, and (2) how does the DNA enter the prohead?

IIIE.1.a. DNA recognition

It is important to understand the options these phages have for modes of replication and packaging. Among those phage that are well understood, there are several overall strategies. (1) Phages lambda and P22 DNA molecules circularize upon injection (by different mechanisms, see below), replicate to give daughter circles for a short time, and then begin rolling circle replication to give rise to large head-to-tail concatemers of phage DNA which are the substrate for packaging. (2) Phage P2 DNA circularizes upon injection and replicates by simply producing circular daughter chromosomes. (3) Mu is thought to replicate by transposition, so it is always covalently integrated into host DNA. (4) T7 DNA does not circularize but replicates to give linear daughter DNA molecules for a time. Later in infection daughter molecules recombine to give rise to concatemers similar to those of lambda and P22, which are the substrate for packaging. (5) ϕ29 replicates its linear chromosome to give linear daughter chromosomes, so the substrate for packaging is monomeric linear DNA. (6) Phage T4 chromosomes are also packaged from concatemers, but it is not known if the DNA ever circularizes. Since all these phage virions contain linear chromosomes, a striking feature of these replication-packaging strategies is that all but ϕ29 require the phage chromosomes to be nucleolytically cut from the product of replication during or before packaging. In these cases the mature chromosome ends do not accumulate unless packaging occurs concomitantly (reviewed by Earnshaw and Casjens, 1980).

The mechanism of cutting mature chromosomes from replicated DNA varies somewhat among the different phage, and it must complement the mode of circularization and/or replication used by the phage (Figure 4). Lambda and P2 chromosomes are cut precisely in a sequence-specific event in which two single-stranded cuts in the DNA (staggered by 12 and 19 bp, respectively) are made to form each end of the chromosome. The two strands between the staggered nicks must be separated to release the DNA being packaged from the concatemer; it is not known if this is an active process. Each virion thus contains exactly one copy of the genetic information of the phage, and all chromosome ends can be generated by a single type of cut. Circularization in the subsequent round of infection occurs by simply annealing the two single-stranded ends together. Since lambda packages its own DNA quite specifically, the DNA *must* somehow be recognized. Hohn (1983) has shown that the recognition sequence may be as large as 100 bp. Feiss et al. (1983) have shown that in addition to the base sequence, called *cosB*, which is responsible for the choice of substrate by the packaging apparatus, a nearby base sequence, called *cosN*, is required for the nucleolytic cleavage of the DNA from the concatemer. Two phage-encoded proteins are responsible for cutting and recognition of these sites. These two proteins purify as

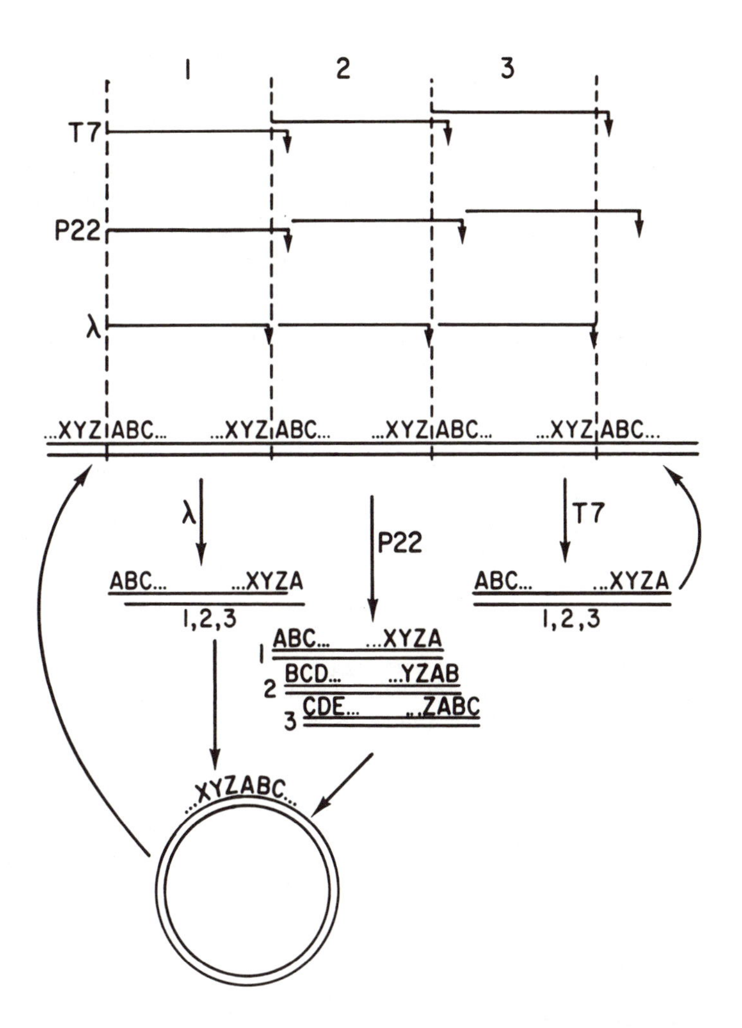

FIGURE 4 DNA cutting and packaging from concatemers. The central double line represents three head-to-tail sequence repeats in a phage DNA concatemer. Above this, three mature phage chromosomes are indicated, and the terminal sequences for each chromosome are shown below. For lambda and P22 the numbers indicate the position in a processive series of packaging events. It is not known if T7 packages DNA in such series (see text). Upon injection, lambda DNA circularizes by annealing the complementary single stranded ends, and P22 circularizes by homologous recombination between the ends. These circles generate concatemers by rolling-circle replication. T7 replicates as linear molecules, and concatemers are thought to be generated by intermolecular homologous recombination between ends.

a complex and are not found in proheads or phage particles (Becker et al., 1977; Gold and Becker, 1983; Feiss and Becker, 1983). In addition, at least one host protein may be required by some lambdoid phages at the recognition stage (Feiss, in Bear et al., 1984). Similar phage proteins are required for P2 packaging (Bowden and Calendar, 1979). It seems clear that packaging of lambda can initiate at an internal recognition site on the concatemer, though studies by Emmons (1974) have shown that the packaging apparatus prefers to initiate at a concatemer end created by a previous packaging event (diagrammed in Figure 4). Thus, packaging in this instance can be said to be processive in the sense that packaging events prefer to continue a series of packaging events to initiating a new series. Packaging series with lambda average about 2 to 3 events.

Phage Mu uses a different type of DNA cutting, which is not sequence specific. Mature Mu chromosomes are thought to be generated by breakage of the host DNA just outside the ends of the integrated phage DNA (Chow and Bukhari, 1977). Again, the DNA must be recognized, probably at some location in the phage DNA but near one end (George and Bukhari, 1981; Castilho et al., 1984). Packaging must then proceed a short distance in one direction and a long distance in the other. The cuts at both ends occur at apparently random sequences in the flanking host DNA.

P22, like lambda, packages from concatemeric DNA, but its DNA circularizes by homologous recombination between direct repeats at the ends of the chromosome. The chromosome ends are likely to be blunt and about 4% terminally redundant (Sgaramella, 1972; Susskind and Botstein, 1978; M. Hayden and S. Casjens, unpublished results). Phage P22 DNA is also more efficiently packaged than host DNA (Ebel-Tsipis et al., 1972), so phage DNA must be recognized by the packaging apparatus. Tye et al. (1974), after finding that the DNA is *partially circularly permuted* (the ends of the chromosomes are scattered over a 20% region of the genome sequence), suggested an elegant model for P22 DNA packaging (see Figure 4) in which (1) a P22 DNA concatemer is recognized by virtue of a *pac* sequence, (2) DNA is cut at or near that site, (3) packaging proceeds unidirectionally from that point until the coat-protein shell is full (104% of the genome length), (4) a second cut is made in the DNA to release the newly filled head from the concatemer, and (5) subsequent packaging events prefer to initiate at the end left by the previous event (instead of recognizing another *pac* site), and proceed in the same direction as the previous event. Further studies have given considerable support to this model and have shown that packaging series average 3-6 events (Jackson et al., 1978; Weaver and Levine, 1978; Kufer et al., 1982; Adams et al., 1983). Thus, a sequence-specific recognition event occurs only once in 3-6 rounds of DNA packaging, with some other feature, perhaps proteins left by the previous event or simply naked DNA ends, being recognized in all other events. P1, T1, and SPP1 are other well-studied phage that utilize this strategy (Bachi and Arber, 1977; MacHattie and Gill, 1977; Morelli et al., 1979). It should be noted that with phage that use this type of packaging strategy it has not yet been shown directly that the first end recognized in a packaging series is generated by simple nucleolytic cleavage by the packaging apparatus. It remains possible that it is a concatemer end generated by replication (Laski and Jackson, 1982; Sternberg and Hoess, 1983; M. Hayden and S. Casjens, unpublished results). Curiously, in the cases of SPP1 and

P22 the first ends created (or recognized) in a series are imprecise, with ends scattered over about 10 and 120 bp, respectively (Diechelbohrer et al., 1982; Casjens and Huang, 1982). If these ends are generated by cleavage, the cutting enzyme does not always cut an exact distance from the recognition site, as with lambda. On the other hand, the analogous phage P1 end is more precise, with such ends occurring mostly at a specific point on the sequence (Sternberg and Hoess, 1983; N. Sternberg, pers. commun.). As with lambda, the generation of the first end in a P22 packaging sequence is caused by a complex of two proteins not found in association with proheads or phage (Casjens and King, 1974; Poteete and Botstein, 1979; Laski and Jackson, 1982). Mutational changes in one of these proteins can cause an altered specificity of packaging (Raj et al., 1974; Jackson et al., 1982), strongly suggesting that they are involved in the substrate-recognition process.

T7 chromosomes are cut from the concatemer so that all have exactly the same end sequences, with a 160-bp direct repeat on each end (i.e., the chromosomes are terminally redundant but not circularly permuted - see Figure 4). It is not known if the T7 packaging apparatus utilizes DNA processively. Phage SP01 may package DNA in a similar manner with a much larger terminal redundancy (Cregg and Stewart, 1978). T7 concatemers, generated by recombination, contain one copy of the terminally redundant sequence between genomes (Dunn and Studier, 1983). How can every phage particle contain two terminally redundant sequences and one genome sequence? Either the packaging process is very wasteful, discarding half of the genomes, which seems unlikely, or the end sequences are regenerated in some way. Watson (1973) suggested that they could be regenerated if cutting were to occur by staggered nicks in opposite strands at the two ends of the terminal redundancy and the single-stranded ends thus generated were used as templates to fill in the other strand. If this is the case, there are several notable differences from lambda cleavage. First, it is difficult to imagine that the 160 bp of complementary DNA between the nicks would spontaneously separate, so perhaps there is an active separation mechanism (this could be the repair synthesis itself). It would seem that either lambda blocks repair synthesis at its ends, or T7 has a mechanism to promote it. Second, the lambda nicks are made at symmetrical positions in a region of rotational symmetry in the sequence, so only one nuclease specificity is required for both nicks. The sequence of T7 shows that the two postulated nicks must occur in apparently unrelated sequences (Dunn and Studier, 1983).

Another type of cutting *may* exist. Several phages that are terminally redundant and circularly permuted, T4, ϕ11 and ϕ42, have not been found to initiate packaging series at particular locations on their genomes (MacHattie et al., 1967; Wilson et al., 1979; Lofdahl et al., 1981; Moynet and DeFilippes, 1982). Only T4 has been studied in detail. It is known to (1) highly modify its DNA, (2) destroy much of the host DNA in the infected cell, and (3) to package its DNA in an oriented manner (Black and Silverman, 1978). Since these infected cells contain only phage-specific DNA, it is possible that these phage do not need to use any base sequence for specific recognition of the correct DNA during packaging. On the other hand, the orientation

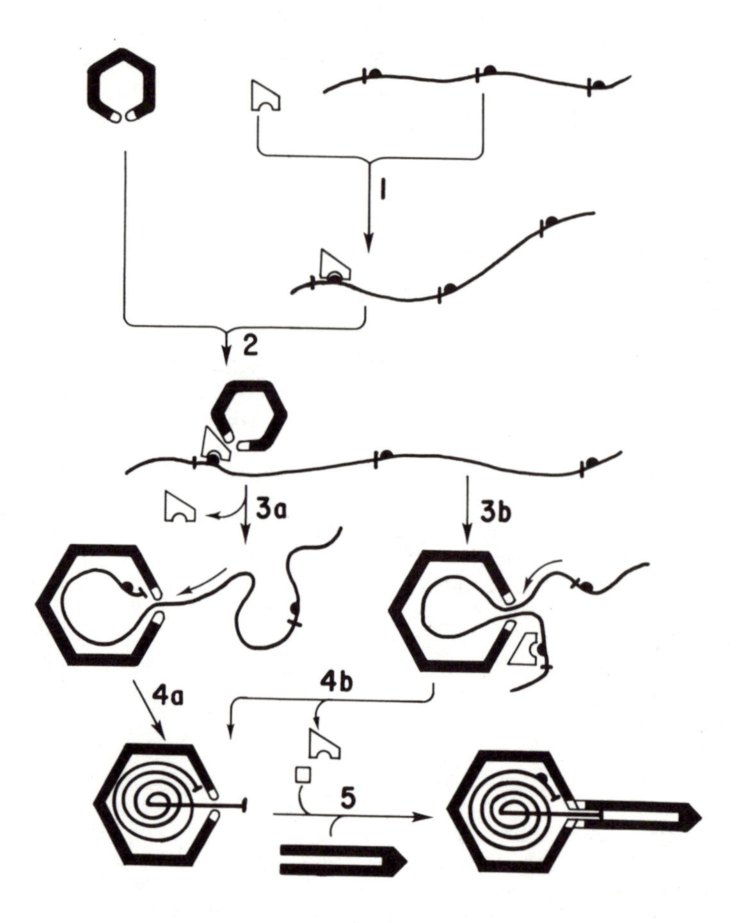

FIGURE 5 Current model for phage dsDNA packaging. Step 1. A specific base sequence on a concatemer of phage DNA is bound by the recognition protein, which is usually a multimer of the products of two phage genes. Step 2. Proheads bind to the recognition protein-DNA complex. Steps 3 and 4. DNA begins to enter the prohead shell, with the first DNA entering occupying positions at greater radius. These are complex steps in which the DNA enters the head in a unidirectional fashion (indicated by arrows), the coat protein shell expands, and packaged DNA is cut from the concatemer. Questions remain about the order of DNA entry and DNA cleavage. It is also not known if the recognition protein leaves the filling head before or after the bulk of the DNA is packaged (alternate pathways a and b in figure). Evidence from the study of P22 indicates that the recognition protein leaves only after the DNA is fully packaged and cut from the concatemer (M. Adams and S. Casjens, unpublished observations). Step 5. Stabilizing proteins are added, and tails join to the completed heads. In several phages, for example lambda, the last end of the DNA to be packaged protrudes into the hollow core of the tail in the finished virion.

of packaging implies some specificity. Possible alternate explanations are that they have scattered, multiple alternate *pac* sites that have been missed in the experiments done to date, or that directional recombination or replication structures orient the packaging apparatus (Mosig et al., 1981).

ϕ29 DNA replicates as a linear monomer without covalently joined ends (summarized in Blanco et al., 1983). A phage coded protein is covalently bound to the 5' end of each strand in virion DNA (Ito, 1978) and is thought to function as a primer during DNA synthesis (Shib et al., 1982). If this protein is removed, the DNA is not packaged *in vitro*, suggesting that it may also play a role in packaging, perhaps as a recognition signal for the packaging apparatus (Bjornsti et al., 1982). However, these workers also showed that packaging is directional. Since both ends have the attached protein, unless there is a subtle difference between the two proteins, there must be an additional signal, most likely base sequence information. The sequences of the two ends of the chromosome are identical for 6 bp (Yoshikawa et al., 1981), so any such information must be internal to these 6 bp.

With all of these dsDNA phages, proteins not physically associated with the prohead first recognize and bind phage-specific DNA; then, the prohead recognizes that complex and packages the DNA (Earnshaw and Casjens, 1980; Bjornsti et al., 1984; Figure 5). Very little is known about the biochemistry of this second recognition process. Only in phage lambda is it known that the large subunit of the protein carrying out both recognition and cutting is responsible for the specificity of this interaction with proheads (Frackman et al., 1984).

IIIE.1.b. DNA entry

Although little is known about the actual mechanism of DNA entry into proheads, sufficient data have accumulated concerning the nature of the prohead and intravirion DNA to put a number of constraints on possible mechanisms. In all cases studied DNA enters the head in a specific direction (Earnshaw and Casjens, 1980). Riemer and Bloomfield (1978) have discussed the energetic considerations that must go into any model for DNA packaging and concluded that DNA is much more likely to assume conformations with gradual bending than with kinking of the DNA. They calculated that the free-energy cost resulting from entropy loss during condensation is small compared to electrostatic repulsion caused by the negative charges of the phosphate backbone, even in the presence of counterions. The notion that charge repulsion is important is supported by the finding of Miller et al. (1983) that phage ϕW14, whose DNA normally contains positively charged putresinyl-thymine, packages an 11% shorter chromosome when its DNA is enriched for unmodified thymine. On the other hand, condensation of DNA by multivalent cations alone questions this conclusion (Thomas and Bloomfield, 1983; Windom and Baldwin, 1980). Riemer and Bloomfield did not consider the possibility that partial dehydration of the DNA may accompany packaging (Serwer, 1975). More recent calculations by Rao et al. (1985) suggest that the work of altering water structure in the vicinity of the DNA during packaging is the predominant "force" that must be overcome during packaging. They calculate 0.1 to 0.4 Kcal must be supplied per mole of base pairs packaged.

Models for DNA entry in which (1) basic or DNA-binding proteins condense the DNA within the coat-protein shell by stoichiometric binding, (2) acidic peptides force condensation by some sort of "exclusion", (3) DNA undergoes a replacement reaction with proteins initially assembled into the prohead interior, and (4) replication forces the DNA inside the coat-protein shell have been proposed. However, in light of subsequent experiments these models all appear quite unlikely (Earnshaw and Casjens, 1980). Three more tenable models are currently under scrutiny:

1. If DNA is able to overcome some initial barrier, presumably with the aid of the recognition-cleavage proteins, will spontaneously enter the prohead interior because of a "favorable interior environment". If this model were true, DNA packaged *in vivo* might be expected to be stably condensed immediately after entry. However, newly packaged P22 and lambda DNA molecules are spontaneously released *in vivo* unless additional proteins are added to stabilize the structure (Lenk et al., 1976; Sternberg and Weisberg, 1977; H. Strauss and J. King, personal communication). It has been suggested that polyamines might be responsible for creating such a favorable interior environment, since they have been found in *some* phage particles (Ames and Dubin, 1960), and they can cause condensation of dsDNA (Gosule and Schellman, 1976; Schellman and Parthasarathy, 1984). However, bacterial cells lacking spermidine and putrescine can serve as hosts for the members of this group tested (Hafner et al., 1979).

2. All of the proheads of this group of phage are smaller in diameter than their respective virions, and an expansion of the coat-protein shell always accompanies DNA packaging. It has been proposed that, if the shell were impermeable to solvent, the negative pressure (osmotic perhaps) could draw the DNA into the shell (Hohn et al., 1974; Serwer, 1980). Calculations show that the volume change upon expansion is too small to draw in all of the DNA, and it has been found by B. Hohn (in Earnshaw and Casjens, 1980) that a short fragment of DNA (11% genome length) can be packaged without expansion by lambda proheads. These observations do not rule out the possibility that part of the chromosome enters by this mechanism, and Serwer and Gope (1983) have shown that osmotic factors do affect P22 and T7 DNA packaging *in vitro*.

3. A "DNA translocase," physically attached to the prohead,

could force the DNA into the coat-protein shell (Hendrix, 1978; Black and Silverman, 1978). This model is attractive because any such translocase would no doubt require energy, and all *in vitro* packaging reactions of these phages absolutely require an energy source. There are are no experimental results which argue directly against this hypothesis at present. Nonetheless, it remains an unproven model. An observation which *might* be used against this model is the finding that the newly discovered phage Erh1 packages DNA that contains long single-stranded regions (Kozloff et al., 1981), and the putative DNA translocase might be expected to be specific for dsDNA or rely on the stiffness of dsDNA for its function (Zachary and Black, 1981). If this mechanism were operative, the recognition proteins could be the translocase, or since all proheads have specific proteins at one corner of the icosahedral coat-protein shell, these corner proteins could be the translocase. Manne et al. (1982) have identified a DNA-dependent ATPase in T4 proheads, but have not demonstrated that it is required for DNA packaging. It is possible that some combination of these three general mechanisms brings the DNA into the coat-protein shell. In some of the few experiments that have attempted to observe DNA directly as it enters the prohead, Bjornsti et al. (1983) found preliminary evidence for pausing of the ϕ29 packaging apparatus on the DNA during packaging, and Gope and Serwer (1983) have measured the rate of DNA entry for P22 to be >3000 bp/sec.

The structure of DNA in the virion should also reflect its manner of entry. In all cases in which it has been studied, both exit of the DNA during injection into susceptible hosts and abortive ejection of the DNA have been found to occur in a specific direction along the DNA. These observations and the findings that DNA is not very mobile within the phage head (Akutsu et al., 1980; Ashikawa et al., 1984) and that the last end of several phage DNA molecules to enter the capsid during packaging is located near the point of tail attachment, suggest that DNA is not randomly oriented in virions (Gillin and Bode, 1972; Earnshaw and Casjens, 1980). This has been supported by electron microscopic and x-ray scattering observations. Richards et al. (1973) and Earnshaw et al. (1978) found that partially disrupted virions release condensed DNA, in the shape of a "ball of yarn" or a "spool". In agreement with this, phages lambda and P22, upon partial disruption, often release loops of supercoiled DNA (Virrankoski-Castrodenza et al., 1982; S. Casjens, unpublished results). Analysis of particles in solution suggests that a considerable amount of order exists in the DNA in phage particles, with many double helices lying quasi-parallel in approximately closest-packed arrays (Earnshaw and Harrison, 1977; Earnshaw et al., 1979; Adrian et al., 1984). Two poorly understood observations with intraphage DNA are that the DNA in phage P4 heads appears to be knotted (Liu et al., 1981), and that in some phages it may have altered chemical reactivity (reviewed by Tikchonenko (1975). More detailed analysis of the intravirion DNA has proven difficult. Haas et al. (1982) and Windom and Baldwin (1983) have used chemical cross-linking to ask if particular regions of the intravirion lambda chromosome are near other particular DNA regions or the coat-protein shell, respectively. Neither study yielded positive results. There are several possible interpretations of these observations. Harrison (1983) has suggested that DNA in different particles assumes different spatial orientations during packaging. This model, which assumes a translocase mechanism, suggests that at any time during packaging the DNA duplexes within the head occupy positions as far as possible from each other owing to both electrostatic repulsion and the inherent stiffness of DNA. Thus, as the head is filled, the whole DNA structure must be compacted, which requires that much of the DNA slide with respect to neighboring double helices. The compaction could take place in different ways in different particles, and so all particles would not have identical DNA arrangements. As long as the first end to enter is tethered and the DNA is rather stiff, tangling of the compacting DNA should not occur (presumably tangling would inhibit subsequent injection through the narrow tail tube). Although at present this model is reasonable, it remains possible that the packaged DNA is identical in all particles, arranged in some undeciphered complex manner.

The structure of mature proheads and the DNA substrate must also place constraints on the packaging mechanism. All complex dsDNA phage proheads have specific proteins present in small numbers (5-20) at the icosahedral vertex to which tails will eventually bind, and circumstantial evidence suggests that DNA enters at this vertex (Earnshaw and Casjens, 1980). Detailed structural analysis of the vertex proteins for phages ϕ29, lambda and T4 has shown that in each case one of these proteins is a cylindrical dodecamer with 12-fold symmetry and a 2-5 nm central hole (Carrascosa et al., 1982; Driedonks et al., 1981; Kochan et al., 1984). It is possible that the DNA enters the prohead through this hole, which would be consistent with the observations that DNA branches resulting from replication and/or recombination block packaging of phage T4, and dsDNA-binding proteins appear to be stripped off by the packaging process (summarized in Black and Showe, 1983; Earnshaw and Casjens, 1980; however, phage T4 might package chromosomes with up to 22 bp heteroduplex deletion loops, Drake, 1966). The 12-fold symmetry of this structure clashes with the 5-fold symmetry of an icosahedral vertex. Hendrix (1978) has suggested that proteins at such a clash could, in theory, rotate more readily than if there were a one-to-one contact between the coat protein and the vertex protein (see also Chapter 5, this volume). Using this idea he proposed an elegant, but speculative model for DNA packaging in which the vertex proteins actively rotate to screw the DNA into the head. Mutations altering the minor P22 and T4 prohead proteins present at the putative point of DNA filling affect packaging, suggesting that these proteins participate in some way in the filling and/or termination reactions (see below; M. Hayden and S. Casjens, unpublished experiments; Hsiao and Black, 1978).

An important unanswered question is whether a single double helix end or a loop of dsDNA enters (Figure 5). The latter situation seems possible, but not proven for lambda and Mu, since chromosome ends do not accumulate unless packaging can occur; however, the possibility remains that these cuts can be made and are rapidly repaired by DNA ligase in the absence of proheads (Laski and Jackson, 1982; Feiss and Becker, 1983; Kobayashi et al., 1983; Murialdo and Fife, 1984). Attempts to look directly at active packaging reactions in this regard have been equivocal (Yamagishi and Okamoto, 1978; Syvanen and Yin, 1978). On the other hand, phage P22 apparently makes the first cut in the DNA in the absence of proheads and DNA packaging (Laski and Jackson, 1982). It could simply initiate packaging at ends and insert the DNA unidirectionally along a single DNA duplex (Strobel et al., 1984).

IIIE.1.c. Termination of packaging

Cutting of the concatemer to phage lengths by the "headful nuclease" always depends upon successful packaging (Figure 5). This "enzyme" presumably recognizes a head that is "full" of DNA and cuts to release the packaged chromosome from the concatemer. The nature of the coupling of capsid fullness to the headful nuclease is not understood in any system. The phage lambda enzyme requires a specific base sequence only in the actual cutting site (*cosN*) portion of the binding-cutting site for this type of "headful" cutting of the DNA (Feiss and Becker, 1983; Feiss et al., 1983). The recognition portion (*cosB*) is not needed. Presumably, the act of packaging brings the nuclease into proximity with the terminal site, so the recognition reaction is not required. Since in lambda the cuts that generate the two ends of the chromosome occur at identical sequences, there is no need to postulate different enzymes for the initial and final cuts. The presence of a *cosN* sequence in the DNA being packaged is, however, not sufficient for cleavage to occur (Emmons, 1974). Feiss et al. (1977) showed that in addition, the capsid must be >75% full for successful headful cleavage by the lambda packaging apparatus, and that efficient packaging does not occur if the chromosome is >5% longer than wildtype. Presumably the upper limit is a result of the limited space within the coat-protein shell. On the other hand, T7 makes cuts in very different specific sequences to form the two ends of the mature chromosome, and hence might use two different nucleases to make the different cuts. The phage P22 and Mu headful nucleases do not require any rare specific sequence for headful cleavage, but appear to cut whenever the capsid is full. P22 chromosomes are cut from concatemers with a precision in length of ±2% (M. Hayden and S. Casjens, unpublished observations). Other headful packaging phage, SPP1 and Mu, have similar precision in length measurement (Humphreys and Trautner, 1981; Chow and Bukhari, 1977). For this type of phage it is not known if the initial cut in a packaging series and subsequent cuts are made by the same enzyme. SPP1 may use more than simply headfullness in selecting the positions of the final cut, though this is not yet understood (Morelli et al., 1979).

After the headful cut, the DNA is very unstably packaged, and in all cases additional proteins must be added to keep the DNA from spontaneously leaving the capsid, both *in vitro* and *in vivo* (Lenk et al., 1976). Even after the addition of these proteins, in some cases the DNA may not be completely inside the capsid. Gillin and Bode (1972) found that several nucleotides of the single-stranded ends of the lambda chromosome are susceptible to nuclease digestion unless tails are added to the filled heads. Finally, at least with phages Mu and lambda, some DNA protrudes into the tail tube (very approximately 100 bp in lambda) in the mature particle (reviewed in Earnshaw and Casjens, 1980). Nothing is known about the process that places it there.

IIIE.2. Lipid-containing dsDNA bacteriophages

DNA packaging by these phages has not been studied in detail. At least two classes of such phages are known. PRD1 and its relatives (tectiviruses) have linear chromosomes with proteins covalently attached to both 5' ends (Bamford et al., 1983). The other class, characterized by PM2, has a closed circular chromosome; these phages have been reviewed by Mindich (1978). PRD1-type phage are interesting with respect to packaging, because they first assemble icosahedral proheads that contain lipid, and these structures package DNA (Mindich et al., 1982; Davis and Cronan, 1983). Thus, the DNA may have to pass through a "pore" in the lipid of the particle, so this packaging process may be a model for the passage of nucleic acid through lipid membranes.

IIIE.3. Complex dsDNA animal viruses

These are very complex viruses, with genomes in excess of 100,000 bp. In no case is packaging of DNA clearly understood; however, it is of interest to discuss what is known in terms of the models developed above for the dsDNA phages. The herpesviruses and iridoviruses are icosahedral and have quite highly compacted DNA in virions. Calculations based on the size of the herpesvirus nucleocapsid (Wildy et al., 1960) and on DNA length (Roizman, 1979) suggest a DNA density comparable to that of the complex dsDNA phages. In iridovirions the DNA is packaged to about one fourth this density (Cuillel et al., 1979). Too little is known about the molecular nature of the packaging processes in the pox- and baculoviruses to warrant discussion here.

IIIE.3.a. Herpesviruses

Herpesviruses are large enveloped viruses with genomes contained in complex icosahedral nucleocapsids (reviewed by Spear and Roizman, 1980). The sequence organization of the chromosome is also more complex than in most other viruses. The linear chromosome of herpes simplex 1 has been studied in greatest detail. It contains two "halves," which rapidly invert relative to each other during growth. On both ends of each half are the same short inverted repeats, called "*a*" sequences, about 300-600 bp long. The two large invertible sections are covalently joined, so the *a* sequences are directly repeated at the termini of the virion DNA. Replicating DNA does not contain the chromosome ends; they have been joined together by either circularization or concatemerization (see, for example, Roizman, 1979; Preston et al., 1983; Marks and Spector, 1984). Work with temperature-sensitive mutants has led to the conclusion that, like the dsDNA phages, formation of the ends of herpesvirus DNA depends on packaging. All evidence is *consistent with* a model in which prohead-type structures are assembled first, and then DNA is inserted into these capsids (Ladin et al., 1980; Roizman and Furlong, 1974). Vlazny et al. (1982) have studied the packaging of very short defective chromosomes and found that only multiple tandem repeats of these short chromosomes are successfully packaged into complete virions. They also found that before budding the nuclear nucleocapsids contain concatemers that vary in length from one defective chromosome up to about 20 repeats, the wildtype length. It appears that short chromosomes can be cut and packaged from concatemeric DNA, but the resulting nucleocapsids cannot be enveloped. These studies and those of Smiley et al. (1981) support the notion that herpesvirus DNA can be packaged from concatemers or multimeric circles, and the cutting enzyme does not obligatorily cut at every potential end generating sequence. Thus, the DNA may be filled and cut by a system with at least some similarity to the headful packaging seen in the dsDNA phages.

The exact nature of the packaging recognition signal is not understood, but a 500-bp right-end fragment appears to be sufficient to direct initiation of packaging (Spaete and Frenkel, 1982). The base sequences of the terminal portions of the genome have been determined and have implications both for replication and for the packaging and cutting of herpesvirus DNA (Mocarski and Roizman, 1982). The chromosome ends have one protruding 3'-nucleotide, and the cutting enzyme appears to cut always at the same sequence, which is reiterated a variable number of times within the terminally redundant regions between genomes on concatemers. The location of the particular reiterated sequence element chosen for cleavage is precise with respect to internal unique sequences at the right end and is variable at the left end. This is consistent with, but does not prove, a three-step process: (1) precise initiation and cutting at the right end, (2) unidirectional packaging to the left, and (3) headful cutting, which is slightly variable with respect to length but precise with respect to the sequence at the left end. There appear to be more copies of the *a* sequence in the sum of two virion chromosome ends than are present between genome ends in replicating DNA. Hence, the same difficulties that apply to the generation of the T7 terminal redundancy (discussed above) apply and are yet to be understood here. A final, perhaps unique, aspect of the packaging of herpesvirus DNA is that although there appears to be no room inside the nucleocapsid for histones or other DNA condensing proteins, and no histones have been found in virions, replicating DNA may be present as nucleosomes (Shaw et al., 1979; St. Jeor et al., 1982). Thus, it is possible that either histones are removed during packaging or that there is some mechanism for excluding histones from DNA destined to be packaged.

IIIE.3.b. Iridoviruses

Several of these very large icosahedral viruses contain circularly permuted and terminally redundant chromosomes (Goorha and Murti, 1982; Darai et al., 1983; Delius et al., 1984). The physical ends of the chromosomes are limited to a 20-30% region of the genome (Murti et al., 1982), so it appears to utilize a sequential headful-packaging mechanism similar to that of phage P22 (Tye et al., 1974; see above). Very little is known about the biochemistry of packaging in this case. Because it is difficult to imagine alternate ways of generating this pattern of chromosome ends, this observation is probably the strongest evidence for any virus of a higher organism that a protein shell measures the amount of DNA to be packaged. Of course, this does not determine whether the DNA enters a preformed shell or co-condenses with coat proteins.

IIIE.4. Simple dsDNA icosahedral viruses

IIIE.4.a. Papovaviruses

These viruses have been studied intensively because of their relatively simple dsDNA circular chromosomes, 5000-6000 bp in length, and their ability to cause tumors (Tooze, 1980). The virions have one major and two minor capsid proteins, VP1, VP2, and VP3 respectively, which are virus-encoded. Intravirion chromosomes of papovaviruses have bound, host-encoded histones (reviewed by Griffin, 1980). Viral DNA, replicating in the nucleus of infected cells is also present as nucleosomes. Although several modifications, such as histone acetylation and loss of histone H1, of this viral "chromatin" are made during the transition from newly replicated DNA to packaged DNA, none of these seems to be responsible for committing chromosomes to be packaged (Roman, 1982; Brady et al., 1981a). Early studies suggested that capsids are assembled first and then filled with DNA (Ozer and Tegtmeyer, 1972; Tan and Sokol, 1974); however, more recent studies have supported models in which the capsid protein subunits are gradually added to the histone-coated virus chromosomes (Garber et al., 1980; Puvion-Dutilleul et al., 1980; Coca-Prados et al., 1982; Milavetz and Hopkins, 1982; Ng and Bina, 1984). This last type of model is attractive because of the possible difficulty of inserting nucleosomes into a preformed protein shell. In fact, histone H1, which is present on papovavirus DNA as it is packaged and is removed later, is capable of condensing nucleosomes to a density approaching that of intravirion DNA (Muller et al., 1978). Although attempts to assemble papovavirions *in vitro* have met with apparent success starting with either empty capsids isolated from infected cells or with coat-protein oligomers from disassembled virions, such studies have not yet clarified the packaging mechanism (Brady et al., 1979; Yuen and Consigli, 1982).

Control of nucleic acid specificity in packaging by papovaviruses is currently difficult to assess. A large number of packageable deletion and substitution mutations of various papovaviruses have been isolated. A striking feature of these viruses is that the chromosomes are never longer than about 105% of the wildtype genome length (Muzyczka, 1980; Wilson et al., 1982) or shorter than about 60% of the wildtype length (Lee et al. 1975; Scott et al., 1976; Carroll and O'Neill, 1981; Yoshiike et al., 1982). The upper limit is understandable in terms of the finite size of the internal virion cavity. One observation suggests that the lower limit may not be simply the failure to isolate such mutants. Deletions larger than 50% have not been recovered in virus preparations as simple deletions. Invariably these are found as multiple tandem repeats of the remaining sequences, sometimes with attached host sequences, so the packaged chromosomes are within the above size limits (Lee et al., 1975; Carroll and O'Neill, 1981; O'Neill et al., 1982). It seems unlikely that multimer formation always occurs concomitantly with the deletion event, so there must be an additional selection such as requirements for packaging size and/or more rapid replication to explain the failure to find packaged deletion monomers.

Examination of the deletion-substitution strains of papovaviruses reveals that in addition to size limits only one region of the sequence is present in all packaged chromosomes. One such mutant retains only about 3% of the virus genome (Lee et al., 1975). *If* there is a packaging recognition sequence on the DNA of papovaviruses, it must be in this region. These chromosomes also must be able to replicate, and the common region retained in all deletions contains the replication origin. This region of the chromosome is nucleosome-free, so that it may be accessible to sequence-specific binding proteins (Saragosti et al., 1982). No experiments published to date directly address the question of the existence of a possible *pac* site. Lee and Nathans (1979) found that a mutant with five repeats of the origin region was, if anything, less efficiently encapsidated than wildtype DNA. However, interpretation of this result is complex, since this mutant did replicate much faster than wild-type. Slilaty et al. (1982) found that isolated empty capsids associate with DNA of apparently any sequence *in vitro* and suggested that sequence may not be responsible for specific encapsidation. However, this conclusion should be viewed with caution, because wrong conditions or missing components could cause loss of specificity.

Even less clear is the role of the protein components in the specificity of packaging. VP2 and VP3, minor capsid proteins, have unknown roles in the structure and assembly of the virion (Griffin, 1980). VP1, the major coat protein, has been found in a disaggregated form bound to DNA from disrupted virions. There is no evidence for sequence specific binding by VP1, but it may affect the structure of the bound histones (Brady et al., 1981b; Moyne et al., 1982). A fourth protein, the product of the agnogene, has been tentatively implicated in virus assembly as well, but it is not found in virions (Mertz et al., 1983; Jay et al., 1981; Jackson and Chaukley, 1981). Finally, a fraction of the virus particles produced in some host cells contain host DNA, suggesting that the host may play a role in the specificity of packaging (Aposhian, 1975).

IIIE.4.b. Adenoviruses

Adenoviruses are moderately complex, naked, icosahedral particles containing 12-14 species of virion proteins and a 35,000-bp linear chromosome (reviewed by Ginsberg, 1979). The chromosome has 103 bp inverted terminal repeats and a protein covalently linked to each 5' end (Shinagawa et al., 1980; Challberg and Kelley, 1981). All daughter strands have the terminal protein, since the terminal protein serves as primer for asymmetric replication of unit-length linear molecules (Challberg et al., 1982; Ikeda et al., 1982). Thus, linear unit-length molecules are likely to be the substrate for packaging. The structure of the virion is quite well understood, and virions can be disrupted to give compact cores containing the DNA molecule with its terminal proteins holding it in a circle, and three additional proteins, V, VII, and mu (Ginsberg,1979; Robinson and Bellet, 1975; Brown et al., 1975; Vayda et al., 1983). Proteins V and mu are present to the extent of about 180 and 125 molecules per core, respectively. They are removed from the cores with salt, leaving quite compact structures containing about 1000 molecules of VII (Black and Center, 1979). Chemical cross-linking studies have found some V protein molecules bound to icosahedral shell corner proteins, so they could be involved in positioning the condensed DNA in the virion (Everitt et al., 1975). The VII proteins are very basic, and appear to contact DNA along its entire length with a regular spacing, possibly in about 180 "beads" analogous to nucleosomes (Corden et al., 1976). Little is known about the three-dimensional arrangement of this viral "chromatin" in the virion, but Corden et al. (1976) found that very brief digestion of cores with nuclease produced discrete fragments about 1/20 the length of the chromosome. On the other hand, Newcomb et al. (1984) found 12 spherical "condensates" in partially disrupted virions.

Analyses of potential assembly intermediates isolated from cells infected with wildtype virus and of structures that accumulate in cells infected with mutants, have usually been consistent with packaging by preformed protein shells (Sundquist et al., 1973; D'Haluin et al., 1978; Edvardsson et al., 1978; Persson et al., 1979; Chee-Sheung and Ginsberg, 1982). However, this hypothesis has not yet been proven, and recent studies have suggested that capsid assembly is coupled to DNA replication (Weber et al., 1985). The role of the basic protein (VII) in DNA packaging is not yet clear, though it seems central. It is made as a larger precursor and is proteolytically shortened after DNA packaging (Ishibashi and Maizel, 1974; Everitt et al., 1977). The cleaved VII appears to bind DNA more tightly than the precursor (Mirza and Weber, 1977). Unpackaged DNA in infected cells may also have bound VII-precursor protein (Brown and Weber, 1980; Daniell et al., 1981), so a more bulky structure than naked DNA may have to be inserted into any putative procapsid. Paradoxically, VII precursor is also found in the putative procapsid

(Ishibashi and Maizel, 1974), so it is not clear whether it enters assembling virions during or before DNA packaging.

Daniell (1976), Tibbetts (1977), Hammarskjold and Winberg (1980) and Chee-Sheung and Ginsberg (1982) have shown that viral DNA molecules are only found in capsids if they contain a sequence found between 290 and 390 bp from the left end of the chromosome. Robinson and Tibbetts (1984) showed that precise positioning of this site with respect to the left end is not required for packaging. Although it remains possible that any DNA molecule can be packaged and that only those containing this sequence are retained in the capsid, the simplest explanation is that this region contains a packaging-recognition site. Systematic studies of packaging-size constraints have not been reported, but deletion and insertion chromosomes from 50 to 105% of genome length can be stably packaged (Jones and Shenk, 1978; Larson, 1982). In addition, defective virions are often made that have apparently packaged broken or faultily replicated molecules containing only the left end of the chromosome. These chromosomes can be as short as 5% genome length, and these particles appear to be able to deliver their DNA into susceptible cells (Hammarskjold et al., 1977), so there may not be a lower limit on the size of a chromosome that can be successfully packaged and unpackaged.

IV. AN OVERVIEW

IVA. Summary

The study of a large number of viruses has shown that the related problems of protecting viral nucleic acid from environmental insult between growth cycles and delivery of the nucleic acid to very specific target cells have been solved in quite different ways by different viruses, so different that it seems that different viruses must have arisen independently during evolution. Two basic methods are used to protect the nucleic acid, which always occupies an internal position:

1. The nucleic acid is coated throughout its length with a specific protein, which also has the ability to interact with itself to form a regular protective structure. Crick and Watson (1956) originally pointed out that by far the simplest way to achieve this is in a helical structure. This arrangement has been elegantly proved for TMV and many of the details of its structure and assembly are now understood. Still, there are variations on this theme. The filamentous phage DNA may not follow the protein helices in the virion, and the manner by which the DNA is held in the particle must be more complex than was envisioned by Crick and Watson, even though the viruses do use the helical structure principle.

2. The alternate strategy is the containment of the nucleic acid within a closed protein shell. Again, Crick and Watson in

> their prophetic 1956 article, showed that the most efficient way to accomplish this is to build a closed shell of repeating subunits arranged with cubic symmetry. Like the helical structure, the closed-shell arrangement has been studied in great detail, and structural details are known in many virus systems. In spite of the intense study, the question of how the nucleic acid becomes condensed within such a shell has not yet been answered in detail in any case. However, the plethora of interesting facts uncovered makes it clear that there are more modes of packaging into shells than into helices. The structures vary from phage that contain only quasi-crystalline DNA, to the papova and adeno virions in which the DNA is essentially completely covered with nonsequence-specific DNA-binding proteins. Packaging mechanisms vary from dsDNA phages that insert naked DNA into quite complex preformed shells to apparent co-condensation of RNA and single species of coat protein in the plant viruses.

One universal feature of packaging of nucleic acid by viruses is that all have devised mechanisms to avoid packaging the wrong nucleic acid. In most, and perhaps all cases this is achieved through recognition of a base sequence specific for viral nucleic acid, referred to here as the *pac* site. In addition, since there is always a single *pac* site per chromosome, this initiation mechanism avoids simultaneous attempts to package the same chromosome by two "packaging apparatuses." The biochemistry of the recognition process varies considerably with the lifestyle of the virus. Currently interesting in this context are the several dsDNA phage and papovaviruses in which no evidence for *pac*-site recognition exists. Have these viruses solved the problem in an alternate way, or has the recognition site been overlooked in these cases?

Many questions remain that pertain to nucleic acid packaging, whose solutions will no doubt surprise and be useful to us as have the past answers to related questions. These problems include the mode of condensation of nucleic acid in icosahedral shells, the energetics of the process, the relationship of the various packaging processes to the mode of penetration and uncoating of the nucleic acid in susceptible cells, how attempts at packaging chromosomes being replicated, transcribed, or translated are avoided, and the arrangement of DNA within icosahedral capsids and the very complex baculo- and poxviruses. An interesting general question is why the packaging processes of the different viruses vary so greatly in complexity. How is it that simple viruses such as the RNA phages require only two phage-encoded proteins to build virions, yet some like phage T4 require ten proteins to package DNA and about 30 more to complete the virion? What does T4 gain by being so complex?

IVB. Nucleic Acid Packaging and Genetic Engineering

Fifteen years ago, most workers in the field of virus assembly saw the area as particularly interesting in terms of the complex biochemical processes involved. They also accepted the fact that at times the field was seen by others as rather esoteric, having only indirect application in the areas of macromolecular superstructure and therapy of virus-caused diseases. The explosive growth of the use of the powerful methods of molecular cloning and genetic engineering have brought the study of nucleic acid packaging new relevance. Bacterial geneticists had long known that phage were very useful as vectors for transferring genetic information from one bacteria to another via generalized and specialized transduction. The latter involves delivering packaged host and phage DNA, joined by *in vivo* processes, into desired host strains by the normal infection process. As techniques were developed to join DNA fragments of interest efficiently to viral vector chromosomes *in vitro*, getting these joined molecules into cells for amplification, screening and study became limiting for the overall efficiency of the process. Since phage can inject nearly every molecule of DNA packaged, it was clear that if newly constructed viral chromosomes could be packaged *in vitro* more efficiently than they entered cells by transfection, the efficiency of recovery of joined molecules could be raised. This is in fact possible with phage lambda vectors (Hohn, 1979; Enquist and Sternberg, 1979), and the *in vitro* packaging of lambda DNA has become a valuable tool in the acquisition of molecular clones of interest. This improvement could not have happened as rapidly (if at all) had there not been a large body of basic research on the morphogenesis of phages, *done without this final goal in mind*, which led to the first successful *in vitro* phage-DNA packaging by Kaiser and Masuda (1973). Other studies made clear what the size limits on packageable DNA are, showed that a *pac* site must be present to allow packaging, and made possible the design of the *in vitro* packaging systems and phage vectors. The potential use of animal and plant virions for the delivery of genetically engineered nucleic acid into susceptible cells should stimulate research on structure and packaging mechanisms in these systems. The phage work stands as a particularly clear monument to the ultimate practical value of pure basic research, and other unpremeditated uses of knowledge from the study of nucleic acid packaging will certainly emerge in the future.

ACKNOWLEDGEMENTS

I would like to thank M. Abouhaidar, R. Baldwin, M. Feiss, W. Fiers, S. Harrison, M. Hayashi, E. Jaspars, J. Kochan, J. King, Y. Okada, V. Parsegian, T. Schuster, N. Sternberg, G. Wengler, and N. Zinder for supplying unpublished observations used in the preparation of this chapter; V. Bloomfield, T. Schuster, and D. Summers for reviewing portions of the manuscript, and M. Hayden, L. Sampson, and E. Wyckoff for proofreading. My interest in this field was initiated and nurtured by Dale Kaiser and Jonathan King, and my research in the area is supported by NIH research grant GM21975 and NSF grant PCM8017177.

REFERENCES

Abad-Zapatero, C., Abdel-Meguid, J. Johnson, A. Leslie, I. Rayment, M. Rossmann, D. Suck, and T. Tsukihara. 1980. *Nature*, 286, 33.

Abou Haidar, M. and J. Bancroft. 1978a. *J. Gen. Virol.*, 39, 559.

Abou Haidar, M. and J. Bancroft. 1978b. *Virology*, 90, 54

Abou Haidar, M. and J. Bancroft. 1979. *Virology*, 93, 253.

Abou Haidar, M. and J. Bancroft. 1980. *Virology*, 107, 202.

Abou Haidar, M. and J. Erickson. 1983. In *Molecular Plant Virology*. J. Davies, ed., CRC Press, Inc. Vol. 3.

Abou Haidar, M. and L. Hirth. 1977. *Virology*, 76, 173.

Abou Haidar, M., P. Pfeiffer, C. Fritsch, and L. Hirth. 1973. *Virology*, 21, 83.

Adams, M., M. Hayden, and S. Casjens. 1983. *J. Virol.*, 46, 673.

Adolph, K. and P. Butler. 1974. *J. Mol. Biol.*, 88, 603.

Adolph, K. and P. Butler. 1977. *J. Mol. Biol.*, 109, 345.

Adrian, M., J. Dubochet, J. Lepault, and A. McDowell. 1984. *Nature*, 308, 32.

Akutsu, H., H. Satake, and R. Franklin. 1980. *Biochemistry*, 19, 5264.

Al Ani, R., P. Pfeiffer, and G. Lebeurier. 1979. *Virology*, 93, 188.

Alberts, B., L. Frey, and H. Delius. 1972. *J. Mol. Biol.*, 68, 139.

Alma, N., B. Harmsen, E. de Jong, J. Ven, and C. Hilbers. 1983. *J. Mol. Biol.*, 163, 47.

Ames, B. and D. Dubin. 1960. *J. Biol. Chem.*, 235, 769.

Amesse, L., C. Pridgen, and D. Kingsbury. 1982. *Virology*, 118, 17.

Antczak, J., R. Chmelo, D. Pickup, and W. Joklik. 1982. *Virology*, 121, 307.

Aoyama, A., M. Hamatake, and M. Hayashi. 1981. *Proc. Nat. Acad. Sci. USA*, 78, 7285.

Aoyama, A., M. Hamatake, and M. Hayashi. 1983a. *Proc. Nat. Acad.*

Sci. USA, 80, 4195.

Aoyama, A., M. Hamatake, R. Mukai, and M. Hayashi. 1983b. *J. Biol. Chem.*, 258, 5798.

Aoyama, A. and M. Hayashi. 1982. *Nature*, 297, 704.

Aposhian, H. 1975. In *Comprehensive Virology*. Vol 5. H. Fraenkel-Conrat and R. Wagner, eds. p. 155. Plenum.

Argos, P. 1981. *Virology*, 110, 55.

Ashikawa, I., T. Furuno, K. Kinosita, A. Ikegami, H. Takahashi, and H. Akutsu. 1984. *J. Biol. Chem.*, 259, 8338.

Armstrong, J., S. Smeekens, and P. Rottier. 1983. *Nucleic Acids Res.*, 11, 883.

Astell, C., M. Thomson, M. Chow, and D. Ward. 1982. *Cold Spring Harbor Symp. Quant. Biol.*, 47, 751.

Astell, C., M. Thomson, M. Merchlinsky, and D. Ward. 1983. *Nucleic Acids Res.*, 11, 999.

Atebekov, J. 1977. In *Comprehensive Virology*. Vol 11. H. Fraenkel-Conrat and R. Wagner, eds. p.143. Plenum.

Auperin, D., K. Dimmock, P. Cash, W. Rawls, W. Leung, and D. Bishop. 1982. *Virology*, 116, 363.

Bachi, B. and W. Arber. 1977. *Mol. Gen. Genet.*, 153, 311.

Bachrach, H. 1964. *J. Mol. Biol.*, 8, 348.

Bader, J., N. Brown, and A. Bader. 1970. *Virology*, 41, 718.

Bamford, D., T. McGraw, G. MacKenzie, and L. Mindich. 1983. *J. Virol.*, 47, 311.

Bancroft, J., M. Abouhaidar, and J. Erickson. 1979. *Virology*, 98, 121.

Bancroft, J., E. Hiebert, and C. Bracker. 1969. *Virology*, 39, 924.

Banner, D., C. Nave, and D. Marvin. 1981. *Nature*, 289, 814.

Bar-Joseph, M. and R. Hull. 1974. *Virology*, 62, 552.

Beachy, R. and M. Zaitlin. 1977. *Virology*, 81, 160.

Bear, S., D. Court, and D. Friedman. 1984. *J. Virol.* 52, 966.

Becker, A., H. Murialdo, and M. Gold. 1977. *Virology*, 78, 277.

Bisaro, D. and A. Siegel. 1982. *Virology*, 118, 411.

Bishop, D., K. Gould, H. Akashi, and M. Clerx-van Haaster. 1982. *Nucleic Acids Res.*, 10, 3703.

Bishop, D. and R. Shope. 1979. In *Comprehensive Virology*. Vol 14. H. Fraenkel-Conrat and R. Wagner, eds. p.1. Plenum.

Bjornsti, M., B. Reilly, and D. Anderson. 1982. *J. Virol.*, 41, 508.

Bjornsti, M., B. Reilly, and D. Anderson. 1983. *J. Virol.*, 45. 383.

Bjornsti, M., B. Reilly, and D. Anderson. 1984. *J. Virol.*, 50, 766.

Black, B. and M. Center. 1979. *Nucleic Acids Res.*, 6, 2339.

Black, L. and M. Showe. 1983. In *Bacteriophage T4*. C. Mathews, E. Kutter, G. Mosig, and P. Berget, eds. p. 219. American Society for Microbiology.

Black, L, and D. Silverman. 1978. *J. Virol.*, 28, 643.

Blanco, L., J. Garcia, M. Penalva, and M. Salas. 1983. *Nucleic Acids Res.*, 11, 1309.

Bloomer, A., J. Champness, G. Bricogne, R. Staden, and A. Klug. 1978. *Nature*, 276, 362.

Blumberg, B., C. Gorgi, and D. Kolakofsky. 1983. *Cell*, 32, 559.

Blumberg, B. and D. Kolakofsky. 1981. *J. Virol.*, 40, 568.

Blumberg, B., M. Leppert, and D. Kolakofsky. 1981. *Cell*, 23, 837.

Bol, J., B. Kraal, and F. Brederode. 1974. *Virology*, 58, 101.

Bol, J, and J. Kruseman. 1969. *Virology*, 37, 485.

Bonner, P. 1974. *J. Virol.*, 14, 1152.

Boublik, M. and R. Dzerniek. 1976. *J. Gen. Virol.*, 31, 447.

Bowden, D. and R. Calendar. 1979. *J. Mol. Biol.*, 129, 1.

Bozarth, R., Y. Koltin, M. Weissman, R, Parker, R. Dalton, and R. Steinlauf. 1981. *Virology*, 113, 492.

Brady, J., J. Kendall, and R. Consigli. 1979. *J. Virol.*, 32, 640.

Brady, J., C. Lavaille, M. Rodonovich, and N. Salzman. 1981a. *J. Virol.*, 39, 432.

Brady, J., M. Rodonovich, C. Lavaille, and N. Salzman. 1981b. *J. Virol.*, 39, 603.

Briand, J., J. Bouley, G. Jonard, J. Witz, and L. Hirth. 1975. *Virology*, 63, 304.

Briand, J., G. Keith, and H. Guilley. 1978. *Proc. Nat. Acad. Sci. USA*, 75, 3168.

Brown, D., M. Westphal, B. Burlingham, U. Winterhoff, and W. Doerfler. 1975. *J. Virol.*, 16, 366.

Brown, M. and J. Weber. 1980. *Virology*, 107, 306.

Burgess, A. 1969. *Proc. Nat. Acad. Sci. USA*, 64, 613.

Butler, P. 1971. *Cold Spring Harbor Symp. Quant. Biol.*, 36, 461.

Butler, P. and A. Durham. 1977. *Adv. Protein Res.*, 31, 187.

Butler, P., A. Durham, and A. Klug. 1972. *J. Mol. Biol.*, 72, 1.

Butler, P., J. Finch, and D. Zimmern. 1977. *Nature*, 265, 216.

Butler, P. and A. Klug. 1971. *Nature New Biol.*, 229, 48.

Butler, P. and G. Lomonossoff. 1978. *J. Mol. Biol.*, 126, 877.

Butler, P. and G. Lomonossoff. 1980. *Biophys. J.*, 32, 295.

Carrascosa, J., E. Vinuela, E. Garcia, and A. Santisteban. 1982. *J. Mol. Biol.*, 154, 311.

Carroll, D. and F. O'Neill. 1981. *Virology*, 112, 461.

Cartwright, B., C. Smale, F. Brown, and R. Hull. 1972. *J. Virol.*, 10, 256.

Casadevall, A. and L. Day. 1982. *Nucleic Acids Res.*, 10, 2467.

Casjens, S. and W. Huang. 1982. *J. Mol. Biol.*, 157, 287.

Casjens, S. and J. King. 1974. *J. Supramol. Struct.*, 2, 202.

Casjens, S. and J. King. 1975. *Ann. Rev. Biochem.*, 44, 555.

Caspar, D. 1963. *Adv. Protein Chem.*, 18, 37.

Caspar, D. 1980. *Biophys. J.*, 32, 103.

Caspar, D. 1965. In *Viral and Rickettsial Infections of Man* 4th Edition, P. Horsfall and I. Tamm, eds. p. 51. Lippincott.

Caspar, D. and A. Klug. 1962. *Cold Spring Harbor Symp. Quant. Biol.*, 27, 1.

Caspar, D. and G. Stubbs. 1983. In *Plant Infectious Agents*. H. Robertson, S. Howell, M. Zaitlin, and R. Malmberg, eds. p. 88. Cold Spring Harbor.

Castilho, B., P. Olfson, and M. Casadaban. 1984. *J. Bacteriol.*, 158, 488.

Challberg, M. and T. Kelly. 1981. *J. Virol.*, 38, 272.

Challberg, M., J. Ostrove, and T. Kelly. 1982. *J. Virol.*, 41, 265.

Chauvin, C., J. Witz, and B. Jarcot. 1978. *J. Mol. Biol.*, 124, 641.

Chee-Sheung, C. and H. Ginsberg. 1982. *J. Virol.*, 42, 932.

Choppin, P. and R. Compans. 1975. In *Comprehensive Virology*. Vol 4. H. Fraenkel-Conrat and R. Wagner, eds. p.95. Plenum.

Chow, L. and A. Bukhari. 1977. In *DNA Insertion Elements, Plasmids and Episomes*. A. Bukhari, J. Shapiro and S. Adhya, eds. p. 295. Cold Spring Harbor.

Clark, S., R. Spendlove, and B. Barnett. 1980. *J. Virol.*, 46. 378.

Coca-Prados, M., H. Yu, and M. Hsu. 1982. *J. Virol.*, 44, 603.

Coffin, J. 1982. In *Molecular Biology of Tumor Viruses*. R. Weiss, N. Teich, H. Varmus, and J. Coffin, eds. p. 261. Cold Spring Harbor.

Cohen, S. and F. McCormick. 1979. In *Adv. Virus Res.* Vol. 24. M. Lauffer, F. Bang, K. Maramorosch, and K. Smith, eds. p. 331. Academic Press.

Collmer, C., V. Vogt, and M. Zaitlin. 1983. *Virology*, 126, 429.

Collmer, C. and M. Zaitlin. 1983. *Virology*, 126, 449.

Compans, R. and P. Choppin. 1975. In *Comprehensive Virology*.. Vol 4. H. Fraenkel-Conrat and R. Wagner, eds. p.179. Plenum.

Compans, R., J. Content, and P. Duesberg. 1972. *J. Virol.*, 10, 795.

Corden, J., H. Engelking, and G. Pearson. 1976. *Proc. Nat. Acad. Sci. USA*, 73, 401.

Cornelissen, B., F. Bredorode, G. Veeneman, J. van Boom, and J. Bol. 1983. *Nucleic Acids Res.*, 11, 3019.

Correia, J., S. Shire, D. Yphantis, and T. Schuster. 1981. *Biophys. J.* 33, 254a.

Crane, H. 1950. *The Scientific Mon.*, 70, 376.

Cregg, J. and C. Stewart. 1978. *Virology*, 86, 530.

Crick, F. and J. Watson. 1956. *Nature*, 177, 473.

Cross, T., S. Opella, G. Stubbs, and D. Caspar. 1983. *J. Mol. Biol.*, 170, 1037.

Crowther, A., L. Amos, and J. Finch. 1975. *J. Mol. Biol.*, 98, 631.

Cuillel, M., M. Herzog, and L. Hirth. *Virology*, 95, 146.

Cuillel, M., F. Tripier, J. Braunwald, and B. Jarcot. 1979. *Virology*, 99, 277.

Cuillel, M., M. Zulauf, and B. Jarcot. 1983. *J. Mol. Biol.*, 164, 589.

Curtiss, L. and R. Krueger. 1974. *J. Virol.*, 14, 503.

Daniell, E. 1976. *J. Virol.*, 19, 685.

Daniell, E., D. Groft, and M. Fedor. 1981. *Mol. Cell Biol.*, 1, 1094.

Darai, G., K. Anders, H. Koch, H. Delius, H. Geldblom, C. Samalecos, and R. Flugel. 1983. *Virology*, 126, 466.

Darlix, J., L. Schwager, P. Spahr, and P. Bromley. 1980. *Nucleic Acids Res.*, 8, 3335.

Darlix, J. and P. Spahr. 1982. *J. Mol. Biol.*, 160, 147.

Davis, T. and J. Cronan. 1983. *Virology*, 126, 600.

Day, L., and R. Wiseman. 1978. In *The Single-Stranded DNA Phages*. D. Dressler and D. Ray, eds. p. 605. Cold Spring Harbor.

Day, L., R. Wiseman, and C. Marzec. 1979. *Nucleic Acids Res.*, 7, 1393.

Deichelbohrer, I., W. Messer, and T. Trautner. 1982. *J. Virol.*, 42, 83.

Delius, H., D. Darai, and R. Flugel. 1984. *J. Virol.*, 49, 609.

Demsey, A., F. Collins, and D. Kawka. 1980. *J. Virol.*, 36, 872.

Denloye, A., R. Homer, and R. Hull. 1978. *J. Gen. Virol.*, 41, 77.

D'Halluin, J., M. Milleville, P. Boulanger, and G. Martin. 1978. *J. Virol.*, 26, 344.

Dickerson, R. 1983. *J. Mol. Biol.*, 166, 419.

Dickson, C., R. Eisenman, H. Fan, E. Hunter, and N. Teich. 1982. In *Molecular Biology of Tumor Viruses*. R. Weiss, N. Teich, H. Varmus, and J. Coffin, eds. p. 513. Cold Spring Harbor.

Diener, T. 1979. *Viroids and Viroid Diseases*. p. 143. Wiley.

DiVerdi, J. and S. Opella. 1981. *Biochemistry*, 20, 380.

Dobbs, P., R. Hallet, D. Kells, O. Sorenson, and D. Rowe. 1977. *J. Virol.*, 22, 150.

Dotto, G., V. Enea, and N. Zinder. 1981. *Virology*, 114, 463.

Drake, J. 1966. *Proc. Nat. Acad. Sci. USA*, 55, 506.

Driedonks, R., A. Engel, B. tenHeggeler, and R. van Driel. 1981. *J. Mol. Biol.*, 152, 641.

Driedonks, R., P. Krigsman, and J. Mellema. 1976. *Phil. Trans. Soc. Lond.*, B276, 131.

Driedonks, R., P. Krigsman, and J. Mellema. 1977. *J. Mol. Biol.*, 113, 123.

Driedonks, R., P. Krigsman, and J. Mellema. 1978a. *Eur. J. Biochem.*, 82, 405.

Driedonks, R., P. Krigsman, and J. Mellema. 1978b. *J. Mol. Biol.*, 124, 713.

Dubra, M., J. La Torre, E. Scodeller, C. Denoya, and C. Vasquez. 1982. *Virology*, 116, 349.

Dunker, K. 1981. In *Progress in Clinical and Biological Research*. Vol. 64. M. DuBow, ed. p. 383. Alan R. Liss, Inc., New York.

Dunker, K. and W. Paranchych. 1975. *Virology*, 67, 297.

Dunn, J. and F. Studier. 1983. *J. Mol. Biol.*, 166, 477.

Durbin, R. and J. Manning. 1982. *Virology*, 116, 31.

Durham, A. 1972. *J. Mol. Biol.*, 67, 289.

Durham, A. and J. Bancroft. 1979. *Virology*, 93, 246.

Durham, A., J. Finch, and A. Klug. 1971. *Nature New Biol.*, 229, 37.

Durham, A. and A. Klug. 1971. *Nature*, 177, 42

Earnshaw, W. and S. Casjens. 1980. *Cell*, 21, 319.

Earnshaw, W. and S. Harrison. 1977. *Nature*, 286, 598.

Earnshaw, W., R. Hendrix, and J. King. 1979. *J. Mol. Biol.*, 134, 575.

Earnshaw, W., J. King, S. Harrison, and F. Eiserling. 1978. *Cell*, 14. 559.

Ebel-Tsipis, J., D. Botstein, and M. Fox. 1972. *J. Mol. Biol.*, 71, 433.

Edvardsson, B., S. Ustacelebi, J. Williams, and L. Philipson. 1978. *J. Virol.*, 25, 641.

Emmons, S. 1974. *J. Mol. Biol.*, 83, 511.

Emori, Y., H. Iba, and Y. Okada. 1982. *J. Mol. Biol.*, 154, 287.

Enea, V. and N. Zinder. 1975. *Virology*, 68, 105.

Enquist, L. and N. Sternberg. 1979. *Methods in Enzym.*, 68, 281. Academic Press.

Enzmann, P. and F. Weiland. 1979. *Virology*, 95, 501.

Erickson, J., M. Abouhaidar, and J. Bancroft. 1978. *Virology*, 90, 60.

Erickson, J. and J. Bancroft. 1978. *Virology*, 90, 36.

Erickson, J. and J. Bancroft. 1981. *Virology*, 108, 235.

Erickson, J. and M. Rossmann. 1982. *Virology*, 116, 128.

Erikson, E., J. Brugge, and R. Erikson. 1977. *Virology*, 80, 177.

Everitt, E., L. Lutter, and L. Philipson. 1975. *Virology*, 67, 197.

Everitt, E., S. Meador, and A. Levine. 1977. *J. Virol.*, 21, 199.

Faust, E. and D. Ward. 1979. *J. Virol.*, 32, 276.

Feiss, M., and A. Becker. 1983. In *Lambda II*. R. Hendrix, J. Roberts, F. Stahl, and R. Weisberg, eds. p. 305. Cold Spring Harbor.

Feiss, M., R. Fisher, D. Siegele, M. Crayton, and C. Egner. 1977. *Virology*, 77, 281.

Feiss, M., I. Kobayashi, and W. Widner. 1983. *Proc. Nat. Acad. Sci. USA*, 80, 955.

Fiers, W. 1979. In *Comprehensive Virology*.. Vol 13. H. Fraenkel-C onrat and R. Wagner, eds. p.69. Plenum.

Finch, J. and A. Gibbs. 1970. *J. Gen. Virol.*, 6, 141.

Frackman, S., D. Siegele, and M. Feiss. 1984. *J. Mol. Biol.* 180, 283.

Fraenkel-Conrat, H. and R. Williams. 1955. Proc. Nat. Acad. Sci. U.S.A., 41, 690.

Franck, A., H. Guilley, G. Jonard, K. Richards, and L. Hirth. 1980. *Cell*, 21, 285.

Franklin, R. 1955. *Biochim. Biophys. Acta*, 18, 313.

Freymeyer, D., P. Shank, M. Edgell, C. Hutchison, and T. Vanaman. 1977. *Biochemistry*, 16, 4550.

Frisby, D., R. Cotter, and B. Richards. 1977. *J. Gen. Virol.*, 37, 311.

Fugisawa, H. and M. Hayashi. 1976. *J. Virol.*, 19, 409.

Fugisawa, H. and M. Hayashi. 1977a. *J. Virol.*, 21, 506.

Fugisawa, H. and M. Hayashi. 1977b. *J. Virol.*, 24, 303.

Fukuda, M., T. Meshi, Y. Okada, Y. Otsuki, and I. Takebe. 1981. *Proc. Nat. Acad. Sci. USA*, 78, 4231.

Fukuda, M., T. Ohno, Y. Okada, Y. Otsuki, and I. Takebe. 1978. *Proc. Nat. Acad. Sci. USA*, 75, 1727.

Fukuda, M. and Y. Okada. 1982. *Proc. Nat. Acad. Sci. USA*, 79, 5836.

Fukuyama, K., S. Abdel-Meguid, and M. Rossmann. 1981. *J. Mol. Biol.*, 150, 33.

Fulton, A. 1982. *Cell*, 30, 345.

Gaillard, R., J. Li, J. Keene, and W. Joklik. 1982. *Virology*, 121, 320.

Gallione, C., J. Greene, L. Iverson, and J. Rose. 1981. *J. Virol.*, 39, 529.

Garber, E., M. Seidman, and A. Levine. 1980. *Virology*, 107, 389.

Garoff, H., A. Frischauf, K. Simons, H. Lehrach, and H. Delius. 1980. *Proc. Nat. Acad. Sci. USA*, 77, 6376.

George, M. and A. Bukhari. 1981. *Nature*, 292, 175.

Georgopoulos, C., K. Tilly, and S. Casjens. 1983. In *Lambda II*. Hendrix, J. Roberts, F. Stahl, and R. Weisberg, eds. p.279. Cold Spring Harbor.

Ghosh, K. and H. Ghosh. 1982. Nucleic Acid Res., 10, 6341.

Gillin, F. and V. Bode. 1972. *J. Virol.*, 10, 863.

Ginsberg, H. 1979. In *Comprehensive Virology*. Vol 13. H. Fraenkel-Conrat and R. Wagner, eds. p.409. Plenum.

Godson, N., J. Fiddes, B. Barrell, and F. Sanger. 1978. In *The Single-Stranaed DNA Phages*. D. Denhardt, D. Dressler and D. Ray, eds. p.51. Cold Spring Harbor.

Goelet, P., G. Lomonossoff, P. Butler, M. Akam, M. Gait, and J. Karn. 1982. *Proc. Nat. Acad. Sci. USA*, 79, 5818.

Gold, M. and M. Becker. 1983. *J. Biol. Chem.*, 258, 14619.

Goodman, R. 1981. *J. Gen. Virol.*, 54, 9.

Goodman, R., J. McDonald, R. Horne, and J. Bancroft. 1976. *Phil. Trans. Roy. Soc. Lond.*, B276, 173.

Goorha, R. and K. Murti. 1982. *Proc. Nat. Acad. Sci. USA*, 79, 248.

Gope, R. and P. Serwer. 1983. *J. Virol.*, 47, 96.

Gordon, K. and R. Symons. 1983. *Nucleic Acids Res.*, 11, 947.

Gosule, L. and J. Schellman. 1976. *Nature*, 259, 333.

Grant, R., T. Lin, W. Kingsberg, and R. Webster. 1981. *J. Biol. Chem.*, 256, 539.

Gray, C., G. Kneale, K. Leonard, H. Siegrist, and D. Marvin. 1982. *Virology*, 116, 40.

Griffin, B. 1980. In *DNA Tumor Viruses*. J. Tooze, ed. p. 61. Cold Spring Harbor.

Gronenborn, B., R. Gardner, S. Schaefer, and R. Shepherd. 1981.

Nature, 294, 773.

Gunther, M. and B. Revet. 1978. In *Replication of Mammalian Parvoviruses*. D. Ward and P. Tattersall, eds. p. 469. Cold Spring Harbor.

Haas, R., R. Murphy, and C. Cantor. 1982. *J. Mol. Biol.*, 159, 71.

Hafner, E., C. Tabor, and H. Tabor. 1979 *J. Biol. Chem.*, 254, 12419.

Hagen, F. and A. Huang. 1981. *J. Virol.*, 37, 363.

Hall, T. 1979. In *Inter. Rev. Cytology*, vol. 60. G. Bourne and J. Danielli, eds. p. 1. Academic Press.

Hammarskjold, M. and G. Winberg. 1980. *Cell*, 20, 787.

Hammarskjold, M., G. Winberg, E. Norrby, and G. Wadell. 1977. *Virology*, 82, 449.

Hari, V. 1981. *Virology*, 112 391.

Harms, E., W. Rhode, R. Friis, and H. Baurer. 1977. *J. Virol.*, 21, 419.

Harrison, B. and D. Robinson. 1978. *Adv. Virus Res.*, 23, 25.

Harrison, S. 1983. *J. Mol. Biol.*, 171, 577.

Harrison, S., A. Jack, D. Goodenough, and B. Sefton. 1974. *J. Supramol. Struct.*, 2, 486.

Harrison, S., A. Olson, C. Schutt, and F. Winkler. 1978. *Nature*, 276, 368.

Harvey, J., A. Bellamy, W. Earnshaw, and C. Schutt. 1981. *Virology*, 112, 240.

Haseloff, J. and R. Symons. 1982. *Nucleic Acids Res.*, 10, 3681.

Hatta, T. and R. Franki. 1979. *Virology*, 92, 428.

Heggeness, M., A. Scheid, and P. Choppin. 1981. *Virology*, 114, 555.

Heggeness, M., P. Smith, I. Ulmanen, R. Krug, and P. Choppin. 1982. *Virology*, 118, 466.

Heisenberg, M. 1966. *J. Mol. Biol.*, 17, 136.

Hendrix, R. 1978. *Proc. Nat. Acad. Sci. USA*, 75, 4779.

Hewlett, M., R. Pettersson, and D. Baltimore. 1977. *J. Virol.*, 21, 1085.

Hill, D. and G. Petersen. 1982. *J. Virol.*, 44, 32.

Hill, V., L. Marnell, and D. Summers. 1981. *Virology*, 113, 109.

Hill, V., C. Simonsen, and D. Summers. 1979. *Virology*, 99, 75.

Hirst, G. 1973. *Virology*, 55, 81.

Hirth, L. 1976. In *Comprehensive Virology*. Vol 6. H. Fraenkel-Conrat and R. Wagner, eds. p.39. Plenum.

Hirth, L. and K. Richards. 1981. *Adv. Virus Res.*, 26, 145.

Hogan, M., J. LeGrange, and B. Austin. 1983. *Nature*, 304, 752.

Hogle, J. 1982. *J. Mol. Biol.*, 160, 663.

Hohn, B. 1979. *Meth. Enzym.*, 68, 299, Academic Press.

Hohn, B. 1983. *Proc. Nat. Acad. Sci. USA*, 80, 7456.

Hohn, B., M. Wurtz, B. Klein, A. Lustig, and T. Hohn. 1974. *J. Supramol. Struct.*, 2, 302.

Hohn, T. 1969. *J. Mol. Biol.*, 43, 191.

Hosaka, Y., H. Kitano, and S. Ikeguchi. 1966. *Virology*, 29, 205.

Houwing, C., and E. Jaspars. 1982. *Biochemistry*, 21, 3408.

Howarth, A., R. Gardner, J. Messing, and R. Shepherd. 1981. *Virology*, 112, 678.

Hsiao, C. and L. Black. 1978. *Virology*, 91, 26.

Hsu, C., D. Kingsbury, and K. Murti. 1979. *J. Virol.*, 32, 304.

Hull, R. 1970. *Virology*, 40, 34.

Hull, R. and S. Covey. 1983a. *Trends Biochem. Sci.*, 8, 119.

Hull, R. and S. Covey. 1983b. *Nucleic Acids Res.*, 11, 1881.

Humphreys, G., and T. Trautner. 1981. *J. Virol.*, 37, 832.

Hunter, E., J. Bennett, A. Bhown, R. Pepinsky, and V. Vogt. 1983. *J. Virol.*, 45, 885.

Iba, H., T. Watanabe, Y. Emori, and Y. Okada. 1982. *FEBS Lett.*, 141, 111.

Ikeda, J., T. Enomoto, and J. Hurwitz. 1982. *Proc. Nat. Acad. Sci. USA*, 79, 2442.

Ishibashi, M. and J. Maizel. 1974. *Virology*, 57, 409.

Ito, J. 1978. *J. Virol.*, 28, 895.

Jackson, E., D. Jackson, and R. Deans. 1978. *J. Mol. Biol.*, 118, 365.

Jackson, E., F. Laski, and C. Andres. 1982. *J. Mol. Biol.*, 154, 551.

Jackson, V. and R. Chaukley. 1981. *Proc. Nat. Acad. Sci. USA*, 78, 6081.

Jamjoon, G., R. Naso, and R. Arlinghaus. 1977. *Virology*, 78, 11.

Jarcot, B., C. Chauvin, and J.Witz. 1977. *Nature*, 266, 417.

Jay, G., S. Nomura, C. Anderson, and G. Khoury. 1981. *Nature*, 291, 346.

Jennings, P., J. Finch, G. Winter, and J. Robertson. 1983. *Cell*, 34, 619.

Jonard, G., K. Richards, H. Guilley and L. Hirth. 1977. *Cell*, 11, 483.

Jonard, G., J. Witz, and L. Hirth. 1972. *J. Mol. Biol.*, 67, 165.

Jones, N., and T. Shenk. 1978. *Cell*, 13, 181.

Joshi, S., R. Joshi, A. Haenni, and F. Chapeville. 1983. *Trends Biochem. Sci.*, 8, 402.

Kaiser, A., and T. Masuda. 1973. *Proc. Nat. Acad. Sci. USA*, 70, 260.

Kaiser, A., M. Syvanen, and T. Masuda. 1975. *J. Mol. Biol.*, 91, 175.

Kaplan, I., Y. Kozlov, E. Pshennikova, M. Taliansky, and J. Atabekov. 1982. *Virology*, 118, 317.

Kayajanian, G. 1968. *Virology*, 36, 30

Kayajanian, G. and A. Campbell. 1966. *Virology*, 30, 482.

Keene, J., G. Thornton, and S. Emerson. 1981. *Proc. Nat. Acad. Sci. USA*, 78, 6191.

Kennedy, S. 1980. In *The Togaviruses*. R. Schlesinger, ed. p. 351.

Academic Press.

Kennell, D. and H. Riezman. 1977. *J. Mol. Biol.*, 114, 1.

Kiley, M. and R. Wagner. 1972. *J. Virol.*, 10, 244.

King, J., 1980. In *Biological Regulation and Development*. Vol. 3. R. Goldberger, ed. p.101. Plenum.

King, J., R. Griffin-Shea, and M. Fuller. 1980. *Quart. Rev. Biol.*, 55. 369.

Kitamura, N., B. Semler, P. Rothberg, G. Larsen, C. Adler, A. Dorner, E. Emini, R. Hanecak, J. Lee, S. van der Werf, C. Anderson, and E. Wimmer. 1981. *Nature*, 291, 547.

Kneale,G., R. Freeman, and D. Marvin. 1982. *J. Mol. Biol.* 156, 279.

Knolle, P. and T. Hohn. 1975. In *RNA Phages*. N. Zinder, ed. p. 147. Cold Spring Harbor.

Kobayashi, I., M. Stahl, D. Leach, and F. Stahl. 1983. *Genetics*, 104, 549.

Kochan, J., J. Carrascosa, and H. Murialdo. 1984. *J. Mol. Biol.* 174, 433.

Koper-Zwarthoff, E., and J. Bol. 1980. *Nucleic Acids Res.*, 8, 3307.

Koper-Zwarthoff, E., F. Brederode, G. Veeneman, J. van Boomn and J. Bol. 1980. *Nucleic Acids Res.*, 8, 5635.

Koths, K., and D. Dressler. 1980. *J. Biol. Chem.*, 255, 4328.

Kozloff, L., V. Chapman, and S. DeLong. 1981. In *Progress In Clinical and Biological Research*. Vol. 64. M. DuBow, ed. p. 253. Alan Liss.

Krahn, P., R. O'Callaghan, and W. Paranchych. 1972. *Virology*, 47, 628.

Krug, M., P. de Haseth, and O. Uhlenbeck. 1982. *Biochemistry*, 21, 4713.

Kruse, J., P. Timmins, and J. Witz. 1982. *Virology*, 119, 42.

Kuchino, Y., S. Nishimura, R. Smith, and Y. Furiuchi. 1982. *J. Virol.*, 44, 538.

Kufer, B., H. Backhaus, and H. Schmieger. 1982. *Mol. Gen. Genet.*, 187, 510.

Ladin, B., M. Blankenship, and T. Ben-Porat. 1980. *J. Virol.*, 33, 1151.

Larson, S. 1982. *Virology*, 116, 573.

Laski, F. and E. Jackson. 1982. *J. Mol. Biol.*, 154, 565.

Lazzarini, R., J. Keene, and M. Schubert. 1981. *Cell*, 26, 145.

Lebeurier, G., H. Fraenkel-Conrat, M. Wurtz, and L. Hirth. 1971. *Virology*, 43, 51.

Lebeurier, G., A. Nicolaieff, and K. Richards. 1977. *Proc. Nat. Acad. Sci. USA*, 74, 149.

Ledneva, R., T. Lanina, G. Teganova and A. Bogdanov. 1980. *Nucleic Acids Res.*, 8, 5129.

Lee, T., W. Brockman, and D. Nathans. 1975. *Virology*, 66, 53.

Lee, T. and D. Nathans. 1979. *Virology*, 92, 291.

Leibowitz, J., K. Wilhelmsen, and C. Bond. 1981. *Virology*, 114, 39.

Leis, L., J. McGinnis, and R. Greene. 1978. *Virology*, 84, 87.

Lenk, E., S. Casjens, J. Weeks, and J. King. 1976. *Virology*, 68, 182.

Lesnaw, J. and R. Reichmann. 1970. *Proc. Nat. Acad. Sci. USA*, 66, 140.

Levin, J. and J. Seidmen. 1979. *J. Virol.*, 29, 328.

Liljas, L., T. Unge, T. Jones, K. Fridborg, S. Lovgren, U. Skoglund, and B. Strandberg. 1982. *J. Mol. Biol.*, 159, 93.

Ling, C., P. Hung, and L. Overby. 1970. *Virology*, 40, 920.

Liu, L., L. Perkocha, R. Calendar, and J. Wang. 1981. *Proc. Nat. Acad. Sci. USA*, 78, 5498.

Lodish, H. 1968. *J. Mol. Biol.*, 32, 681.

Lofdahl, S., J. Zabielski, and L. Philipson. 1981. *J. Virol.*, 37, 784.

Lomonossoff, G. and P. Butler. 1979. *Eur. J. Biochem.*, 93, 157.

Lopez, J. and R. Webster. 1983. *Virology*, 127, 177.

Luftig, R., W. Wood, and R. Okinada. 1971. *J. Mol. Biol.*, 57, 555.

Lundquist, R., M. Sullivan, and J. Maizel. 1979. *Cell*, 18, 759.

Lynch, S. and D. Kolakofsky. 1978. *J. Virol.*, 28, 584.

MacHattie, L. and G. Gill. 1977. *J. Mol. Biol.*, 110, 441.

MacHattie, L., D. Ritchie, and C. Thomas. 1967. *J. Mol. Biol.*, 23, 355.

Makowski, L. and D. Caspar. 1981. *J. Mol. Biol.*, 145, 611.

Makowski, L., D. Caspar, D. Marvin. 1981. *J. Mol. Biol.*, 140, 149.

Mandelkow, E., G. Stubbs, and S. Warren. 1981. *J. Mol. Biol.*, 152, 375.

Mann, R., R. Mulligan, and D. Baltimore. 1983. *Cell*, 33, 153.

Manne, V., V. Rao, and L. Black. 1982. *J. Biol. Chem.*, 257, 13223.

Marini, J., S. Levene, D. Crothers, and P. Englund. 1982. *Proc. Nat. Acad. Sci. USA*, 79, 7668.

Marks, J. and D. Spector. 1984. *J. Virol.*, 52, 24.

Marvin, D. 1978. In *The Single-Stranded DNA Phages*. D. Denhardt, D. Dressler and D. Ray, eds. p. 583. Cold Spring Harbor.

Marvin, D. and E. Wachtel. 1975. *Nature*, 253, 19.

Marvin, D., R. Wiseman, and E. Wachtel. 1974. *J. Mol. Biol.*, 82, 121.

Marzec, C. and L. Day. 1980. *Biophys. J.*, 32, 240.

Matthew, B., D. Ohlendorf, W. Anderson, and Y. Takeda. 1982. *Proc. Nat. Acad. Sci. USA*, 79, 1428.

Matthews, R. 1981. *Plant Virology*, 2nd ed., Academic Press.

Matthews, R. 1982. *Intervirology*, 17, 1.

Mayo, M. and D. DeMarcillac. 1977. *Virology*, 76, 560.

McCain, D., R. Virudachalam, R. Santini, A. Abdel-Meguid, and J. Markley. 1982. *Biochemistry*, 21, 5390.

McCrae, M. 1981. *J. Gen. Virol.*, 55, 393.

McPherson, A., F. Jurnak, A. Wang, F. Kolpak, A. Rich, I. Molineux, and P. Fitzgerald. 1980. *Biophys. J.*, 32, 155.

Mellon, M. and S. Emerson. 1978. *J. Virol.*, 27, 560.

Meric, C., J. Darlix, and P. Spahr. 1984. *J. Mol. Biol.* 173, 531.

Mertz, J., A. Murphy, and A. Barkan. 1983. *J. Virol.*, 45, 36.

Meshi, T., R. Kiyama, T. Ohno, and Y. Okada. 1983. *Virology*, 127, 54.

Meshi, T., T. Ohno, H. Iba, and Y. Okada. 1981. *Molec. Gen. Genet.*, 184, 20.

Metcalf, P. 1982. *J. Ultrastr. Res.*, 78, 292.

Milavetz, B. and T. Hopkins. 1982. *J. Virol.*, 43, 830.

Miller, P., D. Scraba, M. Leyritz-Wills, K. Maltman, and R. Warren. 1983. *J. Virol.*, 47, 399.

Mindich, L. 1978. In *Comprehensive Virology*. Vol 12. H. Fraenkel-Conrat and R. Wagner, eds. p.271. Plenum.

Mindich, L., D. Bamford, T. McGraw, and G. MacKenzie. 1982. *J. Virol.*, 44, 1021.

Mirza, M. and J. Weber. 1977. *Virology*, 80, 83.

Mocarski, E. and B. Roizman. 1982. *Cell*, 31, 89.

Morelli, G., C. Fisseau, B. Behrens, T. Trautner, J. Luh, S. Ratcliff, D. Allison, and A. Ganesan. 1979. *Molec. Gen. Genet.*, 168, 165.

Morris, T. and J. Semancik. 1973. *Virology*, 53, 215.

Mosig, G., D. Ghosal, and S. Bock. 1981. In *Progress In Clinical and Biological Research*. Vol. 64. M DuBow, ed. p. 139. Alan R. Liss.

Moyne, G., F. Harper, S. Saragosti, and M. Yaniv. 1983. *Cell*, 30, 123.

Moynet, D. and F. DeFelippes. 1982. *Virology*, 117, 475.

Mukai, R., R. Hamatake, and M. Hayashi. 1979. *Proc. Nat. Acad. Sci. USA*, 76, 4877.

Muller, U., H. Zentgraf, I. Eicken, and W. Keller. 1978. *Science*, 201, 1406.

Munowitz, M., C. Dobson, R. Griffin, and S. Harrison. 1980. *J. Mol. Biol.*, 141, 327.

Murialdo, H. and W. Fife. 1984. *Gene*, 30, 183.

Murti, K., R. Goorha, and A. Granoff. 1982. *Virology*, 116, 275.

Muzyczka, N. 1980. *Gene*, 11, 63.

Naeve, C., C. Kolakofsky, and D. Summers. 1980. *J. Virol.*, 33, 856.

Nakashima, Y., B. Frangione, R. Wiseman, and W. Konigsberg. 1981. *J. Biol. Chem.*, 256, 5792.

Nakashima, Y. and W. Konigsberg. 1980. *J. Mol. Biol.*, 138. 493.

Nasco, R., L. Stanker, J. Kopchich, V. Ng, W. Karshin, and R. Arlinghaus. 1983. *J. Virol.*, 45, 1200.

Nave, C., R. Brown, A. Fowler, J. Lander, and D. Marvin. 1981. *J. Mol. Biol.*, 149, 675.

Nayak, D. 1980. *Ann. Rev. Microbiol.*, 34, 619.

Nermut, N., H. Frank, and W. Schafer. 1972. *Virology*, 49, 345.

Newcomb, W., J. Boring, and J. Brown. 1984. *J. Virol.*, 51, 52.

Newcomb, W., G. Tobin, J. McGowan, and J. Brown. 1982. *J. Virol.*, 41, 1055.

Newman, J. and F. Brown. 1977. *J. Gen. Virol.*, 38, 83.

Newman, J., L. Day, G. Dalack, and D. Eden. 1982. *Biochemistry*, 21, 3352.

Ng, S. and M. Bina. 1984. *J. Virol.*, 50, 471.

Nonomuna, Y. and K. Kohama. 1974. *J. Mol. Biol.*, 86, 621.

Novikov, V., V. Kimaev, and J. Atabekov. 1972. *Dok. Akad. Nauk SSSR*, 204, 1259.

Obijeski, J., D. Bishop, E. Palmer, and F. Murphy. 1976. *J. Virol.*, 20, 664.

Obijeski, J., J. McCauly, and J. Skehel. 1980. *Nucleic Acids Res.*, 8, 2431.

Odumusu, R., B. Homer, and R. Hull. 1981. *J. Gen. Virol.*, 53, 193.

Ohno, T., M. Takahashi, and Y. Okada. 1977. *Proc. Nat. Acad. Sci. USA*, 74, 552.

O'Neill, F., E. Maryon, and D. Carroll. 1982. *J. Virol.*, 43, 18.

Otsuki, Y., I. Takebe, T. Ohno, M. Fukuda, and Y. Okada. 1977. *Proc. Nat. Acad. Sci. USA*, 74, 1913.

Ozer, H. and P. Tegtmeyer. 1972. *J. Virol.*, 9, 52.

Palmer, E., J. Obijeski, P. Webb, and K. Johnson. 1977. *J. Gen. Virol.*, 36, 541.

Paradiso, P. 1983. *J. Virol.*, 46, 94.

Paranchych, W., P. Krahn, and R. Bradley. 1970. *Virology*, 41, 465.

Patterson, J., C. Cabradilla, B. Holloway, J. Obijeski, and D. Kolakofsky. 1983a. *Cell*, 33, 791.

Patterson, J., D. Kolakofsky, B.Holloway, and J. Obijeski. 1983b. *J. Virol.*, 45, 882.

Patton, J., N. Davis, and G. Wertz. 1983. *J. Virol.*, 45, 155.

Persson, H., B. Mathison, L. Philipson, and U. Petterson. 1979. *Virology*, 93, 198.

Peterson, C., G. Dalack, L. Day, and W. Winter. 1982. *J. Mol. Biol.*, 162, 877.

Pettersson, R. and L. Kaariainen. 1973. *Virology*, 56, 608.

Pettersson, R. and C. von Bonsdorff. 1975. *J. Virol.*, 15, 386.

Pfeiffer, P. and T. Hohn. 1983. *Cell*, 33, 781.

Pons, M. 1970. *Curr. Top. Microbiol. Immunol.*, 52, 142.

Pons, M., I. Schulze, G. Hirst, and R. Hauser. 1969. *Virology*, 39, 250.

Poteete, A., and D. Botstein. 1979. *Virology*, 95, 565.

Pratt, D., P. Laws, J. Griffith. 1974. *J. Mol. Biol.*, 82, 425.

Preston, V., J. Coates, and F. Rixon. 1983. *J. Virol.*, 45, 1056.

Putnak, J. and B. Philips. 1981. *Microbiol. Rev.*, 45, 287.

Puvion-Dutilleul, F., J. Pedron, and M. Lange. 1980. *J. Gen. Virol.*, 51, 15.

Racaniello, V., and D. Baltimore. 1981. *Proc. Nat. Acad. Sci. USA*, 78, 4887.

Raghavendra, K., M. Adams, and T. Schuster. 1985. *Biochemistry*, in press.

Raj, A., A. Raj, and H. Schmieger. 1974. *Molec. Gen. Genet.*, 135, 175.

Rao, C., B. Lee, and V. Parsegian. 1985. *Proc. Nat. Acad. Sci. USA*, In press.

Rao, D. and A. Huang. 1982. *J. Virol.*, 41, 210.

Rawls, W., and W. Leung. 1979. In *Comprehensive Virology*. Vol 14. H. Fraenkel-Conrat and R. Wagner, eds. p.157. Plenum.

Revie, D., B. Tseng, R. Grafstrom, and M. Goulian. 1979. *Proc. Nat. Acad. Sci. USA*, 76, 5539.

Rhode, S. 1976. *J. Virol.*, 17, 659.

Rice, C., and J. Strauss. 1981. *Proc. Nat. Acad. Sci. USA*, 78, 2062.

Richards, J. 1981. *Adv. Protein Chem.*, 34, 167.

Richards, K. and R. Williams. 1972. *Proc. Nat. Acad. Sci. USA*, 69, 1121.

Richards, K. and R. Williams. 1976. In *Comprehensive Virology*. Vol. 6. H. Fraenkel-Conrat and R. Wagner, eds. p.1. Plenum.

Richards, K., R. Williams, and R. Calendar. 1973. *J. Mol. Biol.*, 78, 255.

Richards, R., P. Linsen, and R. Armentrout. 1978. In *Replication of Mammalian Parvoviruses*. D. Ward and P. Tattersall, eds. p. 447. Cold Spring Harbor.

Richardson, J., P. Tollin, and J. Bancroft. 1981. *Virology*, 112, 34.

Riemer, S. and V. Bloomfield. 1978. *Biopolymers*, 17, 784.

Roberts, J. and J. Steitz. 1967. *Proc. Nat. Acad. Sci. USA*, 58, 1416.

Robb, J., and C. Bond. 1979. In *Comprehensive Virology*. Vol 14. H. Fraenkel- Conrat and R. Wagner, eds. p.193. Plenum.

Robinson, A. and A. Bellet, 1975. *Cold Spring Harbor Symp. Quant. Biol.*, 39, 523.

Robinson, C. and C. Tibbetts. 1984. *Virology*, 137, 276.

Robinson, W. 1979. In *Comprehensive Virology*. Vol 14. H. Fraenkel-Conrat and R. Wagner, eds. p.471. Plenum.

Roizman, B. 1979. *Cell*, 16, 481.

Roizman, B. and D. Furlong. 1974. In *Comprehensive Virology*. Vol 3. H. Fraenkel-Conrat and R. Wagner, eds. p.229. Plenum.

Roman, A. 1982. *J. Virol.*, 44, 958.

Romanowski, V. and D. Bishop. 1983. *Virology*, 126, 87.

Rossmann, M., R. Chandrasekaran, C. Abad-Zapatero, J. Erickson, and S. Arnott. 1983. *J. Mol. Biol.*, 166, 73.

Rubio, C., C. Kolakofsky, V. Hill and D. Summers. 1980. *Virology*, 105, 123.

Russel, P. and U. Muller. 1984. *J. Virol.*, 52, 822.

Saksena, K. and P. Lemke. 1978. In *Comprehensive Virology*. Vol 12. H. Fraenkel-Conrat and R. Wagner, eds. p.103. Plenum.

Sanger, F., G. Air, B. Barrell, N. Brown, A. Coulson, J. Fiddes, C. Hutchison, P. Slocombe, and M. Smith. 1977. *Nature*, 265, 687.

Saragosti, S., S. Cereghini, and M. Yaniv. 1982. *J. Mol. Biol.*, 160, 133.

Savithri, H. and J. Erickson. 1983. *Virology*, 126, 328.

Sawyer, R. and H. Hanafusa. 1979. *J. Virol.*, 29, 863.

Schellman, J. and N. Parthasarathy. 1984. *J. Mol. Biol.*, 175, 313.

Schlesinger, M. 1934. *Biochem. Z.*, 264, 306.

Schonlaker-Schwarz, R. and H. Engelberg-Kulka. 1981. *J. Virol.*, 38, 833.

Schuster, T., R. Scheele, M. Adams, S. Shire, J. Steckert, and M. Potschka. 1980. *Biophys. J.*, 32, 313.

Schuster, T., R. Scheele, and L. Khairallah. 1979. *J. Mol. Biol.*, 127, 461.

Scott, W., W. Brockman, and D. Nathans. 1976. *Virology*, 75, 319.

Sen, A., C. Sherr, and G. Todaro. 1977. *Cell*, 10, 489.

Sen, A., C. Sherr, and G. Todaro. 1978. *Virology*, 84, 99.

Sen, A. and G. Todaro. 1977. *Cell*, 10, 91.

Serwer, P. 1975. *J. Mol. Biol.*, 99, 185

Serwer, P. 1980. *J. Mol. Biol.*, 138, 65.

Sgaramella, V. 1972. *Proc. Nat. Acad. Sci. USA*, 69, 3389.

Shank, P. and M. Linial. 1980. *J. Virol.*, 36, 450.

Shaw, J., L. Levinger, and C. Carter. 1979. *J. Virol.*, 29, 657.

Shen, C., J. Ikoku, and J. Schoenmakers. 1979. *J. Mol. Biol.*, 127, 163.

Shib, M., K. Watanabe, and J. Ito. 1982. *Biochem. Biophys. Res. Commun.*, 105, 1031.

Shinagawa, M., R. Padmanabhan, and P. Padmanabhan. 1980. *Gene*, 9, 99.

Shire, S., J. Steckert, and T. Schuster. 1981. *Proc. Nat. Acad. Sci. USA*, 78. 256.

Silverstein, S., J. Christman, and G. Acs. 1976. *Ann. Rev. Biochem.*, 45, 375.

Simonsen, C., S. Batt-Humphries, and D. Summers. 1979. *J. Virol.*, 31, 124.

Simons, G., J. Beintema, F. Duisterwinkel, R. Konigs, and J. Schoenmakers. 1981. In *Progress In Clinical and Biological Research*. Vol. 64. M DuBow, ed. p. 401. Alan R. Liss.

Slilaty, S., K. Berns, and H. Aposhian. 1982. *J. Biol. Chem.*, 257, 6571.

Smiley, J., B. Fong, and W. Leung. 1981. *Virology*, 113, 345.

Sorge, J., W. Ricci, and S. Hughes. 1983. *J. Virol.*, 48, 667.

Spaete, R. and N. Frenkel. 1982. *Cell*, 30, 295.

Spahr, P., M. Farber, and R. Gesteland. 1969. *Nature*, 222, 455.

Spear, P. and B. Roizman. 1980. In *Molecular Biology of Tumor Viruses*. R. Weiss, N. Teich, H. Varmus, and J. Coffin, eds. p. 615. Cold Spring Harbor.

Stanley, J. and M. Gray. 1983. *Nature*, 301, 260.

Steckert, J. and T. Schuster. 1982. *Nature*, 299, 32.

Steitz, J. 1968. *J. Mol. Biol.*, 33, 923.

Steitz, J. 1979. In *Biological Regulation and Development*. R. Goldberger, ed. p. 349. Plenum.

Sternberg, N. and R. Hoess. 1983. *Ann. Rev. Genet.*, 17, 123.

Sternberg, N. and R. Weisberg. 1977. *J. Mol. Biol.*, 117, 733.

St. Jeor, S., C. Hall, C. McGaw, and M. Hall. 1982. *J. Virol.*, 41, 309.

Stoller, V. 1980. In *The Togaviruses*. R. Schlesinger, ed. p. 427. Academic Press.

Strauss, E., C. Rice, and J. Strauss. 1984. *Virology*, 133, 92.

Strazielle, C., H. Beniot, and L. Hirth. 1965. *J. Mol. Biol.*, 13, 735.

Streisinger, G., J. Emrich, and M. Stahl. 1967. *Proc. Nat. Acad. Sci. USA*, 57, 292.

Strobel, E., W. Benisch, and H. Schmieger. 1984. *Virology*, 133, 158.

Stubbs, G. and C. Stauffacher. 1981. *J. Mol. Biol*, 152, 387.

Stubbs, G., S. Warren, and K. Holmes. 1977. *Nature*, 267, 216.

Summers, J., and W. Mason. 1982. *Cell*, 29, 403.

Sundquist, B., E. Everitt, L. Philipson, and S. Hoglund. 1973. *J. Virol.*, 11, 449.

Susskind, M. and D. Botstein. 1978. *Microbiol. Rev.*, 42, 385.

Sykora, K. and K. Moelling. 1981. *J. Gen. Virol.*, 55, 379.

Syvanen, M. and J. Yin. 1978. *J. Mol. Biol.*, 126, 333.

Takamatsu, H. and K. Iso. 1981. *Nature*, 298, 819.

Tan, K.,and F. Sokol. 1974. *J. Gen. Virol.*, 25, 37.

Tattersall, P. 1978. In *Replication of Mammalian Parvoviruses*. D. Ward and P. Tattersall, eds. p. 53. Cold Spring Harbor.

Thomas, T. and V. Bloomfield. 1983. *Biopolymers*, 22, 1097.

Tibbetts, C. 1977. *Cell*, 12, 243.

Tikchonenko, T. 1975. In *Comprehensive Virology*. Vol 5. H. Fraenkel-Conrat and R. Wagner, eds. p. 1. Plenum.

Tollin, P., J. Bancroft, J. Richardson, N. Payne, and T. Beveridge. 1979. *Virology*, 98, 108.

Tooze, J. ed. 1980. *DNA Tumor Viruses*. Cold Spring Harbor Laboratory, N. Y.

Torbet, J., D. Gray, C. Gray, D. Marvin, and H. Siegrist. 1981. *J. Mol. Biol.*, 146, 305.

Tye, B., J. Huberman, and D. Botstein. 1974. *J. Mol. Biol.*, 85, 501.

Van der Ende, A., R. Teertstra, and P. Weisbek. 1982. *Nucleic Acids Res.*, 10, 6849.

Van Etten, J., A. Vidaver, R.Koski, and J. Burnett. 1974. *J. Virol.*, 13, 1254.

Van Vloten-Doting, L., and E. Jaspars. 1977. In *Comprehensive Virology*. Vol 11. H. Fraenkel-Conrat and R. Wagner, eds. p.1. Plenum.

Varma, A., A. Gibbs, R. Woods, and J. Finch. 1968. *J. Gen. Virol.*, 2, 107.

Varmus, H. 1982. *Science*, 216, 812.

Vayda, M., A. Rogers, and S. Flint. 1983. *Nucleic Acids Res.*, 11, 441.

Virrankoski-Castrodeza, V., M. Fraser, and J. Parish. 1982. *J. Gen. Virol.*, 58, 181.

Vlanzy, D., A. Kwong, and N. Frenkel. 1982. *Proc. Nat. Acad. Sci. USA*, 79, 1432.

Wachtel, E., R. Wiseman, W. Pigram, D. Marvin, and L. Manuelidis. 1974. *J. Mol. Biol.*, 88, 601.

Wagner, R. 1975. In *Comprehensive Virology*. Vol 4. H. Fraenkel-Conrat and R. Wagner, eds. p.1. Plenum.

Watanabe, S. and H. Temin. 1982. *Proc. Nat. Acad. Sci. USA*, 79, 5986.

Watson, J. 1973. *Nature New Biol.*, 239, 197.

Weaver, S. and M. Levine. 1978. *J. Mol. Biol.*, 118, 389.

Weber, J., C. Dery, M. Mirza, and J. Horvath. 1985. *Virology*, 140, 351.

Webster, R., and J. Cashman. 1978. In *The Single-Stranded DNA Phages*. D. Denhardt, D. Dressler and D. Ray, eds. p.557. Cold Spring Harbor.

Webster, R., R. Grant, and L. Hamilton. 1981. *J. Mol. Biol.*, 152, 357.

Weisbeek, P. and R. Sinsheimer. 1974. *Proc. Nat. Acad. Sci. USA*, 74, 3054.

Wengler, G., U. Boege, G. Wengler, H. Bischoff, and K. Wahn. 1982. *Virology*, 118, 401.

Wengler, G., G. Wengler, U. Boege, and K. Wahn. 1984. *Virology*, 132, 401.

Wetz, K. and K. Habermehl. 1982. *J. Gen. Virol.*, 59, 397.

Wildy, P., W. Russell and R. Horne. 1960. *Virology*, 12, 204.

Wilson, G., R. Neve, G. Edlin, and W. Konigsberg. 1979. *Genetics*, 93, 285.

Wilson, J., P. Berget, and J. Pipas. 1982. *Mol. Cell Biol.*, 2, 1258.

Wimmer, E. 1982. *Cell*, 28, 199.

Windom, J. and R. Baldwin. 1980. *J. Mol. Biol.*, 144, 431.

Windom, J. and R. Baldwin. 1983. *J. Mol. Biol.*, 171, 419.

Winter, G. and S. Fields. 1981. *Virology*, 114, 423.

Wiseman, R. and L. Day. 1977. *J. Mol. Biol.*, 116, 607.

Wood, W., and J. King. 1979. In *Comprehensive Virology*. Vol 13. H. Fraenkel-Conrat and R. Wagner, eds. p.581. Plenum.

Wu, H. and D. Crothers. 1984. *Nature*, 308, 509. Yamagishi, H. and M. Okamoto. 1978. *Proc. Nat. Acad. Sci. USA*, 75, 3206.

Yoshiike, K., T. Miyamura, H.Chan, and K. Takemoto. 1982. *J. Virol.*, 42, 395.

Yoshikawa, H., T. Friedmann, and J. Ito. 1981. *Proc. Nat. Acad. Sci. USA*, 78, 1336.

Yoshinaka, Y. and R. Luftig. 1977. *Proc. Nat. Acad. Sci. USA*, 74, 3446.

Yoshinaka, Y. and R. Luftig. 1982. *Virology*, 116, 181.

Ysebaert, M., J. van Emmelo, and W. Fiers. 1980. *J. Mol. Biol.*, 143,273.

Yuen, L. and R. Consigli. 1982. *J. Virol.*, 43, 337.

Zachary, A. and L. Black. 1981. *J. Mol. Biol.*, 149, 641.

Zimmermann, R., C. Watts, and W. Wickner. 1982. *J. Biol. Chem.*, 257, 6529.

Zimmern, D. 1975. *Nucleic Acids Res.*, 2, 1189.

Zimmern, D. 1977. *Cell*, 11, 463.

Zimmern, D. and P. Butler. 1977. *Cell*, 11, 455.

Zuidema, D., R. Cool, and E. Jaspars. 1984. *Virology*, 136, 282.

Zuidema, D., M. Bierhuizen, B. Cornelissen, J. Bol, and E. Jaspars. 1983a. *Virology*, 125, 361.

Zuidema, D., M. Bierhuizen, and E. Jaspars. 1983b. *Virology*, 129, 255.

4
Pathways in Viral Morphogenesis

Peter B. Berget
University of Texas
Health Science Center

INTRODUCTION

Over the last 30 years, the study of viral morphogenesis has provided the paradigm for the study of the genetic control of structural development. Early in these studies the observation was made that bacteriophage development is similar to the pathways of intermediary metabolism in that bacteriophage morphogenesis occurs through controlled pathways. It has been recently stated that the rapid success of investigators in identifying the existence of and elucidating the individual steps in assembly pathways in phage morphogenesis has led to their incorporation into textbooks in an oversimplified or misleading form (King, 1980). It is the purpose of this chapter to clarify some of the thoughts concerning the roles of pathways in biological construction, and to provide a detailed description of a controlled-assembly pathway in bacteriophage morphogenesis.

REQUIREMENTS OF BIOLOGICAL CONSTRUCTION

Supramolecular assemblies or subcellular organelles are built from a variety of biological components. For most bacteriophages only two types of biological molecules, protein and nucleic acid, are used in any great amount. The relative mass proportions of protein and nucleic acid can vary over a substantial range. For example, T4 phage is approximately 50% protein and the filamentous phages, such as fd or M13, are approximately 80% protein by weight. Furthermore, the number of protein molecules involved can vary from 60 in the simplest icosahedral viruses to several thousand in the helical and large icosahedral viruses. We now know that nucleic acid is the genetic component of viruses and that in the simplest of interpretations the protein component provides the packaging and delivery system for this information into the next host. Although the building blocks for phage structures are limited in their types, the variation that can exist in different protein molecules makes the actual number of kinds of building blocks effectively infinite. In this chapter I shall mainly address the problem of biological construction from protein molecules.

The primary requirement of biological construction is that the need of the organism for the structure should be met by the final product. This may seem like an obvious statement but all other requirements stem from it; clearly, if a bacteriophage constructs a DNA-injection organelle as part of its overall structure, it must in fact work as a DNA-injection organelle. The secondary requirements of biological construction are that the structure must be constructed in an appropriate time frame for the organism and that the construction should be carried out efficiently. It is my contention that the use of pathways and subassemblies allow these requirements of accuracy and efficiency to be met.

THEORETICAL CONSIDERATIONS OF PATHWAYS

Consider a very simple assembly system in which only three components—A, B, and C—interact to form an ABC complex. As shown in Figure 1, there are three possible orders in which ABC can be built: joining either A and B, A and C, or B and C first, and then adding the final component. Each of these is a potential pathway for the assembly of ABC. As I shall describe below, bacteriophages are invariably assembled from their protein and nucleic acid components according to very specific pathways; that is, the building blocks assemble in a strict order. In this simple example there are three possible assembly pathways. In Figure 1, a particular pathway, A + B ⟶ AB + C ⟶ ABC, is boxed. If only the boxed pathway is functional, A and B will bind one another, but C will bind neither A nor B alone. In a real structure, as the number of components increases, the number of possible pathways becomes very large. One

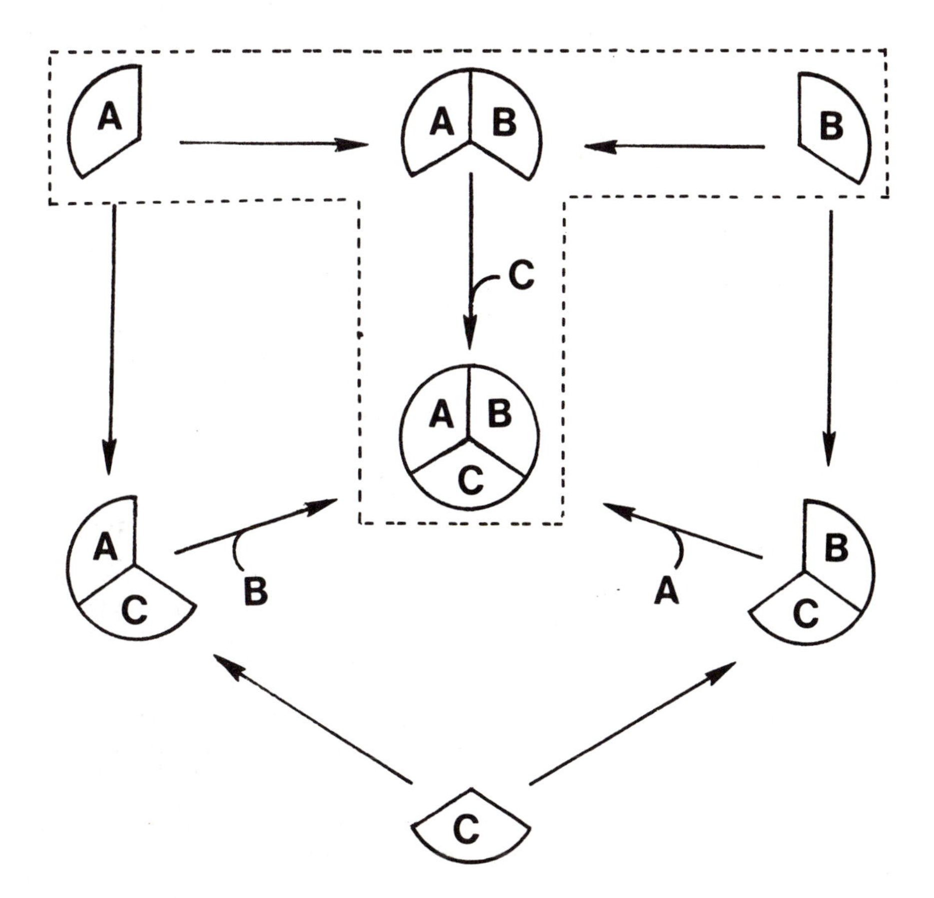

FIGURE 1 Potential assembly pathways for a three-component system. The subunits A, B, and C of the ABC complex may interact in three possible ordered pathways of assembly by initiating with unique heterodimer formation, as described in the text. The outlined pathway may be selected as the unique pathway by "nucleation control" of the formation of the AB heterodimer.

great advantage of assembly by specific pathways is the ease by which control can be exerted over the assembly pathway. One such control is the regulation of the number of structures made. Again referring to the example in Figure 1, in order to regulate the entire assembly process control need only be exerted on the A + B step, rather than all three heterodimerization steps. This has often been referred to as control of the "nucleation" of an assembly process, and such regulation requires that assembly occur by a specific pathway. A second advantage of assembly by a specific pathway is that "spatial" control can be exerted . Imagine that a closed protein shell (as many viral capsids are thought to be), which has internal and external components, were being constructed. If the shell were built and closed before the internal components were added, the resulting structure would no doubt be nonfunctional, since the internal components could not be added after the outer shell were closed. Thus, if assembly were required to occur by a specific pathway in which the internal components were inserted before the shell were closed, the assembly of nonfunctional structures would be completely avoided. This type of control is no doubt very important in the assembly of structures, such as viruses, that contain a large number of protein molecules. (See Chapter 5, this volume, for details.)

Inherent in the concept of pathways is that specific "intermediates" or subassemblies are generated. It is through the consideration of subassemblies and "assembly-line" construction that useful thoughts on this concept have been generated. One of the first discussions on the benefits of subassemblies and assembly-lines in biology was put forth in "Principles and Problems of Biological Growth" (Crane, 1950). Several of Crane's thoughts are expanded upon in the remainder of this section, in which it will be seen that accuracy and efficiency are direct results of the use of subassemblies.

A useful approach to understanding the roles pathways and subassemblies play in the accuracy of structure formation is to consider the probability that a given structure will be formed successfully under a defined set of rules. For example, let us consider the assembly of a supramolecular structure having 1000 subunits that may or may not all be the same. In this example we make the assumptions: (1) the probability of making a mistake in the assembly of one subunit into a structure or in the interaction of two subassemblies is 0.1%; and (2) one incorrect assembly reaction either makes the structure unusable or causes the substructure to become assembly-incompetent for further reactions. The probability $P(f)$ of successfully finishing the structure by adding one subunit at a time is the product of the probabilities of a correct reaction at each step, or $P(f) = (1 - x)^{n - 1}$, in which n = the number of subunits and x = the probability of incorrect association at each step. Thus, $P(f) = (0.999)^{999} = 0.368$, which means that there is only a 36.8% chance of correctly completing such a structure by using this mode of assembly. Let us explore alternate means of assembling this structure by determining the efficiency of subunit utilization.

Assume now that we are given 1,000,000 subunits to start with, that any structure consumes its normal complement of subunits even if a mistake is made during its assembly, and that the error frequency is 0.1%. Furthermore, assume that if a mistake is made in the assembly of a component that is an integral part of the whole, it will be incapable of entering into succeeding rounds of assembly. How many subunits will reside in perfect structures using a given assembly pathway? If we use an assembly scheme in which each 1000-mer is made by 999 successive additions of subunits, the probability of successful assembly of each 1000-mer is 0.368; thus, 368 out of a possible thousand 1000-mers will be perfect, or 368,000 of the 1,000,000 subunits will reside in perfect structures. If we make the structure by forming 500-mers and then assembling two of them into a 1000-mer as the last step, a different result is obtained. The probability of successfully assembling a 500-mer is $(0.999)^{499}$ or 0.607; thus, 606,985 monomers will exist in perfect 500-mers (using only integer values, 1,213 perfect 500-mers will be formed consuming 606,500 monomers). Thus, 606 1000-mers could be assembled from the 1,213 500-mers, but only (606)(0.999) = 605 perfect 1000-mers will form. Using two 500-mer subassemblies to form 1000-mers has increased the efficiency of usage of monomers from 368,000 to 605,000. In Figure 2 the results of calculating efficiencies by this method are presented for forming 1000-mers by different combinations of subunits and subassemblies.

According to this logic, the most successful way of assembling this structure is from 25 subassemblies with 40 subunits, or or 40 subassemblies with 25 subunits, in which almost every start results in a perfect structure ($P(f) = 0.939$). In general, the most successful way of assembling a multisubunit supramolecular structure using one type of subassembly is to maximize both the total number of subassemblies per final structure and the number of subunits per subassembly. The "ideal" number of subunits per subassembly is the integer factor nearest the square root of the total number of subunits in the structure.

The probability of successful assembly at each step has been termed the "degree of selectivity" (Crane, 1950). We can investigate the theoretical "degree of selectivity" that may be required in such a biological process. By varying the "degree of selectivity", DS, one can calculate the probability of successful assembly of a 1000-mer using the best pathway of 25 40-mers with the following equation: $P(f) = (DS)^{24}(DS)^{39}$. Figure 3 shows the plot of the probability of successful assembly of a 1000-mer using this best pathway as a function of degree of selectivity. Clearly the success of assembly depends quite strongly on the "degree of selectivity" at each step in the assembly pathway. If this model assembly system is to create structures with at least a 50% chance of the final structure being correct or a 50% chance of successful completion, then each step in the assembly pathway needs to be carried out with an accuracy of about 99%. This demonstrates that an acceptable degree of successful assembly can be achieved without impossibly stringent requirements at each step of this assembly reaction. Compare this to the acceptability of a genetic replication system operating with a 0.1% error rate at each step in the "assembly" of the nucleic acid polymer!

How accurate is the process of assembly for bacteriophages? One possible measure of the total accuracy of the process is the relative

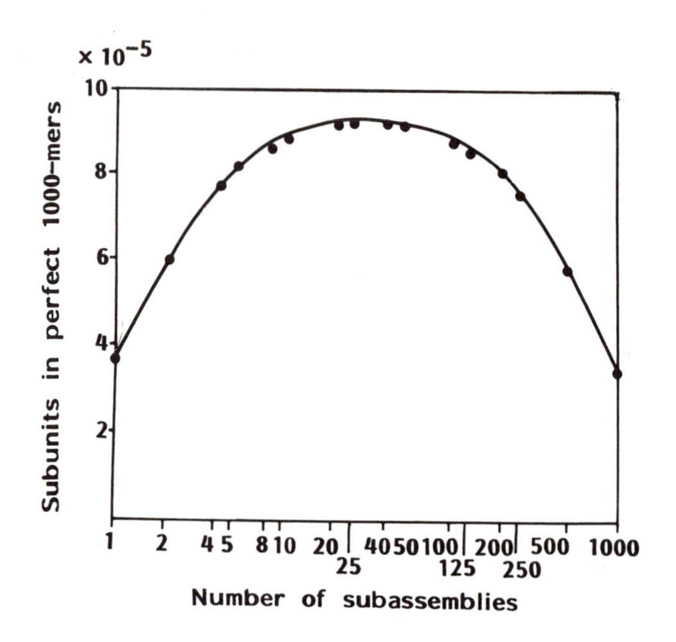

FIGURE 2 Efficiencies of alternate modes of assembly. The efficieny of utilizing subunits is plotted against the number of subassemblies used to form the final structure, as described in the text. Starting with 1,000,000 subunits, the number of subunits found in perfectly assembled 1000-mers is plotted against the number of subassemblies used to make the 1000-mer. For example, 368,000 subunits reside in perfect 1000-mers if the 1000-mers are made by 999 successive monomer additions (i.e., 1 or 1000 subassemblies). If two 500-mers (or 500 dimers) are used to form 1000-mers, then 605,000 of the 1,000,000 starting subunits end up in perfect 1000-mers, etc.

infectivity of the final particles, that is, what percent of the final structures are capable of initiating a subsequent round of infection in a new host? For *Escherichia coli* bacteriophage T4 and *Salmonella typhimurium* bacteriophage P22 this number is between 50% and 100%; however, for many animal viruses the number is less than 10%. Surely these numbers reflect both the accuracy of the final structure and the efficiency of the process of viral infection. However, it is clear that for bacteriophages such as T4, which contain more than 1000 protein molecules and have an infectious particle to physical particle ratio near 1, the overall assembly process must indeed be very accurate.

A further result of the subassembly principle is the increased speed with which a complicated structure can be made. If each assembly reaction, which may occur between subunits or subassemblies, takes a certain amount of time, the subassembly scheme can be shown to be more efficient. If each reaction in the previous example were to consume 1 second, the assembly of a 1000-mer by

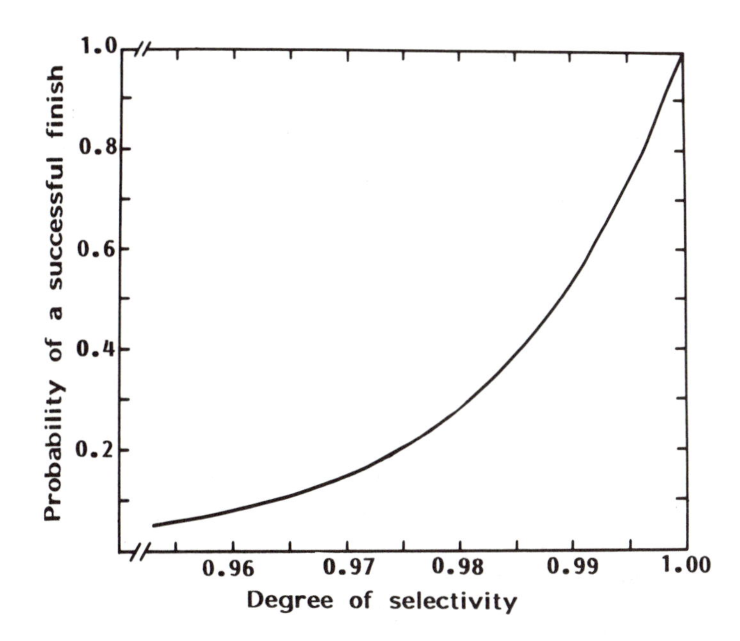

FIGURE 3 The effect of the degree of selectivity on the probability of successful assembly. The probability of successfully assembling a 1000-mer from 25 40-mers (or 40 25-mers) is plotted against the probability of a correct assembly reaction at each step of the assembly process.

successive monomer additions would take 999 seconds to complete. However, assembling the same structure from 40 25-mers would take 39 + 24 = 83 seconds to complete, which is a direct result of the simultaneous construction of subassemblies.

The reader should dwell for a moment on the thought of assembling a 1000-mer from two different types of intermediate subassemblies (a logical extension of the discussion above). Suppose that 10-mers are initially formed by monomer additions and that ten of these 10-mers subsequently assemble to form 100-mers, then ten of these 100-mers assemble to form the final 1000-mer. This can be shown by the same arguments made above to be an even more efficient pathway of assembling the desired structure. Nevertheless, the important points about accuracy and efficiency have been made. Using only probabilistic considerations, the use of subassemblies and "rules" or pathways by which they are brought together make the process of assembly possible using relatively relaxed "degrees of selectivity". Again, in contrast, consider the "degree of selectivity" required to replicate genetic information faithfully; this is on the order of 0.9999999999 (Drake, 1969).

H. R. Crane's discussion of "degree of selectivity" closes with a story too good to resist re-telling some 34 years later. "...During the early part of the war I had occasion to go through a vacuum tube factory, where the miniature tubes for proximity fuses were being made. There I saw rows of women assembling the most minute parts under magnifying glasses, and giving the every appearance of being highly skilled jewelers. I asked if the training of these women did not present a great problem. To my surprise my guide told me that a completely green person could be put on the production line four hours after walking in the door. "Of course," he said, "they make mistakes, but each worker does only a small thing, and the pieces he spoils are, of course, not used in the next subassembly operation." Here a highly complex and precise product was being made entirely by unskilled hands, not only by virtue of the fact that each worker had only a small job to learn to do, but also by virtue of the fact that the system provided for casting aside the imperfect units at many levels of assembly. Therefore no individual was required to make a high score on accuracy. It was at this time that I was struck by the apparent parallels that existed between this method of factory production and many processes in nature, from evolution and natural selection down to the relations of the smallest molecules. The problem of how the fantastically complex and precise 'products' could be manufactured in nature without the requirement of an excessively high degree of selectivity or accuracy at any one stage of development seemed slightly more understandable."

Now, in addition to the rules of assembly put forth for the example system described above, we impose an additional rule: namely, as soon as a mistake is made in the assembly process of any component, the component is discarded and we start again. With that rule, the total number of monomers available for incorporation into perfect final structures increases. Casting aside imperfect structures at any point in assembly provides increased accuracy and efficiency.

The advantage of assembling supramolecular structures from subassemblies should now seem obvious. However, in the biological realm, where is the "blueprint" for the final structure, where are the skilled or unskilled laborers that assemble the desired structure, i.e., what determines the pathways through which subassemblies are formed and subsequently brought together? These questions impinge on the central problems in developmental biology, and they are the questions that investigators studying viral morphogenesis ask.

The specificity or degree of selectivity that gives rise to the interaction of subassemblies to form the final structure and to form the subassemblies themselves must somehow reside in both the protein molecules that comprise the final structure and any other proteins that are utilized as "catalytic" factors in the overall assembly scheme. All would agree that proteins seem ideally suited for these roles; however, the fact that some proteins must simultaneously serve as structural components and catalytic factors for assembly and must perform some highly specific function in the final structure is indeed remarkable.

The analysis of morphogenetic assembly pathways in bacteriophages has proceeded through two basic approaches. The first is a genetic approach through which the developmental phenomena are observed. The second is a biochemical approach through which the molecular mechanisms of the phenomena are elucidated. The success of the former approach in identifying many pathways of morphogenesis and our current lack of understanding of the underlying molecular mechanisms indicate both the power of genetics and the complex nature of the problem.

As alluded to in the introduction, pathways of morphogenesis in viral development were observed early in the general studies of bacteriophages, especially T4. These observations were made possible by the development of the powerful genetic system of temperature-sensitive and amber conditionally lethal mutations in genes necessary for T4 growth (Epstein et al., 1963; Edgar, 1969). A virus, carrying a conditionally lethal mutation in an essential gene, grows under one set of permissive conditions but fails to grow under another defined set of restrictive conditions. At high temperatures or in the nonpermissive host, bacteriophage conditionally lethal mutants make all the active proteins normally found in the wildtype-infected cell, except for one. Using physical and immunological tools and a large collection of these conditionally lethal mutations in almost all of the essential T4 genes, investigators cataloged the defects in viral development that resulted from the absence of various gene functions (Epstein et al., 1963). It was clear from this study that gene products interacted with each other through specific pathways rather than in "random" fashion (Edgar and Lielausis, 1968; King and Wood, 1969). For example, defects in any of a set of seven genes resulted in the formation of T4 virus particles that lacked the characteristic T4 tail fibers. Defects in a larger different set of genes (ca. 20) resulted in the formation of free T4-head structures and tail fibers but no tail structures. Defects in a third set of genes (ca. 20) resulted in the formation of free tail structures and tail fibers, but no head structures. Such observations allowed these investigators to propose a model for the final steps in the assembly of the T4 phage particle. Three major subassemblies, the T4 head, the T4 tail, and the T4 tail fibers are each assembled through independent pathways and then joined together in two distinct assembly reactions. The T4 head and tail combine to form a fiberless particle; then and only then, the T4 tail fibers attach to finish the assembly of the virus particle. Thus, it is clear that the assembly of bacteriophage T4 utilizes the principles of subassembly and assembly-line structure development.

Further research has provided detailed descriptions of the morphogenetic pathways that operate in the assembly of these phage T4 subassemblies. A recent review that appears in the book *Bacteriophage T4* (Berget and King, 1983) provides details that supplements the material presented in the following section.

ASSEMBLY OF THE MAJOR T4 BASEPLATE SUBASSEMBLY, A PROTEIN-DETERMINED, ORDERED ASSEMBLY PATHWAY

The baseplate of T4 is a structure located at the distal end of the tail. It is the adsorption organelle of the phage and a complex molecular machine that triggers tail contraction and DNA ejection. This structure is composed of 18 different protein gene products, which are present in specific stoichiometries. The baseplate itself is assembled from two unique subassemblies: the "wedge", which contains eight types of structural proteins and the "hub", which contains 5 different structural proteins (Kikuchi and King, 1975a,b,c; Mosher and Mathews, 1979; Kozloff, 1981). The hub and six wedge subassemblies co-assemble to form the hexagonally symmetrical baseplate. Then, five additional species of structural proteins then assemble onto this structure to finish the baseplate and to provide the nucleation sites for the polymerization of the T4 tail tube and sheath (King and Mykolajewycz, 1973; Berget and Warner, 1975; Meezan and Wood, 1971; King, 1971).

The assembly of the "wedge" structure provides an excellent example of the ordered, specific interaction of protein subunits to form a biological structure that is itself an intermediate in the assembly of a more complex structure. As depicted in Figure 4, the wedge is formed from the protein products of genes *11*, *10*, *7*, *8*, *6*, *frd*, *53*, and *25*. No catalytic protein factors are required for the association of these proteins. Thus, the information or "blueprint" for their

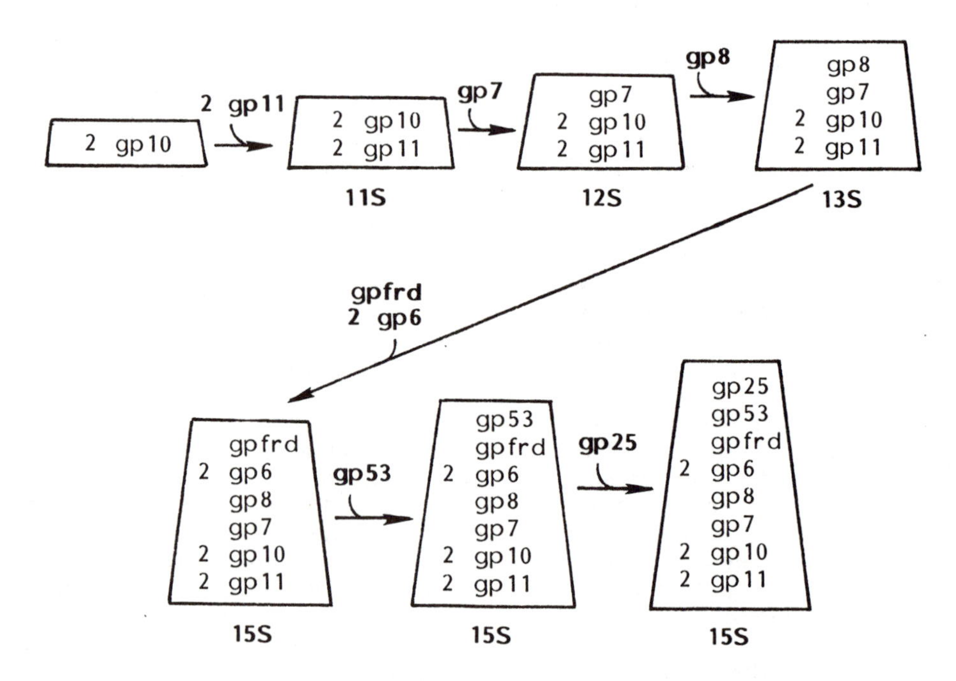

FIGURE 4 The morphogenetic pathway of the "wedge," the major subassembly of the T4 baseplate. The data used to construct this pathway come from Kikuchi and King (1975a, b, c); Plishker, Chidambaram, and Berget (1983; unpublished data); and Kozloff (1981; personal communication). The abbreviation "gp" indicates the protein product of an identified T4 gene; i.e., gp10 refers to the protein product of T4 gene *10*. When known, the stoichiometries of the various protein components are indicated. Note that the point of addition of gpfrd is not exactly known; its point of addition is shown according to the data of Mosher and Mathews (1979).

interaction must be contained in the structural proteins themselves, and the final structure must be entirely determined by specific protein-protein interactions.

The assembly pathway of the wedge is unique and linear. The pathway was elucidated by Kikuchi and King (1975a,b, c) in a set of experiments utilizing a genetic approach. These investigators showed by sucrose-gradient sedimentation analysis and *in vitro* complementation tests (Edgar and Wood, 1966; Wood et al., 1968) that in infected cells, missing only the product of gene *10* (gp10), all of the other wedge components are found free in the cell in an unassociated, unassembled, active form. This observation pointed out the "keystone" role of gp10 in the assembly pathway. In infected cells missing only gp7, the structural components gp10 and gp11 are found associated with each other in a 10/11 complex, whereas the rest of the wedge components remain unassembled. In similar fashion, in infected cells lacking only gp8, an assembly complex consisting of gp10, gp11, and gp7 (10/11/7 complex) is found, while the rest of the wedge components remain unassembled. The complete set of experiments examined the state of assembly of all of the wedge components in infected cells lacking each structural element one at a time. Edgar and Lielausis (1968) had previously shown that in infected cells missing only gp11, defective phage particles accumulate that lack only gp11 and gp12, a component added later in the assembly of the baseplate. Thus, gp11 is not required for the assembly pathway depicted in Figure 4 to operate; in fact, gp11 can be added much later after the tail is completed, showing that in not all cases must there be an obligate order of addition. Nevertheless, in infected cells missing gp7, all gp11 is found associated with gp10 in a 10/11 complex (Berget and King, 1978). In T4 assembly the vast majority of assembly steps are obligately ordered.

Thus, the assembly pathway of the wedge shown in Figure 4 is initiated by the interaction of gp7 with the 10/11 complex to form the 10/11/7 complex. This complex is the only molecule in the infected cell with which gp8 can form a stable complex, resulting in the 10/11/7/8 complex. Likewise, this latter complex is the only molecule in the infected cell that can form a stable complex with gp6, etc. To put it another way, the only reactive site for accepting each structural protein into the growing wedge structure is a complex of proteins made from all the previous proteins in the wedge pathway. The linear assembly pathway of the wedge has two consequences: first, the reactive sites for protein-protein interaction are limited to the growing structure; second, all the proteins that comprise the structural elements of the wedge (except gp10, gp11, and gp7) are synthesized in forms unreactive towards each other, and these require incorporation into a growing wedge structure to become reactive. The overall effect is to limit the number of possible assembly reactions occurring in the cell and to generate only the desired final structure.

Possible Schemes for Control of Assembly

In molecular terms, how is this linear assembly pathway generated? One possible control mechanism which is most certainly used in eukaryotic development, is temporal control of synthesis of component parts. In T4 development, as in most virus systems, all components of the virion are synthesized simultaneously during the late stage of viral development. Thus, temporal control is not applied to the ordering of the assembly pathways in T4 morphogenesis. In analytical terms, the problem of ordering biological assembly reactions in the absence of temporal control can be simplified to one in which there are three protein molecules A, B, and C that combine to form an ABC complex via the pathway A + B ⟶ AB, AB + C ⟶ ABC. That is, two proteins spontaneously interact to form the reactive site for a third protein.

The molecular mechanisms that may control such a reaction should be thought of as hypotheses to be experimentally tested. Three possible mechanisms are illustrated in Figure 5. The first scheme incorporates covalent modification of one or both of the proteins A and B after their interaction, thereby making available the complementary site for C. As depicted, proteolysis of A catalyzed by the association of A and B could uncover the reactive site for C. Any other type of covalent modification of either structural protein would serve as well as the one depicted. In the second scheme, the complementary reactive site for C is found separated as two half-sites, each half on A and B, respectively. By their interaction to form the AB complex, A and B bring together these half-sites to generate the complementary region with which C reacts. In this scheme, C cannot form a stable complex with either A or B as separate molecules, though C may interact transiently with A or B in a nonproductive manner. In the third scheme, neither A nor B display any portion of their structure as a complementary site for C; however, after or during their assembly they interact in such a way that a conformational rearrangement or refolding of one or both proteins occurs, generating or exposing the complementary site for C. Wood (1980) coined the name "heterocooperativity" for this mechanism. In contrast to the second scheme, C would not be capable of transient, nonproductive interactions with either A or B, because neither protein displays a part of the complementary site for C until after their assembly.

FIGURE 5 Three schematic models (I, II, III) for molecular sorting of three proteins, A, B, and C, into an ordered reaction sequence. All models depict the assembly of an ABC complex through the ordered pathway A + B ⟶ AB, AB + C ⟶ ABC, where the complementary assembly site for subunit C (CSC) is formed after the AB complex. Model I: formation of CSC by covalent modification (proteolysis) of subunit A catalyzed by its assembly with subunit B. Model II: formation of the CSC from two half-sites found on A and B. Model III: formation of the complementary site for C by conformational rearrangement of subunits A and B after they have formed the AB complex, i.e., "heterocooperative" interaction (Wood, 1980). 1 and 2 refer to the first and second assembly steps; italicized letters label the complexes; dotted lines surround the CSC; 1/2 denotes a region that contributes to the CSC.

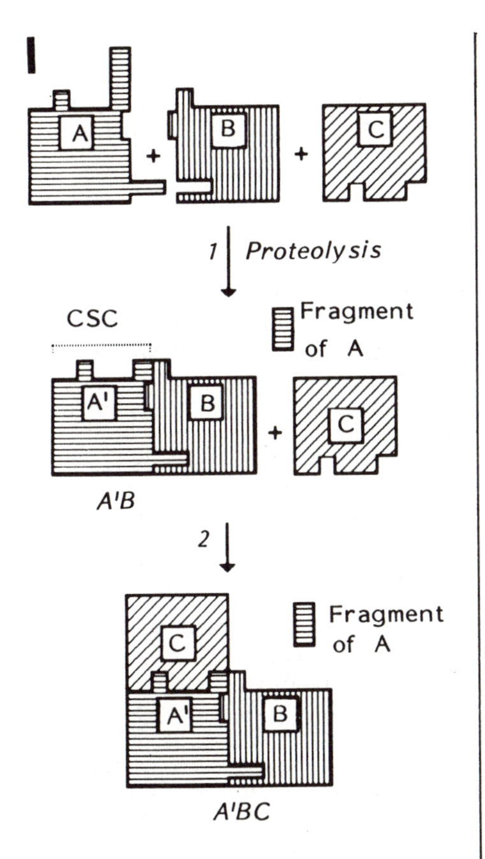
I
A
+
B
+
C
1
Proteolysis
CSC
Fragment of A
A'
B
+
C
A'B
2
C
Fragment of A
A'
B
A'BC

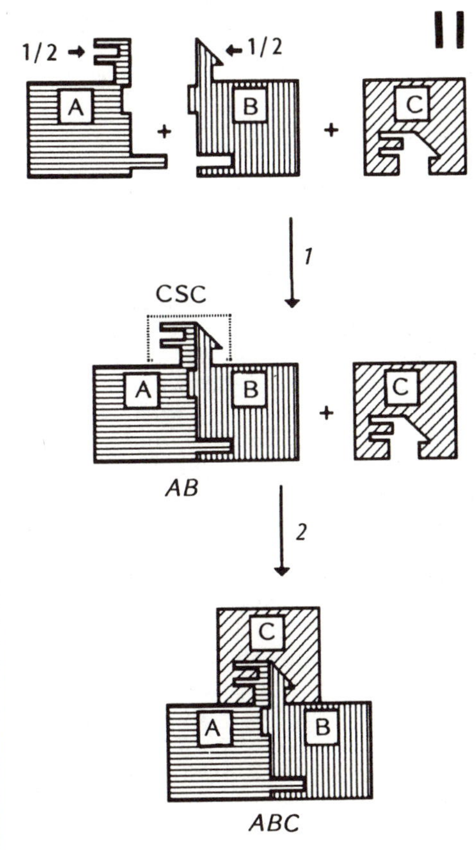
II
1/2
1/2
A
+
B
+
C
1
CSC
A
B
+
C
AB
2
C
A
B
ABC

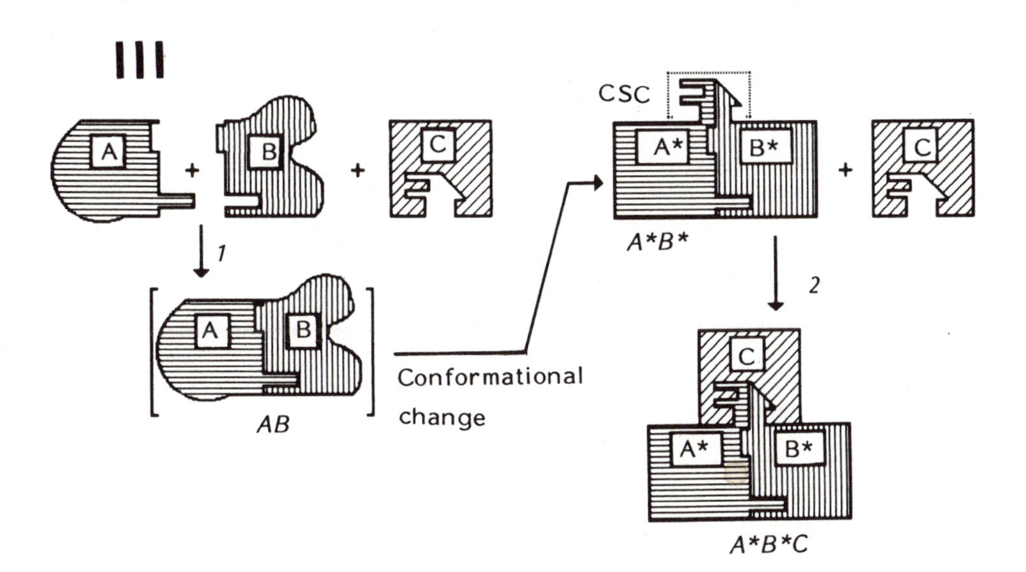
III
A
+
B
+
C
1
A
B
AB
Conformational change
CSC
A*
B*
+
C
A*B*
2
C
A*
B*
A*B*C

Each one of these hypothetical assembly schemes accomplishes the ordering of the interactions of A, B, and C. In addition, if the assembly of C into the ABC complex also utilizes one of these proposed schemes, then the reaction of a fourth molecule could be ordered, and so on.

Experimental Approaches for Distinguishing Assembly Schemes

Experimental verification of any of these schemes will most certainly require the purification of assembly-competent proteins and the careful analysis of assembly reactions carried out *in vitro* using these purified components. Considerable effort has gone into such an endeavor in my laboratory, as we have focused on the initial assembly reactions in the T4 wedge pathway. We have purified the first four structural proteins that interact with each other in the wedge pathway (Plishker et al., 1983; Plishker and Berget, 1984; unpublished results). The protein products of genes *11*, *10*, *7*, and *8* have been isolated either from cells infected with T4 carrying mutations that prevent wedge assembly or from cells carrying cloned genes that encode these proteins. Thus, these proteins are isolated in a form which I call "assembly-naive"; that is, they have never been assembled into a complex with other wedge proteins. Each of these proteins, which may exist in some multimeric form, is designated as P11, P10, P7, and P8 to differentiate them from the primary gene product (gp11, gp10, etc.). The polypeptide chains in these "assembly-naive" proteins have the same molecular weight (within the confidence limits of SDS gel electrophoresis) as the assembled forms found in mature phage or completed baseplates. This result is in accordance with the observations made by Laemmli (1970) and seemingly rules out all but the most subtle involvement of the covalent modification scheme described above as a possible control mechanism for the ordered interaction of these structural proteins. Although substantial proteolytic modification of structural proteins occurs in the morphogenesis of the T4 head subassembly, none seems to occur in the morphogenesis of the wedge.

We have developed a native gel-electrophoresis assay that detects the formation of complexes between the purified wedge proteins (Plishker et al., 1983). Our preliminary *in vitro* assembly experiments have shown that the order of interaction predicted from the assembly pathway proposed by Kikuchi and King (1975) and shown in Figure 4 can be verified using purified proteins. Using this assay system, we have demonstrated the *in vitro* formation of the following complexes: (10/11), (10/7), (10/11/7), (10/7/8), and (10/11/7/8). No interactions have been detected between P11 and any other protein other than P10. P10 does not interact to form a complex with P8 except in the presence of P7, again as predicted from the assembly pathway. However, P8 does interact inefficiently with P7 to form a complex that can be seen in our gel assay system. This interesting result could be taken as support for the second scheme for ordering shown in Figure 5.

By measuring both the affinities of the wedge structural proteins for each other and the kinetics of the *in vitro* assembly reactions we should be able to distinguish between the second and third schemes described above. Although no definitive confirmation of any of the proposed ordering schemes has yet resulted, the reagents for the analysis of this part of an ordered pathway of assembly determined by protein-protein interaction are at hand.

CATALYSIS OF PROTEIN-DETERMINED ORDERED ASSEMBLY STEPS

The possibility exists that protein-determined ordered assembly steps might be catalyzed by other proteins. There are numerous examples of proteins that are bound to particular intermediates of an assembly pathway, but are not found in the final product. In at least one well-documented case, that of the scaffolding protein of phage P22, such a protein is released from the structure in an active form and is reused in subsequent rounds of assembly (King and Casjens, 1974; Chapter 5, this volume). Thus, the P22 scaffolding protein, which is required for proper head assembly, can be considered to catalyze P22 head assembly. Because the scaffolding protein is found as a stably bound component of an intermediate in the head-assembly pathway, it is possible to view its addition and release from head-assembly intermediate structures as two separate protein-determined ordered steps of the kind discussed above (fig. 5). In this sense, scaffolding protein does not catalyze an individual assembly step.

Two T4 tail-fiber assembly proteins, gp38 and gp63, are also candidates for such "assembly enzymes". Gp38 is required for proper assembly of the distal portion of the tail fiber, but it is not found stably bound to the components of that structure. This distal fiber is a dimer of gp37 in which the two molecules of gp37 are likely to be interwound (Earnshaw et al., 1978). It has been speculated that gp38 forms a template for the initiation of the proper interaction between the two (possibly unfolded) gp37 polypeptides (Wood, 1979; Wood, 1980). In the reaction promoted by gp63, the completed tail fiber is joined to the otherwise completed phage particle. In this case, the components of the reaction (fibers and fiberless phage) are likely to be fully folded. The reaction is somewhat more complex than might have been expected in that the fiber appears to have to interact with two separated sites on the fiber-less phage, its attachment site on the edge of the baseplate and the whisker (a fiber extending from the tail-proximal portion of the head) (Coombs and Eiserling, 1977; Wood and Conely, 1979; Terzahgi et al., 1979). The mechanism by which gp63 catalyzes tail-fiber attachment is unknown. However, Wood (1979) has speculated that it may help one of the components to occupy transiently an energetically unfavorable conformation needed for assembly. On the other hand, it is possible that gp63 forms a "template" that transiently binds the two components in relative orientations that allow their rapid association.

Thus, both gp38 and gp63 are excellent candidates for "enzymes" that catalyze the noncovalent association of other protein molecules; however, we do not yet understand the mechanism by which they operate or the reason why these particular steps require catalysis.

FURTHER COMMENTS ON PROTEIN-DETERMINED ORDERED ASSEMBLY MECHANISMS

The building of the T4 tail also shows that ordered assembly occurs during polymerization of a large number of identical protein subunits into a supramolecular structure. After the baseplate is completed, 144 molecules of gp19 are added to form the tail tube, a long hollow structure around which the tail sheath assembles (see Berget and King, 1983). During the assembly of the tail tube, after the first annulus of six gp19 subunits is bound to the baseplate, subsequent subunits use previously incorporated subunits as their active site of polymerization. This process repeats until 144 molecules of gp19 are bound. The addition of gp19 monomers stops at this point. It is likely that this stop is controlled by the T4 baseplate protein gp48, which may act as a "tape measure" for determination of the tail tube length (Duda and Eiserling, 1982; Chapter 5, this volume). Assembly-naive P19 has been purified, and it shows no tendency to self-aggregate in the absence of baseplates (Wagenknecht and Bloomfield, 1977). Thus, it is clear that the assembly of multimers of identical subunits can be controlled in a manner analogous to that discussed above for nonidentical subunits, in that after each subunit binds to the growing structure, it creates a binding site for another similar subunit (in Figure 5, subunits A, B, and C would be identical). This type of controlled assembly of multimers from identical subunits has been discussed by King (1980) and Caspar (1980) and called "self-regulated" and "autosterically controlled" assembly, respectively, by these authors.

Assembly of tobacco mosaic virus (see Chapter 3, this volume) also exhibits "self-regulated" assembly of its single species of coat protein. In this case, atomic resolution structures have been determined for the helical virus and for the "free" coat protein (in a nonhelical assembly without RNA), so a detailed model for changes in the coat protein during the assembly process can be constructed. In this case, the coat-protein monomer by itself does show some self-affinity and is found in fairly small oligomers or double ring structures, called disks, containing 34 monomers. However, in the presence of TMV RNA the 34-mer appears to bind to a particular internal location on the RNA, and additional coat-protein molecules (probably in the form of small oligomers) add to the structure in an ordered fashion in both directions (see Chapter 3, this volume for discussion of the initation process). The "elongation" phase of this assembly is an excellent example of ordered assembly, in which the coat protein requires both the RNA and previously assembled coat protein to act as a site upon which to bind.

The atomic resolution structure of the TMV coat protein in the 34-mer and in the virion, where it is bound to RNA, shows that several alpha helices are very flexible in the 34-mer but are rigidly positioned when bound to RNA (Stubbs et al., 1977; Champness et al., 1976; Bloomer et al., 1978). In addition, there appear to be very slight changes in the relative positions of some helices with respect to one another (Stubbs, 1984). It is not known whether the ordering of some helices or the shifting of overall position of some helices is responsible for the formation of the binding site for additional coat protein; however, it certainly shows that some type of conformational change (as in model 3, Figure 5) in TMV coat protein might be responsible for "self-regulated" assembly.

Several nonviral homopolymerization systems clearly appear to utilize this type of control of assembly. *Salmonella typhimurium* flagellin appears to assemble in a similar "self-regulated" manner. In this case, circular dichroism measurements have documented a conformational change upon assembly (Uratani et al., 1972). The kinetics of flagellin assembly suggest that this conformational change after each subunit assembles is rate-limiting for the assembly process under laboratory conditions. Similar arguments have been made for actin polymerization (Wegner, 1976) and microtubule assembly (Erickson, 1974; Kirschner and Williams, 1974).

The molecular nature of the types of conformational changes that might occur within protein subunits upon their assembly have been discussed by Rossmann and colleagues (Chapter 2, this volume; Rossmann, 1984). In spite of the small changes in helix packing observed upon assembly of the TMV coat protein into virions, they argue rather persuasively that most protein structures cannot undergo major "elastic deformations", and they have catalogued the types of differences that have been seen between atomic resolution structures of free proteins and the same proteins bound to some ligand (small molecule, another protein, or nucleic acid). These are (1) order-disorder changes, (2) hinge movements, (3) conformational changes in loop or beta-bend regions and (4) alteration of binding surfaces between one subunit and another. In addition, very slight alterations in the overall folding would alter the positions of surface amino acid side chains, thereby changing the surface properties of the protein (Harrison, 1983). Presumably any of these, and perhaps other types of conformational changes not yet documented, could be responsible for the creation of the binding site for protein C in Figure 5, thus controlling the assembly process.

CONCLUSION

For those who have studied or worked on the morphogenesis of bacteriophages, the concept of assembly pathways seems natural. It has always seemed that pathways of assembly must represent the most efficient mechanism of building a structure. We have all seen examples of the assembly-line concept in day-to-day life, from the way cars are built to the way buildings are constructed. As I have described in this review, the simplest of probability calculations bears out this feeling.

There is a wealth of descriptions of morphological pathways in the scientific literature. Most of the construction problems faced by organisms are dealt with by using pathways of assembly. However, a description on the molecular level of how pathways are controlled is blatantly missing from our knowledge. This then represents the challenge to investigators studying morphogenesis to determine the molecular basis for sorting protein molecules into a reaction sequence.

ACKNOWLEDGEMENTS

I wish to thank my co-workers Mary F. Plishker, Monjula Chidambaram, Shaukat H. Rangwala, Karen Pollack, Roger Kylberg, Colleen Sari, John Schwarz, and Harryette Davis for their hard work, cheerfulness, and patience during the course of our studies and in the preparation of this article.

The preparation of this chapter was supported by Grant No. PCM 83-14241 and PCM 81-04523 from the National Science Foundation and GM28952 from the National Institutes of General Medical Sciences.

REFERENCES

Berget, P. B. and H. R. Warner. 1975. *J. Virol.*, 16, 1669.

Berget, P. B. and J. King. 1978. *J. Mol. Biol.*, 124, 469.

Berget, P. B. and J. King. 1983. In *Bacteriophage T4*. C. K. Mathews, E. M. Kutter, G. Mosig, and P. B. Berget, eds. p. 246. American Society of Microbiology, Washington, D. C.

Bloomer, A., J. Champness, G. Bricogne, R. Staden, and A. Klug. 1978. *Nature*, 276, 362.

Caspar, D. L. D. 1980. *Biophys. J.*, 32, 103.

Champness, J. H., A. C. Bloomer, G. Bricogne, P.J. G. Butler, and A. Klug. 1976. *Nature*, 259, 20.

Coombs, D. and F. A. Eiserling. 1977. *J. Mol. Biol.*, 116, 375.

Crane, H. R. 1950. *The Scientific Mon.*, 70, 376.

Drake, J. W. 1969. *Nature*, 221, 1132.

Duda, R. and F. A. Eiserling. 1982. *J. Virol.*, 43, 714.

Earnshaw, W. C., E. B. Goldberg, and R. A. Crowther. 1979. *J. Mol. Biol.*, 132, 101.

Edgar, R. S. 1969. *Harvey Lect.*, 63, 263.

Edgar, R. S. and I. Lielausis. 1968. *J. Mol. Biol.*, 32, 263.

Edgar, R. S. and W. B. Wood. 1966. *Proc. Nat. Acad. Sci. (USA)*, 55, 498.

Epstein, R. H., A. Bolle, C. M. Steinberg, E. Kellenberger, E. Boy de la Tour, R. Chevalley, R. S. Edgar, M. Sussman, G. H. Denhardt, and I. Leilausis. 1963. *Cold Spring Harbor Symp. Quant. Biol.*, 28, 375.

Erickson, H. P. 1974. *J. Supramol. Struct.*, 2, 393.

Harrison, S. C. 1983. *Adv. Virus Res.*, 28, 175.

Kikuchi, Y. and J. King. 1975a. *J. Mol. Biol.*, 99, 645.

Kikuchi, Y. and J. King. 1975b. *J. Mol. Biol.*, 99, 673.

Kikuchi, Y. and J. King. 1975c. *J. Mol. Biol.*, 99, 695.

King, J. 1971. *J. Mol. Biol.*, 58, 693.

King, J. 1980. In *Biological Regulation and Development*. Volume 2. R. Goldberg, ed. p. 101. Plenum.

King, J. and S. R. Casjens. 1974. *Nature*, 251, 112.

King, J. and N. Mykolajewycz. 1973. *J. Mol. Biol.*, 75, 339.

King, J. and W. B. Wood. 1969. *J. Mol. Biol.*, 39, 583.

Kirschner, M. W. and R. C. Williams. 1974. *J. Supramol. Struct.*, 2, 412.

Kozloff, L. M. 1981. In *Bacteriophage Assembly*. M. DuBow, ed. p. 327. Alan R. Liss.

Laemmli, U. K. 1970. *Nature*, 227, 680.

Meezan, E. and W. B. Wood. 1971. *J. Mol. Biol.*, 58, 685.

Mosher, R. A. and C. K. Mathews. 1979. *J. Virol.*, 31, 94.

Plishker, M. F., M. Chidambaram, and P. B. Berget. 1983. *J. Mol. Biol.*, 170, 119.

Plishker, M. F. and P. B. Berget. 1984. *J. Mol. Biol.*, 178, 699.

Rossmann, M.G. 1984. *Virol.*, 134, 1.

Stubbs, G. 1984. In *Biological Macromolecules and Assemblies*, Vol. 1. F. A. Jurnak and A. McPherson eds. p. 149. Wiley.

Stubbs, G., S. Warren, and K. Holmes. 1977. *Nature*, 267, 216.

Terzaghi, B. E., E. Terzaghi, and D. Coombs. 1979. *J. Mol. Biol.*, 127, 1.

Uratani, Y., S. Asakura, and K. Imahori. 1972. *J. Mol. Biol.*, 76, 85.

Wagenknecht, T. and V. A. Bloomfield. 1977. *J. Mol. Biol.*, 116, 347.

Wegner, A. 1976. *J. Mol. Biol.*, 108, 139.

Wood, W. B. 1979. *Harvey Lect.*, 73, 203.

Wood, W. B. 1980. *Quart. Rev. Biol.*, 55, 353

Wood, W. B. and M. P. Conley. 1979. *J. Mol. Biol.*, 127, 15.

Wood, W. B., R. S. Edgar, M. Henninger, and I. Lielausis. 1968. *Fed. Proc.*, 27, 1160.

5
Shape Determination in Virus Assembly: The Bacteriophage Example

Roger W. Hendrix
University of Pittsburgh

INTRODUCTION

Any organism, as it grows and develops, generates new biological structure. To accomplish this feat, it must be able to convert its genetic information from the one-dimensional form that it has in nucleic acid into the three dimensions of biological structures. Bacteriophage assembly is a particularly tractable experimental system for investigating many features of how this can be done. In this chapter I discuss how the three-dimensional shapes of the tailed phages are regulated, particularly the large-scale features of the phage: How are the shape and size of the head determined? What specifies the length of the tail? It will be apparent that we are only beginning to scratch the surfaces of these questions; nevertheless, principles about how biological form is determined are emerging, and there is good reason to believe that they will apply to areas of biology beyond phage biology.

The simplest aspect of the problem of converting one-dimensional biological information into three dimensions is that of folding a polypeptide chain into a protein. Although the literature makes it clear that protein folding is an extremely complex process, it is probably fair to say that we understand protein folding in principle even if we are far from understanding it in detail. To wit, the amino acid residues of the polypeptide interact with each other and with the solvent as they are polymerized by the ribosome, and spontaneously fold into a precise three-dimensional structure. In other words, the three dimensional form of the protein is inherent in the properties of the polypeptide chain.

Once a protein is folded, it may assemble with other protein subunits into small oligomeric structures, as exemplified by hemoglobin or the oligomeric enzymes. We can also understand assembly of these structures in terms of the intrinsic properties of the subunits themselves. Thus, the subunits bind to each other in relative orientations that are determined by the details of their shapes and surface properties. In these cases, the process is self-limiting, because the oligomeric structure is closed; thus, when four identical subunits assemble into a tetrahedral oligomer, all of their specific binding sites are mutually satisfied and polymerization stops. Many parts of phage assembly exhibit this type of organization; Peter Berget discusses how these processes are controlled in Chapter 4.

The problem takes on a new aspect when we consider how a structure is assembled if its components are potentially capable of assembling in a variety of ways. Stated another way, this is the problem of how to assemble a specific structure when the form of that structure is not fully specified by the properties of its major subunits. That this is true of phages will become apparent when we consider the structures produced in mutant infections or by in vitro assembly of purified components. We will ask how the versatility of the protein subunits of phages is channeled; that is, how assembly is directed toward the correct structure and away from the many possible incorrect ones. The most general principle that is emerging is that there are proteins that direct polymerization of other proteins. Some of these proteins remain as components of the mature structure, but others do not, and they appear in some cases to have other roles in addition to directing assembly. The detailed mechanisms by which these proteins direct assembly are only beginning to be revealed, but it is already clear that no single mechanism is used in all cases.

This review of the literature on shape determination in phages will deal mostly with the tailed phages, all of which have dsDNA genomes. Brief mention will be made of other viruses, but with a few exceptions these questions have not been approached in comparable detail. A number of other reviews on various aspects of phage assembly and shape determination are available. Among these, the reviews by Black and Showe (1983), Murialdo and Becker (1978), and Wood and King (1979) are especially useful and relevant to the topics considered in this chapter.

A striking outcome of the large number of studies on assembly of tailed phages is that the assembly pathways for the different phages are very similar in overall form, despite numerous differences in detail. This strongly suggests that shape determination, which all of these phages accomplish successfully, is carried out by the same basic mechanisms by all of them. I will assume for most of the discussion that this is true. This assumption has distinct advantages in trying to analyze which features of phage assembly may be fundamental to shape determination and which are variations on the basic theme. It also has obvious pitfalls if the mechanisms are not as similar as we suppose, and this should be borne in mind by the reader.

SHAPE DETERMINATION IN PHAGE HEADS

Structure of Phage Heads

The phages we are considering have head shells made of several hundred copies of a major protein subunit arranged with icosahedral symmetry according to the scheme elaborated by Caspar and Klug (1962) (for a detailed discussion of this topic, see Chapters 1 and 2, this volume). The shells are either isometric (i.e., quasi-spherical), as with λ, P22, T7, T3, P2, and P4, or prolate, as with T4 and ϕ29. The prolate heads are modified icosahedra that are elongated along one axis by virtue of extra subunits inserted around an equator. One of the 12 vertices of the icosahedral shell is the unique site to which the tail attaches; this is called the proximal vertex. In prolate shells, the proximal vertex is located on the long axis of the shell.

In the isometric phage heads that have been examined in sufficient detail, the pentameric rings of protein subunits at the vertices of the icosahedron are made of the same protein as are the hexameric rings that make up the rest of the shell. However, in some phages, for example, T4, a separate protein forms the pentamers at the 11 nonproximal vertices. This may be an evolutionary elaboration on the simpler scheme in which both roles are played by the same protein: the vertex protein of T4, gp24, is very similar to the major shell subunit, gp23, and gene *24* is thought to have arisen from gene *23* by gene duplication (Showe and Onorato, 1978). In fact, amber mutations in gene *24* can be suppressed by missense mutations in gene *23* that allow gp23 to assume both roles (McNicol, Simon, and Black, 1977). Showe and Onorato (1978) suggested that the acquisition of a separate vertex protein by T4 may have accompanied the change from an ancestral isometric head to a prolate head. This is an attractive idea, since gp24 may have some role in determining the length of the prolate head (see below), but whether there is a more general correlation between prolate heads and the presence of a specialized vertex protein will not be clear until more data are available. The other well-studied phage with a prolate head, ø29, does not have a separate vertex protein. However, ø29 should perhaps be regarded as a special case, since its shell is structurally simpler than those of most other phages. ø29 is thought to have a triangulation number of 1 and may have as few as ca. 85 subunits in its shell (Vinuela et al., 1976). In contrast, T4 has a triangulation number of 13 and about 1000 subunits of gp23 and gp24 (Aebi et al., 1974; Branton and Klug, 1975; Moody, 1965). Another virus with a separate vertex protein, adenovirus, is large ($T = 25$) and isometric. In this case, the requirement for the vertex protein may derive from the fact that the adenovirus "hexons" are made not of six subunits but of three, each occupying the positions of two proteins in a conventional Caspar-Klug style shell (Cornick, Sigler, and Ginsberg, 1971; Crowther and Franklin, 1972). To make a quasi-equivalent penton from this same protein would require 2-1/2 subunits, clearly a structurally implausible possibility.

In addition to the subunit(s) that make up the shell proper, phage heads always have several quantitatively minor proteins, present in amounts ranging from a few to a few tens of copies per virion. For the most part, these "minor" proteins are located at the proximal vertex, and most of these make up a knoblike structure called the head-tail connector (also variously known as the neck or upper collar), which is the part of the head to which the tail attaches. The head-tail connector plays central roles in several phage functions, including DNA packaging and injection, and head-tail joining. As I will discuss below, it also appears to participate in the initiation of shell assembly.

Some phages have other proteins in the mature heads, but these proteins have no role in determining head shape and will therefore be mentioned only briefly. These include gpD of λ and the Soc and Hoc of T4, which are major head proteins that add to the surface of the shell several steps in the assembly pathway after the basic shell is completed. They play a role in stabilizing the head (Sternberg and Weisberg, 1977; Ishii and Yanagida, 1977) and can be thought of as auxiliary domains of the main shell protein that are added (noncovalently) to the shell at the time they are needed. T4 also has a number of small internal proteins and peptides that may facilitate DNA packing. These have no essential role in shape determination as far as is known, although two of the peptides are fragments of proteins (gp22 and gp67) that play important roles in shape determination earlier in the assembly pathway (see below). Finally, some viruses, including ø29 and adenovirus, have simple appendages ("fibers") attached to the outside of the shell (Anderson, Hickman, and Reilly, 1966). These also do not participate in determination of shell shape, at least for ϕ29, since mutations that remove the appendage proteins do not affect shell shape (Reilly, Nelson, and Anderson, 1977).

As Caspar and Klug (1962) first realized, the protein subunits of a phage head gain the ability to construct large icosahedral structures ($T > 1$) by being somewhat flexible in terms of the angles at which they can bond to their neighbors. Thus, for a phage with a triangulation number of 7 (420 subunits), there are 7 similar but conformationally distinct ("quasi-equivalent") classes of positions that identical subunits must fit into in the shell. (One of these positions corresponds to the subunits in pentamers at the vertices, and the others to subunits which are at various positions on the faces as members of hexamers; see Chapter 1, this volume.) Inherent in the flexibility (or versatility) that the subunits must have in order to build a large shell is the possibility of building a variety of different, incorrect structures. In wildtype infections, such aberrant structures are rare, but in certain mutant infections they can account for most of the major head subunits in the cell. These "monsters" can take various forms (see Figure 1); most can be described either as tubes made of a cylindrical folding of a hexameric lattice of shell subunits (which may or may not have caps on the ends), or as irregular structures in which pentameric vertices are apparently inserted into a hexagonal lattice, as in normal shells, but with irregular spacing. The tubular structures are usually called "polyheads;" I will refer to the irregular structures as "irregular monsters." The production of aberrant forms in mutant infections both confirms the ability of the subunits to assemble incorrectly and gives us a genetic tool for understanding how normal assembly is controlled.

Pathway for Head Assembly

Figure 2 shows a generalized assembly pathway for the heads of dsDNA phages, including features that are shared by all the well-studied phages in this class, plus a few features (in parentheses) that are seen in more than one phage but are absent in others.

For all the dsDNA phages for which we have appropriate information, head assembly involves assembling an empty protein shell into which DNA is subsequently packaged. This structure, known as a prohead, procapsid, or prehead, is preceded by a structure that has the same shell but also contains an inner protein core or shell, called a scaffold. The scaffold is made from a different protein subunit, the scaffolding protein, and plays an important role in directing assembly of the major shell protein, as will be discussed below.

The scaffolding protein leaves the prohead after shell construction is complete. In some phages (P22, T7, ϕ29) the protein leaves intact, and for P22 it has been shown that the scaffolding protein is recycled for participation in repeated rounds of shell assembly (King and Casjens, 1974). Therefore, it is acting in a formal sense as a catalyst of shell assembly. For other phages (T4, λ) the scaffolding protein is degraded proteolytically at the time it disappears from the prohead. For some phages, exit of the scaffolding protein has not been separated experimentally from DNA packaging. However, in the cases

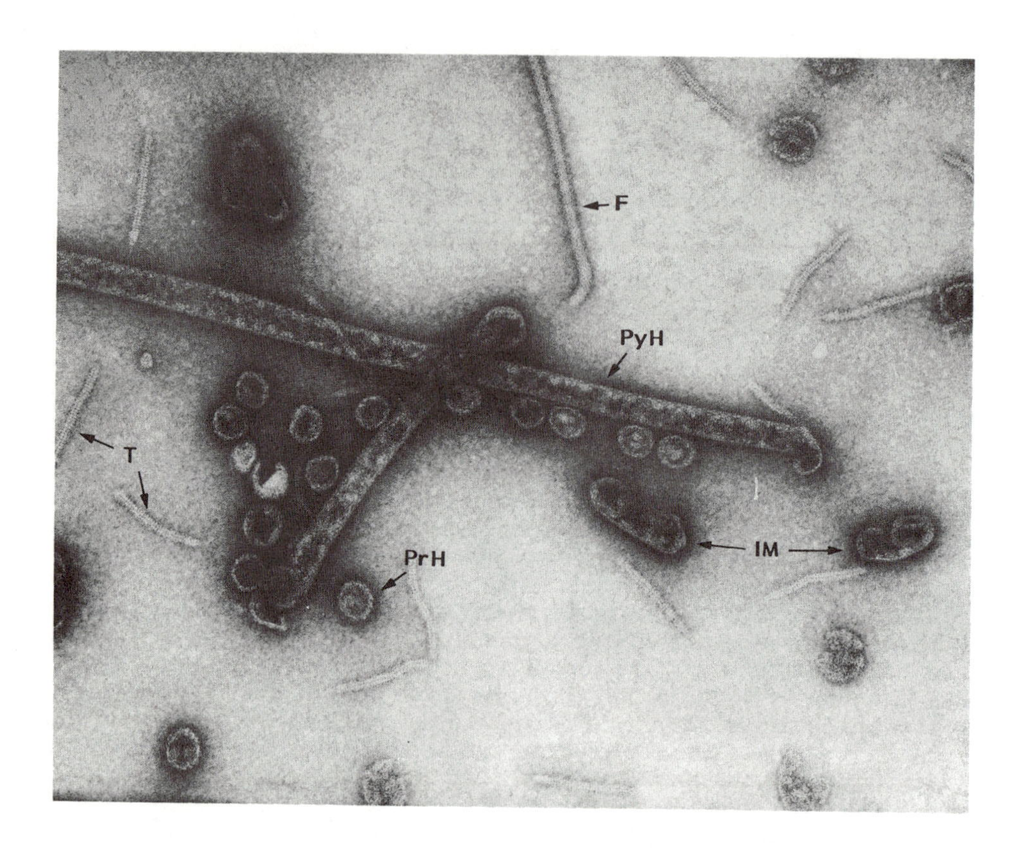

FIGURE 1 Aberrant λ head-related structures. Arrows indicate polyheads (PyH), proheads (PrH), and irregular monsters (IM); tails (T) and a bacterial flagellum (F) are also visible. This lysate was produced by infecting a *groE* mutant of *E. coli* with wildtype λ. Initiation of head assembly is defective in this host.

of λ and T4, scaffoldless proheads can package DNA efficiently *in vitro*, arguing against an early suggestion that exit of the scaffolding protein might be coupled obligatorily to DNA entrance.

By the time the scaffolded prohead is made, the ultimate dimensions, and therefore the shape, of the head are determined (though not yet attained). Thus, proheads of phages with isometric heads are isometric and those of phages with prolate heads are prolate. Furthermore, even though the dimensions of the prohead are smaller than those of the mature head, the prohead shell has the same number of subunits as the head shell. The transition to the final dimensions is accomplished by a conformational change in the shell subunits that causes the shell to expand. Shell expansion occurs at about the time of DNA entry.

In many phages some of the head proteins undergo proteolytic cleavage after they are assembled. In T4 at least eight head proteins, including gp23 (major shell protein) and gp24 (vertex protein) are

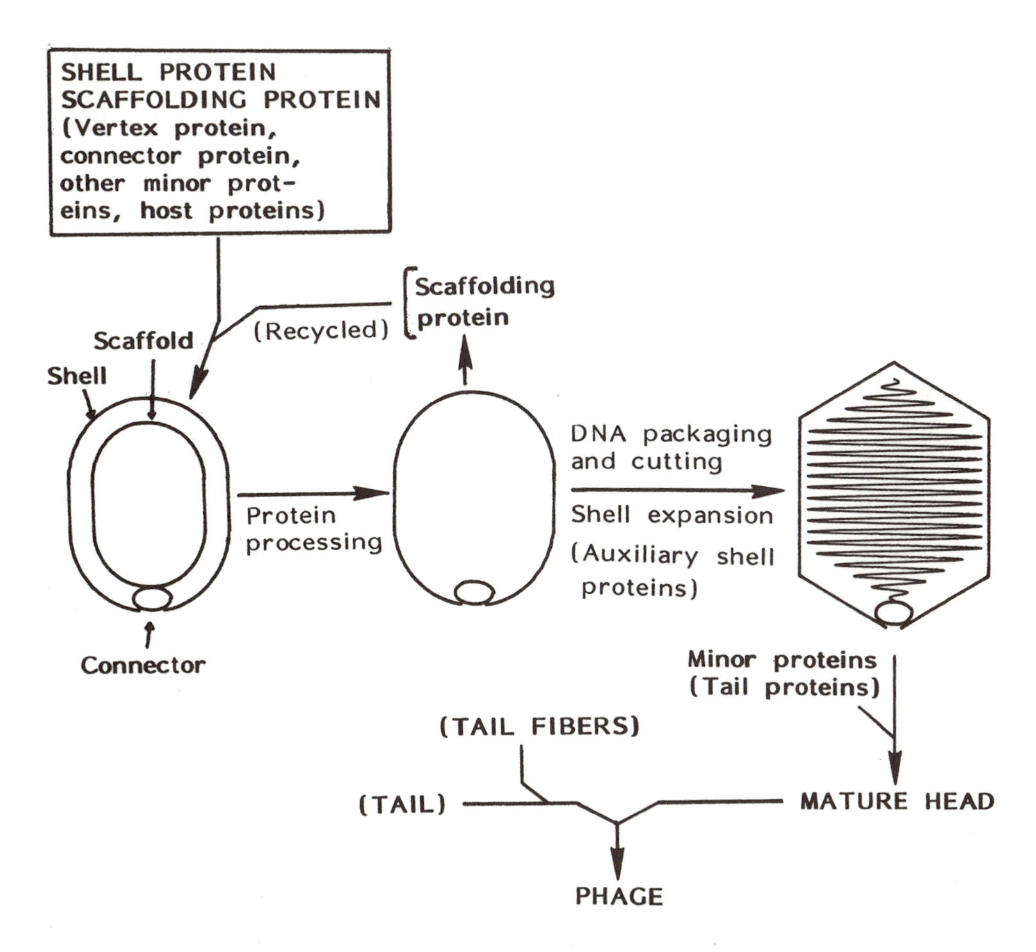

FIGURE 2 Generalized head assembly pathway for the dsDNA bacteriophages. Features in parentheses are found for some phages but not for others; all other features appear to be universal among these phages. The head and proheads are shown as being prolate.

cleaved (Black and Showe, 1983). For λ most of the major shell protein, gpE, is untouched, but gpB, the head-tail connector protein has 22 amino acid residues removed from its amino terminus (Hendrix and Casjens, 1975; Walker et al., 1982). Also, the ca. 10 copies of the minor λ head protein gpC each pair with a copy of gpE, and the two are covalently joined and trimmed (Hendrix and Casjens, 1974a). There are many other examples of protein processing available from other phages. However, some phages, for example P22 and T7, do not show any detectable processing. This argues that protein processing is not central to the mechanism of determination of head shape that we suppose all of these phages share. This conclusion is corroborated by the results of infection by a *21⁻* amber mutant of T4. Gp21 is the phage-coded protease that catalyzes the T4 protein cleavages. In a *21⁻* infection, scaffolded proheads accumulate that are prolate and of normal dimensions (Onorato, Stirmer, and Showe, 1978). Apparently further assembly is blocked as a result of the failure of cleavage, but by the time this happens the correct prohead shape has been attained.

Shell Assembly and the Role of the Scaffold

Experiments on shell assembly indicate that there are three major contributors to correct assembly: the shell protein itself, the scaffolding protein, and the initiation process by which their assembly is started. Leaving initiation aside until the following section, I will first consider the roles of the shell and scafffolding proteins and their interactions.

Isometric Heads

The central importance of the scaffolding protein in assembly of isometric shells can be seen by considering what happens in mutant infections in which the phage carries an amber mutation in the scaffolding-protein gene. In a P22 8^- (scaffold minus) infection, most of the gp5 (shell protein) fails to assemble into recognizable head-related structures (Earnshaw and King, 1978). Of the few structures that are made, some are irregular monsters and some are empty isometric shells. Most of the isometric shells can be shown to have the diameter of normal ($T = 7$) proheads, but some are smaller and probably have the next smaller triangulation number, 4. These results imply that scaffolding protein is necessary for efficient assembly of gp5. The normal-sized isometric shells indicate that gp5 does have the capacity to assemble into a regular closed shell of the correct size in the absence of scaffolding protein, but the abundance of irregular monsters and smaller isometric shells shows that the fidelity of assembly is poor. A λ *Nu3*$^-$ (scaffold minus) infection gives much the same results as seen in the P22 case (Ray and Murialdo, 1975), though the number of structures may be somewhat higher. Apparently a significant amount of the shell subunit gpE remains unassembled, since a *Nu3*$^-$ lysate (unlike other mutant lysates) can provide gpE in a low-molecular-weight form for *in vitro* assembly of functional proheads (Murialdo and Becker, 1977).

The shell protein itself plays an important role in determining shell size. Katsura (1983) isolated a series of missense mutants in the gene for the λ shell subunit, *E*, which make heads that are smaller than normal. The head shells have a triangulation number of 4 instead of the normal $T = 7$. The proheads produced by these mutants are also proportionally smaller than wildtype proheads. Furthermore, if assembly of these small structures is blocked at a stage when the scaffold is still associated with the prohead (a *groE*$^-$ infection), the number of copies of the scaffolding protein (which is itself wildtype) is also reduced about threefold. Therefore, in these mutants the properties of gpE determine not only the size of the shell but also the size of the scaffold. (These proheads also have a normal complement of minor proteins such as the head-tail connector protein, and they are capable of DNA packaging, tail attachment, and subsequent injection of their DNA.)

The ability of the shell subunit to determine a particular shell size is taken to reflect an intrinsic curvature of the lattices into which it assembles. This curvature of the shell protein lattice would force the insertion of pentameric vertices at the appropriate positions as shell assembly proceeds. Mutants such as those described above are interpreted as having an increased intrinsic curvature, resulting in more frequent insertion of vertices, and therefore smaller shells. In isometric shells, the role of the scaffold is apparently to improve the efficiency and precision of this assembly process, but not to specify the size. The fact that in Katsura's mutants the altered gpE imposes a smaller size on the scaffold argues that the scaffold does not assemble in advance of shell assembly, since if it did, it should not be smaller than normal. Thus, it seems likely that the scaffold and shell co-polymerize, starting from the initiator at the proximal vertex (see next section).

Fuller and King (1982) carried out *in vitro* assembly experiments with purified shell (gp5) and scaffolding (gp8) proteins of P22 that agree with the view that the shell and the scaffold influence each other as they copolymerize. Neither gp5 or gp8 alone assembled into large structures, except for a few gp5 structures similar to those seen in an *8*⁻ infection. However, if the two proteins were mixed they assembled rather efficiently into structures with the overall morphology of normal scaffolded proheads.

Although the roles of the shell and scaffolding proteins in assembly of isometric shells are starting to become clear in a general sense, the details of how they interact with each other, and how that interaction promotes correct assembly are still essentially mysterious. A particularly important area of ignorance blocking more detailed understanding concerns the structure of the scaffold; its size and shape are known at low resolution (Earnshaw, Casjens, and Harrison, 1976; Earnshaw, Hendrix, and King, 1979; Kunzler and Berger, 1981), but detailed evidence about the spatial arrangement of individual scaffolding protein subunits within that structure is completely lacking. This means that it is not even clear whether coat and scaffolding proteins make regular repeating contacts with each other, or whether their contacts are substantially different from one part of the structure to another. The question of the structure of the scaffold becomes even more interesting in light of the evidence described above that the wildtype λ scaffolding protein can make a fully functional small scaffold with less than half the usual number of subunits.

A possible counter-example to the idea that the shell subunit has the primary role in determining head size is provided by the P2/P4 system. P2 is a dsDNA phage with an isometric head of triangulation number 9; P4 is a satellite phage of P2 which must use the head and tail genes of P2. In a P2/P4 coinfection, P4 directs the assembly of P2 head proteins into P4 heads, which are smaller than P2 heads, probably $T = 4$ (Geisselsoder et al., 1982). P4 mutations, called *sid*, that affect the P4 head size have been isolated, but they are not well understood and appear to act rather indirectly (Shore et al., 1978; Sauer et al., 1981).

P4 codes for two late proteins, protein 3 and protein 4, and these have been considered as possible candidates for size-determining proteins. Protein 4 is found in mature P4 heads, but apparently is not essential to size determination since a mutant that fails to make detectable protein 4 still makes small heads (Sauer et al., 1981). Protein 3, though not present in mature heads, has been detected in structures that may be proheads, and it could be acting (for example) as a size-determining scaffold (Barrett, Marsh, and Calendar, 1976).

Prolate Heads: T4

Assembly of prolate heads presents two distinct problems of size determination, namely, how to determine the width of the head, which is analogous to determining the diameter (triangulation number) of an isometric head, and how to determine its length. The best-studied phage with a prolate head, T4, is structurally more complex than the phages discussed above. The T4 scaffold has several different components, of which gp22 (the major scaffolding protein) and gp67 are both essential for correct shell assembly, and a third component, IPIII, is classified as "semi-essential". Instead of one shell subunit there are two, the major subunit gp23 and the subunit that makes the pentamers at the eleven nonproximal vertices, gp24.

Mutants of either of the essential scaffolding proteins of T4 produce proheads or polyheads of abnormal width. For a *22⁻* infection the structures have a smaller diameter than normal heads (Yanagida et al., 1970); for a *67⁻* infection they have abnormal, variable diameters (Volker et al., 1982). Since mutations in other genes have not been seen to affect the width of head structures, these results are taken to mean that the width of the head is determined, at least in part, by the scaffold.

There is more information available about the detailed structure of the T4 scaffold than for those of other phages, and it puts interesting constraints on the process of shell assembly. The scaffold has sixfold rotational symmetry (Paulson and Laemmli, 1977; Engel, van Driel, and Driedonks, 1982), with the symmetry axis coincident with the long fivefold axis of the shell. The presence of these two different rotational symmetries aligned on the same axis, usually referred to as a symmetry mismatch, means that the contacts between the scaffold subunits and the shell subunits cannot all be the same. This argues against the type of assembly model in which shell and scaffolding subunits first complex with each other and then add to the growing structure as a unit. It also argues against a scheme in which shell subunits (or groups of shell subunits) find their place in the shell by recognizing and attaching to a particular site on the scaffold. Rather, the difference in symmetry seems more compatible with the view that the shell subunits see the scaffold as a more or less featureless surface against which they can polymerize but with which they do not make specific interactions in the usual sense. Returning to the question the determination of T4 shell width, we might then imagine that the scaffold can assemble by itself to the correct diameter, and that it acts as a mandrel around which a shell of the correct diameter can polymerize.

Much of our understanding of the determination of T4 head length has relied upon studying head-length variants. These variants include "petits," which have isometric shells of normal diameter, and "giants," which also have normal diameter but are elongated, sometimes by a factor of 10 or more over the normal length. Petits can be produced by missense mutations in genes *23* (shell) or *22* (scaffold); giants are seen with mutations in *23*, *22*, or *24*, or by growing wildtype phage in the presence of the amino acid analog, canavanine. A *24*⁻ amber mutant produces a related structure, a polyhead of normal diameter in which the top of the tube has failed to close. The scaffolds in the proheads of petits or giants are themselves petit or giant. Thus, a point mutation in any one of genes *22*, *23*, or *24* can alter the length of both the shell and the scaffold. This argues that neither the shell nor the scaffold dominates the other in determining length; rather, they collaborate to determine length. The theme of interaction among the scaffold, major shell, and corner proteins has received support from a number of experiments.

Doherty (1982) studied the second site revertants of gene *23* missense mutants that make abnormal length capsids. Such reversion requires two separate mutations, one in gene *22* and one in gene *24*. Since the revertants were allele specific, this argues that the three proteins, gp22, gp23, and gp24, interact directly with each other to produce a prohead of the correct length.

Showe and Onorato (1978) obtained different sort of evidence for interactions among these proteins by varying their relative concentrations during assembly. This was accomplished by coinfecting cells with wildtype phage and phage with an amber mutation in one of these genes. Reduction of the relative concentration of gp23 increased the frequency of giant heads, while reduction of gp22 caused an increase in the frequency of petits. Showe and Onorato interpreted these results in terms of a kinetic model in which the relative rates of polymerization of the shell and the scaffold determine how soon the shell is able to close off.

Another series of experiments involving assembly of purified head proteins has given an interesting and somewhat different perspective to these studies. Van Driel and Couture (1978) purified gp23, gp22, and gp20, the head-tail connector protein, following low-salt dissociation of scaffolded proheads. Neither gp23 nor gp22 alone would polymerize into large structures, nor would a mixture of gp22 and gp23, but if gp20 was added to a mixture of gp22 and gp23, they assembled efficiently into normal length scaffolded proheads. Most likely, gp20, which normally plays a role in initiation of shell assembly (see below) acts as an initiator *in vitro*. Remarkably, when gp22 and gp20 were mixed in the absence of gp23, prolate scaffolds (without shells) of apparently normal dimensions assembled. This argues that gp22, in the absence of gp23, contains all the information necessary to specify head length. Van Driel and Couture accordingly proposed that the scaffold assembles to the correct dimensions in advance of the shell and provides a template for assembly of the shell. In its simplest form, this model is difficult to reconcile with the *in vivo* genetic data cited above. We can perhaps best make sense of the results by saying that gp22 has a *preference* for the correct length but that aberrant interactions with gp23 and gp24 as the shell and scaffold grow (as when the conformations or relative amounts of the proteins are altered by mutation) can override that preference and produce structures of aberrant lengths. In this view of shell assembly, the scaffold is the primary determinant of length, but it cannot carry out its role successfully if the shell proteins interact with it incorrectly.

It must be borne in mind that the proteins used in the *in vitro* assembly experiments were obtained by dissociation of assembled structures; that is, they were not "assembly-naive." This leaves open the possibility that their behavior in the *in vitro* assembly reaction is influenced by a "memory" of the structure that they were part of prior to dissociation. However, the fact that the scaffolding protein used in the *in vitro* experiments was not assembly-naive apparently does not explain its ability to polymerize to the correct size; Traub and Maeder (1984) detected similar naked scaffolds of normal dimensions in thin sections of cells infected with *23⁻* mutants. This means that scaffolding proteins that have never seen gp23 can assemble into scaffolds of the correct size, but of course does not necessarily mean that scaffold assembly occurs independently of shell assembly when the shell proteins are present. These experiments lend strong support to the argument made from the *in vitro* experiments that the scaffold has a primary role in determining length.

Traub and Maeder also found the surprising result that naked scaffolds will form in a *23⁻20⁻* infection, that is, in the absence of the head-tail connector protein. This is in apparent contradiction to the results obtained with the *in vitro* assembly system, in which gp20 appeared to be necessary for initiation of scaffold assembly. Traub and Maeder postulate that some other unidentified component of the "neck" region is responsible for scaffold initiation, and that the role of gp20 is to initiate shell polymerization.

The question of how head shape is determined in T4 heads is clearly not yet solved. However, it begins to appear that it may be largely explained by the simple hypothesis that the scaffold determines shell shape by polymerizing into a correctly shaped template that guides polymerization of gp23 and gp24. If this is the case, then the question becomes how the shape of the scaffold is determined. Little information is available on this topic, but Engel, van Driel, and Driedonks (1982) have provided an interesting start in the form of a model based on structural studies of gp22 scaffolds assembled *in vitro*. These authors propose that the scaffold is an assemblage of six ribbonlike polymers of gp22. Purified gp22 assembles into such ribbons under appropriate conditions (Van Driel, 1980); these ribbons look curly in electron micrographs, suggesting that they have an intrinsic curvature. Engel, van Driel, and Driedonks argue that the ribbons are the product of side-to-side polymerization of cigar-shaped gp22 monomers.

In the model of scaffold shape proposed by these authors, each of the ribbons of gp22 has one end attached to the head-tail connector, and they wind from there toward the other end of the scaffold in a six-stranded distorted helix. The ribbons start at a small radius where they are attached to the connector, then move out to a larger radius at the equator of the scaffold, and finally back to a small radius at the top, at which point further addition of subunits is prevented by steric interference among the ribbons. In essence, Engel, van Driel, and Driedonks argue that once the first gp22 subunits are fixed in position on the head-tail connector, then the detailed physical and chemical properties of the ribbons (such as their intrinsic curvature and a postulated propensity to keep their edges touching) may be sufficient to completely specify the course of scaffold polymerization and therefore to determine the shape of the scaffold. Although the usefulness of the model is limited by the lack of information about the actual properties of the ribbons (or for that matter, information about what properties they would need to have for the model to work), it does give an interesting general view of how shape might be determined by a combination of the properties of the scaffolding protein and the initiation process. In this sort of model the dimensions of the shell are determined primarily by the properties of the scaffolding protein, but it seems plausible that mutant shell proteins copolymerizing with the scaffold might well perturb the process enough to produce heads of aberrant lengths, as are seen with some shell protein mutants.

Prolate Heads: ϕ29

ϕ29, like T4, has a prolate head but, as mentioned earlier, it is structurally simpler than T4. Shell assembly for ϕ29 also appears to be somewhat simpler than for T4. As for the phages already considered, assembly of the ϕ29 major shell subunit (there is no separate vertex subunit) requires the correct function of the scaffolding protein (gp7) and the head-tail connector protein (gp10). In a *7*⁻ (scaffold-minus) infection irregular monsters are formed that have much the same appearance as the λ or P22 structures seen in a scaffold-minus infection (Hagen et al., 1976). Thus, it again appears that the scaffold is essential for accurate shell assembly.

A *10*⁻ (head-tail connector-minus) infection produces a most interesting result, namely, isometric shells (Hagen et al., 1976). This would seem to imply that although the diameter (triangulation number) of the shell is determined accurately with just the shell subunit and the scaffolding protein, the prolate shape of the shell is imposed on it by the head-tail connector protein. We can get some insight into how the connector might accomplish this by considering the structure of the shell, which is thought to have a triangulation number of 1 (Vinuela et al., 1976).■ In such a shell, the head-tail connector,

■ There is some uncertainty about the structure of the ϕ29 shell. Carrascosa et al. (1981) made careful measurements of the mass of the head and concluded that the number of shell subunits is about twice as large as the number (85) reported previously. They proposed a model in which the shell is prolate with a triangulation number of 1 (as described in the text and illustrated in Figure 3), but with each structural subunit of the shell lattice consisting of a dimer of protein subunits. Thus, each letter Q in Figure 3 would consist of two identical protein subunits in non-equivalent environments. (This would make the ϕ29 shell subunit the structural equivalent of an enzyme that has two separate catalytic activities.) The description of ϕ29 shell assembly given in the text assumes the simpler (85-subunit) model for the shell structure, but it would also apply if the more complex model is correct.

which occupies the proximal vertex, is immediately surrounded by five pentamers of the shell subunit, each forming one of the five adjacent vertices. In an isometric $T = 1$ shell, these five pentamers would be followed by another five pentameric vertices and finally by the last vertex at a position directly opposite the head-tail connector. However, in the prolate ϕ29 head, the first five pentamers are followed by five hexamers, after which come the 5 + 1 pentamers that close the shell (see Figure 3). If we assume that assembly starts from the head-tail connector (as seems to be the case for other phages; see below), we are led to the view (Murialdo and Becker, 1978a) that the head-tail connector distorts the geometry of the adjacent pentameric vertices in such a way that the next subunits to polymerize into the shell do so in such a way as to form hexamers, thus leading to a prolate shell.

In this view of ϕ29 shell assembly, we imagine that the ultimate shape of the head is determined by a combination of two factors: the first is the angles at which the subunits of the first five pentamers are held by the head-tail connector, and the second is the detailed properties of the shell subunits (limited conformational flexibility, intrinsic curvature), which allow positional information to be transmitted along the shell as it assembles. The role of the scaffold in this scheme is a secondary one of preventing assembly errors rather than specifying dimensions. For a fairly simple structure like the ϕ29 shell, it seems plausible that such a scheme could work well. However, for a structure like T4 ($T = 13$), a scheme like this as the sole determinant of shell shape produces two conflicting demands on the properties of the shell subunit. First, since positional information would have to transmitted over a large number of subunits in the T4 shell, the subunits would have to be relatively inflexible to prevent the accumulation of angular errors. On the other hand, the large triangulation number of T4 requires that the shell subunit be relatively flexible in order to accommodate the large number of distinct quasi-equivalent positions in the shell. Consequently, it should not be considered surprising that T4 appears to have features of shell assembly that are not seen in ϕ29, such as a separate protein for making the vertices (which would reduce the amount of conformational flexibility required of gp23) and a more obviously important role for the scaffold in determining shell shape.

Initiation of Shell Assembly

It has been suggested (e.g., Black and Showe (1983)) that, in a wildtype infection, the formation of incorrectly assembled structures is prevented not because their assembly is blocked directly but because the correct assembly pathway is strongly favored kinetically. By this view, one way that head subunits are directed into the correct assembly pathway is by incorporating an efficient initiation process into head assembly. By this means, head protein subunits can begin to assemble while they are still at low concentration, and the unassembled proteins are thereby prevented from accumulating to the relatively high concentrations that allow aberrant structures to initiate

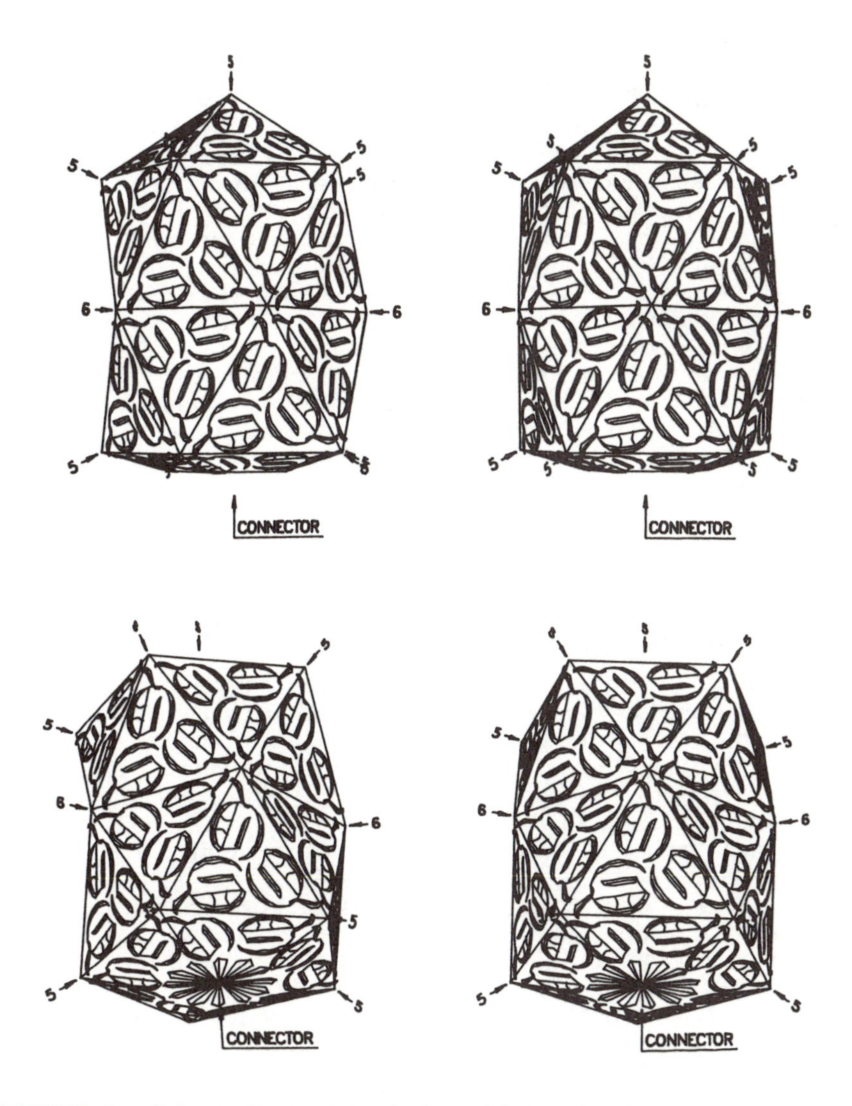

FIGURE 3 Schematic model of the ϕ29 head. Two views are shown, each as a stereo pair. The shell protein subunits, represented by the letter Q, can be thought of as being arranged in hexamers and pentamers, as described in the text.

and assemble. This version of how incorrect shell assembly is prevented has not been completely documented, since the early steps of head assembly, and particularly their kinetics, are very difficult to analyze experimentally. However, it is supported by the observation that the aberrant, head-related structures seen in mutant infections appear later than do the normal head-related structures in a wildtype infection (Laemmli and Eiserling, 1968; Laemmli et al., 1970). An exception to this rule is the variety of structures made in those mutant infections in which initiation is thought to be normal (for example, a T4 *24*⁻ infection), which form with wildtype kinetics.

The details of shell initiation are not well understood for any phage. The process has been investigated most thoroughly for λ and T4, and some common features are emerging. One of these is the requirement for a host function, GroE. There are two *groE* genes in *E. coli*, *groEL* and *groES*. Both genes are required for λ head assembly and at least groEL is required for T4 head assembly. Although similar requirements have not been demonstrated for other phages, it is worth noting that GroE-like proteins are probably present in all eubacterial species, among them *B. subtilis*, for which a protein with the antigenic and morphological properties of GroEL of *E. coli* was shown to associate with head-precursor structures of ø29 (Carrascosa, Garcia, and Salas, 1982). Another similarity between λ and T4 initiation is that the phage proteins involved include those found at the proximal vertex of the assembled shell, prominent among them the head-tail connector protein. These observations lead to the picture that assembly starts at the proximal vertex, assisted in some fashion by the GroE proteins. If shell assembly only occurs following initiation on a head-tail connector, then the problem of how to insure that each head has exactly one head-tail connector is neatly solved.

For λ, *in vitro* complementation experiments suggest that the first step in head assembly involves an interaction between GroE and the λ head-tail connector protein, gpB (Murialdo and Becker, 1978b). In agreement with this, there is both genetic and physical evidence for a direct interaction between GroEL and gpB (Georgopoulos et al., 1973; Murialdo, 1979). Furthermore, in a *groE*⁻ infection, head-tail connectors are not seen by electron microscopy under conditions in which they would be seen in a *groE*⁺ infection, suggesting that the GroE proteins are required for the assembly of gpB into the oligomeric structure of the connector (Tsui and Hendrix, 1980a). A plausible interpretation of these observations is that the GroE proteins act as a surface or template on which the connector is assembled. The GroE proteins are not part of the mature prohead or head, but some mutant proheads, for example, those from a λ *C*⁻ infection (Kar, 1983) or from a λ wildtype infection of some *groE* missense mutants (Hohn et al., 1979), appear to have GroEL loosely associated.

The idea that GroEL acts as a template for connector assembly raises interesting questions about the structural details of the interaction, given the known structures of GroEL and connectors. GroEL is a 14-subunit oligomer arranged with 7-fold rotational symmetry (Hendrix, 1979; Hohn et al., 1979). On the other hand, the λ connector is a dodecamer of gpB with 12-fold rotational symmetry (Kochan, Carrascosa, and Murialdo, 1984), which means that in a complex between GroEL and a connector the different subunits of the connector could not make equivalent contacts with GroEL subunits. Possible implications of this symmetry mismatch are discussed later. The 12-fold symmetry of connectors is conserved in the three phages (λ, T4, ø29) for which data are available (Driedonks et al., 1981;

Carrascosa et al., 1982).

In vitro complementation experiments also implicate the λ scaffolding protein, gpNu3, as well as a minor head protein, gpC, in the earliest steps of head assembly, possibly initiation. It has been proposed that gpC occupies the edge of the shell lattice that lies in contact with the connector (Murialdo and Ray, 1975). Shaw and Murialdo (1980) showed that the *C* and *Nu3* genes overlap in such a way that gpNu3 is identical to the C-terminal half of gpC. It seems likely that this structural relation between gpC and gpNu3 plays some significant role in initiation, but what this might be is still obscure.

Initiation of shell assembly in T4 has the additional feature that it occurs in association with the inner membrane of the cell, with the result that proheads are initially found attached to the membrane through their proximal vertex. Four phage genes, *22*, *20*, *40*, *31*, and the host gene *groEL*, are thought to be required for initiation of correct gp23 assembly (Laemmli et al., 1970; Laemmli, Beguin, and Gujer-Kellenberger, 1970; Georgopoulos et al., 1972; Revel et al., 1980). Gp20 is the head-tail connector subunit; in its absence the shell protein gp23 fails to form the icosahedral caps that constitute the proximal end of a normal head shell. Instead it forms open-ended polyheads (Laemmli et al., 1970). Gp40 is a membrane protein and appears to have a role in anchoring gp20 to the membrane (Hsiao and Black, 1978). *40*⁻ mutants make polyheads of the same sort as seen in a *20*⁻ infection (Laemmli et al., 1970). Genetic evidence argues that gp31 and GroEL act together. In the absence of either, gp23 fails to assemble into regular structures and is found instead in amorphous "lumps" associated with the membrane (Laemmli, Beguin, and Gujer-Kellenberger, 1970; Georgopoulos et al., 1972). It has been proposed that the role of GroEL and gp31 may be to transport or modify newly synthesized gp23, allowing it to assemble into regular lattices. Consistent with this possibility, gp31 has been found to be dispensable in the presence of certain missense mutants of gene *23* (L. D. Simon and B. Randolf, cited in Black and Showe (1983)).

In the purified *in vitro* system, as mentioned earlier, gp20 alone is sufficient to allow assembly of gp23 and gp22 into a morphologically normal scaffolded prohead. We can ask why the other proteins required for *in vivo* shell assembly, namely, gp31, gp40, and GroEL, are not required *in vitro*. Recall that the components used in the *in vitro* assembly were obtained by dissociating assembled prohead; that is, they are not assembly-naive. Any modification of gp23, gp20, or gp22 that might be caused by gp31, gp40, or GroEL would already have occurred. Similarly, any role these proteins have in preventing gp23 from associating with the membrane would very likely be dispensable in the *in vitro* system, which has no membranes. Finally, gp20 is supplied to the *in vitro* system in the form of assembled connectors. If the role of GroEL is to cause assembly of the connectors, as was suggested above for the λ case, it is not surprising that it is not needed for *in vitro* assembly. This sort of comparison of the *in vivo* and *in vitro* cases suggests some possibilities for the biochemical roles of the proteins involved in initiation, but as with λ, the structural and kinetic details of initiation are still quite obscure.

The Problem of Symmetry Mismatches

It was realized nearly twenty years ago that the symmetries of the heads and tails of the tailed phages do not match (Moody, 1965). Tails have sixfold rotational symmetry (threefold in the case of the *Caulobacter* phage øCbK (Papadopoulos and Smith, 1982; Leonard, Kleinschmidt, and Lake, 1973), and they attach to the proximal vertex of the icosahedral head, which lies on a fivefold rotational axis. This mismatch in symmetries poses interesting questions both with regard to the detailed structure of the phage and with regard to the mechanisms by which such a structure is assembled. There are now other examples of symmetry mismatches during phage-head assembly, some of which have been alluded to above. The head-tail connector has 12-fold symmetry, which matches the 6-fold symmetry of the tail but not the 5-fold symmetry of the head shell. The scaffold has 6-fold symmetry, at least in T4 (Paulson and Laemmli, 1977; Engel, van Driel, and Driedonks, 1982), and it is aligned along the same 5-fold axis of the shell as is the head-tail connector. The host GroEL protein molecule, which is thought to form a complex with gpB of λ and is probably responsible, at least in part, for assembling it into the 12-fold head-tail connector, has 7-fold symmetry (Hendrix, 1979; Hohn et al, 1979).

There are at least two different (but not mutually exclusive) explanations for the existence of the symmetry mismatches. First, symmetry mismatch might in itself be useful for the assembly process. That is, the use of mismatched symmetries during assembly may make possible certain ways of constructing large shells or other unwieldy structures that could not otherwise be accomplished as well or at all. By this view, the symmetry mismatches that we observe in phage assembly may be pointing to a general principle of biological assembly that has so far eluded description. Second, the significance of the mismatched symmetries might lie not with the assembly process but with the assembled product. In this view, the mismatch in symmetry during shell assembly does not necessarily make assembly easier but is required in order to produce a finished product with a (functionally important) symmetry mismatch. I have proposed elsewhere (Hendrix, 1978) that the symmetry mismatch between the shell and the head-tail connector that exists in the assembled prohead may play an important role in DNA packaging.

What role symmetry mismatch might play in the process of assembly remains to be seen, but it may be significant that two of the three symmetry clashes seen in shell assembly (GroEL vs. head-tail connector; scaffold vs. shell) involve transient associations that are lost by the time shell assembly is complete. Perhaps symmetry mismatch is a means of assuring that those associations are transient. In such a scheme, one protein (e.g., the scaffolding protein) could provide a template on which a second protein (e.g., the shell protein) could assemble, but without becoming too tightly associated with it through the repeated bonding interactions that characterize many stable structures, such as mature phage heads and tails. The third symmetry clash that we observe (head-tail connector vs shell) is not transient; it survives to be found in the mature phage (where the 12-fold symmetry of the connector does match the tail that has joined to it). In this case there is direct evidence that the bonding across the symmetry mismatch boundary is relatively weak; that is, a variety of chemical and physical treatments that separate heads from tails leave the head-tail connector attached to the tail (Coombs and Eiserling, 1977; Tsui and Hendrix, 1980). In fact, it may be this absence of rigid bonding between the connector and the shell that is functionally significant, if the connector must be able to rotate in its socket during DNA packaging, as Hendrix (1978) suggested.

Shell Expansion

The last step before the shell attains its final shape is expansion. Before expansion the shell has a rounded shape, nearly spherical for the isometric phages, and afterward it is much more angular, with flat triangular faces and sharp corners and edges. The diameter of the shell increases about 20% during expansion, resulting in an approximately two fold increase in internal volume. The expansion is the result of a conformational change in the shell subunits rather than addition of new subunits; the thickness of the shell decreases and the spacing between subunits in the surface lattice increases sufficiently to compensate for the increase in surface area. Expansion is apparently an energetically favorable reaction, since it can be made to occur spontaneously in purified proheads, with the help of 4 *M* urea or 0.8% sodium dodecyl sulfate in λ and P22, respectively (Hohn, Wurtz, and Hohn, 1976; Casjens and King, 1974). Shell contraction has never been observed.

Expansion is accompanied by changes in the surface of the shell. As seen by electron microscopy, the surface texture changes (Wurtz, Kistler, and Hohn, 1976; Yanagida et al, 1970). Antibody probes show that new antigens appear on the surface of the T4 head following expansion (Kistler et al. 1978), and in phages that have auxiliary shell proteins, such as λ and T4, the binding sites for those proteins become available (Imber et al, 1980; Ishii and Yanagida, 1975). Since high resolution structures of shell proteins, either expanded or not, have not yet been determined, the details of the conformational change in the subunits are not known. However, on the basis of electron microscopic studies Casjens (1979) proposed that gp5 of P22 consists of two domains joined by a hinge, and that expansion is accomplished by a by a flexing of the hinge.

Expansion occurs at about the same time as DNA packaging. B. Hohn (cited by Hohn (1983)) found that λ proheads could package a 5.5-kb restriction fragment derived from the left end of λ DNA, without expanding; in contrast, if a 21-kb fragment was used instead, the prohead expanded. This result suggests that expansion is triggered by DNA packaging after a certain amount of DNA has entered the shell, more than 5.5 kb and less than 21 kb in this case.

The only structures that have been caught in the act of expanding are T4 polyheads. An electron micrograph published by Steven et al (1976) shows a polyhead with one half of the lattice unexpanded and the other half expanded. The expansion is apparently proceeding along the polyhead in a wave, with a sharp boundary between the two regions. It seems reasonable to suppose that expansion sweeps across a prohead in a similar manner. Although P22 shells containing only gp5 do expand *in vitro* in response to mildly disruptive conditions, showing that the ability to expand is a property of gp5 alone (Casjens and King, 1974), expansion under normal conditions *in vivo* would presumably be initiated at some point on the shell; the proximal vertex seems the most plausible candidate. Earnshaw, Hendrix, and King (1979) observed that λ C^{-} proheads were more prone to spontaneous expansion than the other proheads they examined, and they suggested that gpC, which is missing in these proheads, may participate in regulating or triggering expansion. The covalently processed forms of gpC that are present at the time of expansion are most probably located at the proximal vertex and may provide a structural transition between the gpE lattice of the shell and the proteins that make up the DNA-packaging machinery (Murialdo and Ray, 1975).

What is the function of shell expansion? At this point we can only speculate. It has been proposed (Hohn et al, 1974; Serwer, 1980; Gope and Serwer, 1983) that expansion might produce a pressure differential across the shell that could be used to suck the DNA into the head like a piece of spaghetti. However, the experiments with *in vitro* packaging of restriction fragments cited above argue that at least the first part of DNA packaging can proceed without expansion. Also, Hsiao and Black (1977) have reported evidence that expanded T4 proheads can still package DNA. Thus, if expansion has a role in DNA packaging, it seems likely to be a secondary one. It has been noted that expansion substantially increases the internal volume of the head, and therefore the amount of DNA it can hold, but if this were the sole purpose of expansion, it is difficult to understand why the shell is not simply made bigger initially. An alternative view is that it is not expansion *per se* that is functionally significant, but rather the conformational change in the subunits. By this view, the properties that the shell subunit must have for efficient shell assembly within the cell are not fully compatible with the properties it needs in its role as a container and protector of the DNA after cell lysis. The conformational change would then be a transformation from a protein suited to do the first job into one suited to do the second. Some support for this view can be inferred from the observation that expansion of the T4 shell is accompanied by a dramatic decrease in the ease with which the shell subunits can be dissociated (Steven et al, 1976). Expansion may simply be an incidental manifestation of tightening the contacts that hold each subunit to its neighbors.

Expansion of the shells of the simpler icosahedral viruses has also been studied. In these cases, much more detailed structural information is available, and it is possible to infer more of the mechanism than is known for the dsDNA phages. With Tomato Bushy Stunt Virus (TBSV), removal of bound Ca^{2+} ions combined with lowering the pH causes the shell radius to expand 10%. Comparison of the structures of the normal and expanded shells, which are known from X-ray diffraction analysis (Robinson and Harrison, 1982), suggests that expansion results from repulsion between pairs of carboxyl groups whose charges are normally masked by the Ca^{2+} ions. These neighboring carboxyl groups lie on particular pairs of adjacent quasi-equivalent shell subunits, and their mutual repulsion forces the subunits apart at this interface. The reversible expansion process is highly cooperative, perhaps like the expansion of phage shell. There is substantial reorientation of the two domains (one in the shell and one protruding outward) of the coat protein with respect to each other during expansion; however, Robinson and Harrison (1982) argued that this is not important in causing expansion, but instead is a consequence of expansion. The intersubunit contact that is disrupted during expansion is one between pentamers and hexamers of the T = 3 structure. This contact is broken and remade in an apparently new manner (nonequivalent to the original contact) in response to the carboxylate repulsion, so that the pentamers and hexamers become separated by about 2 nm, with a branched 2 x 8 nm hole forming in the shell. This process differs from that with the dsDNA phages in which the expansion appears, if anything, to tighten up the structure. The role of the expansion of TBSV and related viruses is not known. One suggestion is that it may be a step in uncoating the viral RNA (Robinson and Harrison, 1982; Durham, Hendry, and von Wechmar, 1977). Based on the data currently available, it seems unlikely that plant virus expansion carries out a similar function to expansion of the phage head shells, but it does illustrate a possible molecular mechanism for the expansion of an icosahedral protein shell.

LENGTH DETERMINATION IN PHAGE TAILS

Tail Morphologies

Phage tails can be classified into three morphological types. The simplest, exemplified by the tails of T3, T7, and P22, is a small, quite elaborate structure extending out a short distance from the proximal vertex of the head. These tails consist of a small number of subunit types, typically present in a few copies each. Together with the head-tail connector, to which they are attached, they constitute a complex molecular machine capable of a number of distinct functions, including attaching the phage to its cell surface receptors and triggering and mediating passage of the DNA out of the head and into the cell. Other phages, of which λ and T4 are the prototypes, have a tail shaft inserted into the middle of this machine, so that the apparatus that contacts the surface of the host cell is held at some distance from the head. For our discussion of shape determination in tails we will limit our focus to how the length of the shaft is determined. Since this is a problem that the phages with the simplest type of tail do not face, they will not be discussed further.

The λ tail shaft is a simple structure made of a stack of about 32 protein disks, each a hexameric ring of the major tail subunit, gpV (Casjens and Hendrix, 1974; Katsura and Kuhl, 1974). The shaft is about 140 nm long. Why some phages, like λ, have a long tail while others get by perfectly well with no tail shaft at all is puzzling. However, Schwartz (1976) calculated that the long, flexible tail of λ greatly increases the efficiency with which the phage can find a receptor on the surface of its host. He argued that this is the reason that λ has such a tail, and he suggested that phages with long flexible tails may be ones whose receptors are quite rare on the cell surface.

The T4-type, or contractile, tail shaft consists of two concentrically arranged tubes. For present purposes, the central tube, called the core or tail tube, can be thought of as homologous to the shaft of λ-type tails, since these structures occupy homologous positions in their assembly pathways (see below), and since it is the assembly of these structures on which length determination operates. In T4 the tail tube is a helical array of gp19, 100 nm long. Around the outside of the tail tube is the sheath, made of gp18. In T2, and presumably also in T4 and other phages with this type of tail, the helical parameters of the sheath are the same as for the tail tube (Moody and Makowski, 1981); thus, each sheath subunit is paired with a tail-tube subunit. The subunit of the T4 tail shaft is evidently different from that of the λ tail: upon adsorption to the host, the sheath contracts and drives the tail tube through the cell wall, as part of the mechanism by which the DNA is transported into the cytoplasm of the host.

Pathway for Tail Assembly

For both flexible-tailed phages and phages with contractile tails, tail assembly occurs by a separate pathway from head assembly. The process begins with the multistep assembly of the structure known in different phages as the baseplate or basal structure. This structure, which will form the distal tip of the tail, then serves as a nucleation point, or initiator, on which the subunit of the tail shaft (tail tube) polymerizes. Polymerization proceeds until the shaft is the right length; then it stops. The central problem in understanding tail length determination is why polymerization stops at this particular point. Following shaft polymerization one or more proteins add to the top of the shaft to stabilize it, and 2-3 additional steps take place before the tails are ready to join to heads. For λ-type tails, one of these steps is the proteolytic trimming of one of the minor components (gpH in λ) to a form ca. 10 kD smaller than the intact protein (Hendrix and Casjens, 1974b; Tsui and Hendrix, 1983; Zweig and Cummings, 1973). Experiments described below implicate this protein in length determination; the function of the cleavage is not known. For T4-type tails the tail sheath polymerizes around the tail tube, probably commencing before the tail tube has finished. The sheath has no role in length determination, since the tail tube will polymerize to normal length even in a mutant lacking the sheath subunit (King, 1971). Length of the sheath is apparently determined by the length of the tail tube (King, 1971). For T4-type tails there is no proteolysis associated with completion of assembly, as is found for the flexible-tailed phages.

Determination of Tail Length

The precision with which most phages determine their tail length is such that no variation in tail length can be detected experimentally.

It is usually assumed that tail length is exact to the level of the number of polypeptide subunits in the shaft. It has been pointed out that in the case of T4, precision in tail length is probably necessary to produce viable phages. This is because the tail fibers, as they are being attached by one end to the baseplate, must be held at the center by a protein (gpWac) located near the opposite end of the tail. One presumes that a change in the length of the tail would disrupt this interaction and prevent tail fiber attachment. In the case of phages with λ-like tails, there is no analogous constraint known that would demand absolute precision in tail length. In fact, λ mutants with tails slightly shorter than wildtype are viable (see below). The uniformity of length that is nonetheless observed in these cases may simply reflect an intrinsic precision in the terminating mechanism rather than a functional requirement for precision.

Apparently tail length cannot be completely unregulated, even for λ-like tails. Katsura and Kuhl (1975) have produced λ virions with extra long tails by *in vitro* complementation between appropriate mutant lysates. These phages are infective, but with a very much reduced efficiency compared to wildtype phages. MB78, a *Salmonella* phage with a λ style tail, has the unusual property that about 3-4% of the virions produced in a wildtype infection have tails longer than normal (Joshi et al, 1982) (Figure 4). Like the long-tailed λ virions, the long-tailed MB78 virions have a very low infectivity (Hendrix, unpublished results). The length distribution of MB78 tails is instructive. The 97% of the tails that are normal length have as narrow a length distribution as seen in other phages; the long-tailed phages have a smooth length distribution (Hendrix, unpublished results). This argues that the mechanism determining length terminates polymerization at a precise position, but that in this phage it misses occasionally. Once it has missed, there do not appear to be any preferred lengths (such as integral multiples of the normal length) at which polymerization subsequently stops.

Three types of models have been proposed for how tail length is determined (summarized by Kellenberger (1976)). In the first, the length of the tail is intrinsically coded in the main protein subunit of the shaft. In one formulation of this proposal (the "cumulated strain" model), it is hypothesized that the shaft subunits must be slightly distorted in order to fit onto the growing shaft. Each succeeding layer of subunits must distort a bit more than the layer that precedes it, until finally the energy required for distortion is greater than the energy gained by bonding, and polymerization stops. Wagenknecht and Bloomfield (1975) calculated that this could be a thermodynamically plausible way to obtain the precision in length that is observed. However, two lines of evidence argue against any model of length determination in which length is specified by the major shaft subunit. First, the existence of the long-tailed variants of MB78 and λ cited above shows that the shaft subunit can polymerize to arbitrarily long dimensions; similar results have been seen with other phages as well. Second, genetic and electron microscopic evidence from λ and ø80, which is described below, also effectively rules out the hypothesis that shaft length is intrinsic to the shaft subunit.

The second model for length determination proposes that length is measured by a vernier mechanism. In this model it is imagined that the shaft subunit and some other subunit copolymerize into adjacent structures with slightly different repeat periods. Because of the difference in period, the two structures will be "out of phase" until they reach a certain length at which they are back "in phase." This would be a signal for a termination event. Although there is no strong evidence against this model, it is not generally favored because there is no direct evidence to support it. A weak argument against the vernier model is provided by the data cited above on MB78, the phage that makes 3-4% of its particles with long tails. If length were determined by a vernier mechanism, it might be expected that the long tails would preferentially be integral multiples of normal length. This is not observed.

The third model states that tail length is specified by a template or tape measure that determines tail length by physically spanning the distance. Until recently the evidence supporting the template model, though strongly suggestive, was somewhat circumstantial. A new experiment described below, together with the previous evidence,

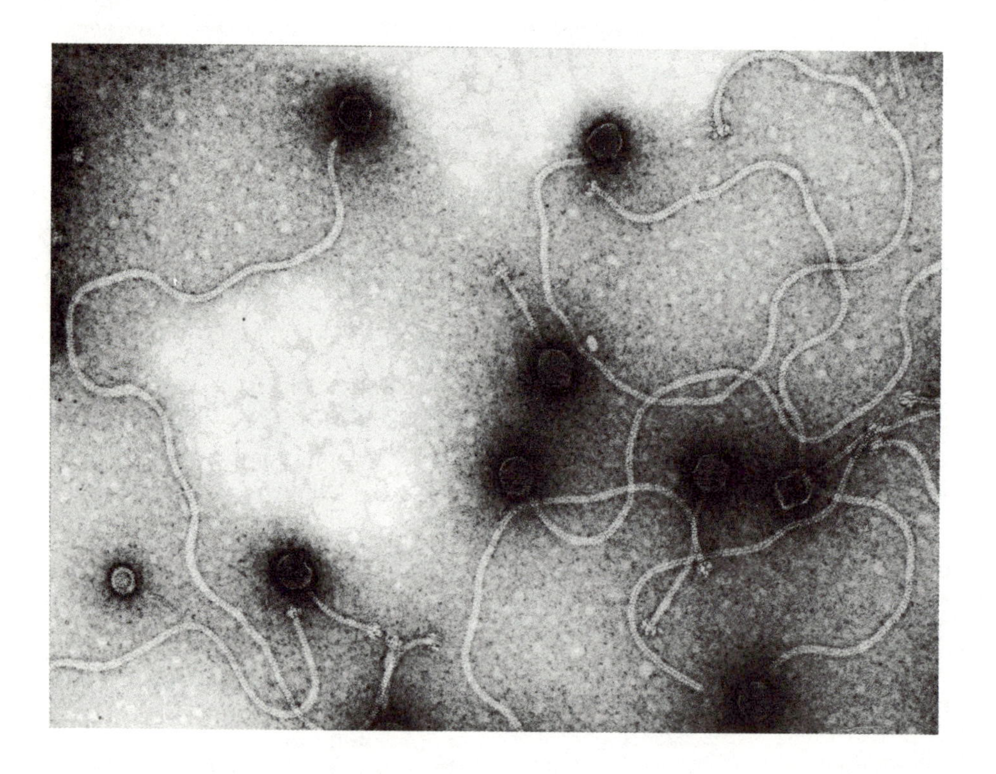

FIGURE 4 Long-tailed variants of MB78. A wildtype lysate, which contains mostly phage with normal length tails, was centrifuged to equilibrium in a CsCl density gradient. This micrograph shows material from the low-density shoulder of the phage band, which is highly enriched for phage with exceptionally long tails. Some phage with normal tails are also visible.

now makes it very likely that it is correct.

The first definitive experiments to provide support for the template model were done by P. Youderian and J. King (Youderian and King, personal communication; Youderian, 1978) using λ and its relative ø80, which has a tail about 17% longer than that of λ. λ and ø80 are similar enough that many of their tail proteins work interchangeably. Youderian and King asked what lengths of tails were produced by various combinations of λ and ø80 proteins. This was done either by constructing recombinational hybrid phages, with some tail genes from λ and some from ø80, or by coinfecting cells with λ and ø80 derivatives in which selected genes in each phage were inactivated by mutation. By these means they were able to map tail length determination to a region that includes only tail genes *T*, *H*, *M*, and *L*. Since these experiments show that the major tail-shaft proteins of λ or ø80 (gpV and p28,000) will both assemble to either λ or ø80 length, depending on which other proteins are present, they strengthen the argument that tail length is not intrinsic to the shaft proteins. All four of the proteins coded in the region that specifies tail length are involved in basal structure assembly (Katsura and Kuhl, 1975b; Kar, 1983). This implies that whichever of these genes specifies tail length, the basal structure must contain the information for how long the shaft will be. Youderian and King proposed that the basal structure contains a length template, and that the template is gpH. They favored this assignment because gpH is the only protein of this group big enough to reach from one end of the tail to the other. In addition, they showed that the ø80 homologue of gpH has a slightly higher molecular weight than gpH and would therefore be an appropriate template for the slightly longer ø80 tail.

Hendrix and Popa (in preparation) extended this type of study by examining the protein compositions of a series of phages with λ-like morphology. In addition to λ and ø80, these included T5 and two newly isolated lambdoid phages, HK022 and HK97 (Dhillon et al, 1980; Dhillon, Dhillon, and Lai, 1981). Each of these phages has a protein that is homologous to the λ gpH by the criteria that it is a minor tail protein, present in ca. 6 copies/tail and that it loses ca. 10 kD to proteolysis during tail assembly. In agreement with the template hypothesis, these "gpH homologues" have molecular weights that are roughly proportional to the lengths of the corresponding tails. Interestingly, the proportionality constant that relates tail length to template size is about 0.15 nm/amino acid, or about the same as the corresponding constant for an α helix. A secondary structure calculation (Chou and Fasman, 1978) for the gpH amino acid sequence predicts, in fact, that gpH is unusually rich in α helix (Hendrix, unpublished results).

Recently, the template model of tail length determination has received strong support from experiments with two lambda mutants, each carrying a small deletion in H (Katsura and Hendrix, 1984). These in-frame deletions remove 246 and 219 bp from the middle of gene H and produce correspondingly smaller gpH molecules. The tails are shorter respectively by 3 and 2 out of the usual 32 discs, which are the differences predicted if gpH is a largely @-helical tape measure. Thus, the available evidence suggests quite strongly that the length of λ tails, and by analogy the lengths of other phage tails, are determined by a length template, which in λ appears to be the protein gpH. On the more interesting question of the detailed mechanism by which the template protein specifies tail length there is little evidence available.

If phages determine tail length with a tape-measure protein, we might expect to find this protein spanning the length of the tail in the mature phage. Several lines of evidence suggest that this may be the case, though none is conclusive. For T4, Duda and Eiserling (1982) used guanidinium hydrochloride to disrupt baseplate + tail-tube structures in such a way that the tail tubes came off the baseplates. These tail tubes had the expected tail-tube protein subunits, but also contained most of the gp48, a baseplate protein (Berget and Warner, 1975). By electron microscopy, a fiber could be seen protruding from one end of many tail tubes, and negative stain penetrated about halfway down the lumen of the tail tube from the other end. Duda and Eiserling suggested that the fiber is gp48, that can be imagined to be the T4 tape measure that runs the length of the tail tube but slides partway out under these conditions. Parker and Eiserling (1983) reported experiments with the *B. subtilis* phage SP01 that suggest that this phage also has stain-excluding material (possibly a tape-measure protein) in the lumen of the tail tube. These authors examined by electron microscopy overlength tails obtained from an SP01 mutant infection and found that negative stain penetrates the lumen of the tail at the end away from the baseplate but stops at the position where the head-proximal end of a normal tail would be.

In the *Caulobacter* phage ϕCbK, Papadopoulos and Smith (1982) have seen a few examples of virions with fractured tails in which the parts of the tail appear to be held together by a thin fiber (Figure 5), as might be expected if a protein runs up the lumen. This phage is an interesting case because it has an unusually long tail of 287 nm (Papadopoulos and Smith, 1982). Given the relationship between tail length and tape-measure size observed with other phages, ϕCbK would be expected to have a very large tape-measure protein in the range 170-190 kD. In fact, virions do have a minor protein of ca. 175 kD; however, this protein has not yet been assigned a location in the structure.

Roughly six copies of the tape-measure proteins are present in the tail of phages like λ (Casjens and Hendrix, 1974). The current view is that the six polypeptides run along the lumen of the tail shaft in parallel. Six α-helical polypeptides would just fit into the lumen, which has a diameter of roughly 3-4 nm. However, this would not leave enough space for the DNA to pass through during injection, which suggests that the tape-measure proteins must depart before the DNA can be injected into the cell. In experiments in which changes in the protease sensitivity of gpH upon DNA injection either into lipid vesicles or into the medium were monitored Roessner and Ihler (1984) obtained evidence that is consistent with the idea that the λ tape measure, gpH, is injected ahead of the DNA. In accord with this view, there is also genetic evidence that argues that gpH is involved in mediating passage of the phage DNA across the inner membrane of the cell during injection (Scandella and Arber, 1976; Elliot and Arber,

1978). If these suggestions about the functions of gpH are correct, then it would provide an interesting example of a bifunctional protein for which the act of carrying out the first function (length determination) positions it correctly in the structure for carrying out the second (leading the DNA into the cell).

Another well-studied virus, tobacco mosaic virus (TMV), shows a clear example of length determination by a template mechanism. TMV has a simple virion with a single type of protein subunit helically arrayed about its ssRNA ohromosome. The ssRNA chromosome acts as the template for determination of the virus length. Other helical ssRNA and ssDNA viruses use their nucleic acids similarly to determine the length of the virion or nucleocapsids (see for example Chapters 3, 6, and 7, this volume). As the details of TMV assembly are currently understood, the coat protein is in a delicately balanced equilibrium between a state in which it cannot assemble into the helical virion and one in which it can. The switch between these two states is controlled by pH and the presence or absence of RNA bound to previously added coat proteins (the initiation process is not relevant to this discussion; see Chapter 3, this volume, for more details). Thus, at the proper pH, unassembled coat protein cannot assemble

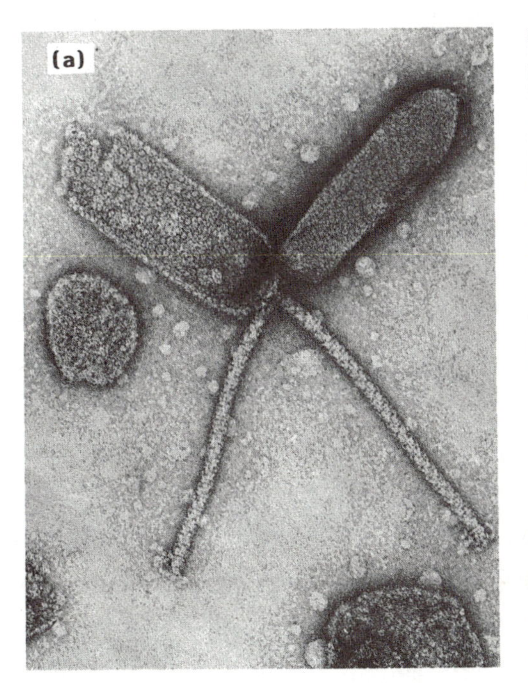

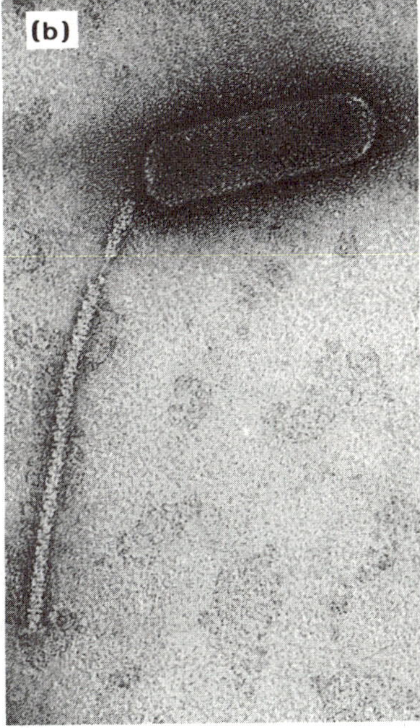

FIGURE 5 (a) øCbK virions. One particle has lost its DNA. (b) øCbK virion with fractured tail. The arrow indicates the fiber that may be the length tape measure. (Courtesy of Dr. P. R. Smith, New York University Medical Center.)

onto RNA without associated protein or onto protein without associated RNA. The three-dimensional structures of these states are at least partially understood (Stubbs, Warren, and Holmes, 1977; Bloomer, 1978). The switch involves the electrostatic interactions among six carboxyls, five arginines, and three RNA phosphate groups. In the "nonhelical" state of the protein, the carboxyls and arginines are arranged to give mutual repulsion and to keep two α helices separated. The "helical assembly" state exists either at low pH (unionized carboxyls) or in the presence of the negatively charged phosphates of RNA (neutralized arginines?); both conditions allow the two α helices to move closer together, which appears to be required for helical assembly (Stubbs, Warren, and Homes, 1977). Therefore, under normal assembly conditions coat protein can only assemble onto itself if the RNA is present, and in addition, it does not assemble onto the RNA at random but rather in a cooperative fashion. Thus, there is an orderly assembly of coat proteins onto the elongating structure until the RNA is covered, at which point assembly stops.

A characteristic feature of the TMV assembly just described is that each e template, specifically, with three phosphates in the RNA backbone. This of course depends on the fact that the RNA has a periodic repeat in the relevant feature of its structure, the phosphates. Whether a similar mechanism operates in the case of phage tails is not yet clear; however, a preliminary examination of the amino acid sequence of gpH of λ does not reveal any periodicity that might correspond to the repeat period of the tail shaft. Thus, it seems possible that the determination of the length of the phage tail may involve a template mechanism which is quite different from the template mechanism used by TMV.

OVERVIEW AND PROSPECTS

Our understanding of phage assembly is now to the point that we can specify the general outlines of the processes that lead to determination of virion shape. The head and tail are both made of major subunits that contain in their own structures much of the information about how they are to assemble; evidence is the formation of aberrant, but nonetheless headlike or taillike structures that they can make by themselves. However, in order to polymerize both correctly and efficiently, the major subunits need the assistance of other proteins. Trying to understand the ways in which this collaboration among proteins brings about correct assembly has become the central theme of studies of how virion shape is determined.

For head assembly, the most striking example of this phenomenon is the participation of a scaffold, which directs polymerization of the shell. It is clear in all cases that both the scaffolding protein and the shell protein have roles in determining the size and shape of the shell, though different sorts of experiments and different phages give a range of impressions about the relative importance of the shell and scaffold in the process. The picture of shell assembly that has emerged is that the shell and scaffold copolymerize, starting from an initiation complex that will become the proximal vertex of the head. The initiation complex appears to have at least two roles: first, to position the initial shell and scaffold subunits correctly, and second, to exercise kinetic control over assembly by providing an efficient entry point into the correct assembly pathway.

A completely unexpected feature of assembly of the tailed phages is the existence of symmetry mismatches between parts of the structure during assembly. Thus, for example, the shell and scaffold have different symmetries around a common axis, which guarantees that the geometry of shell-scaffold contacts changes from one pair of subunits to another. The well-known fact that proteins can make very specific interactions with other molecules has led most biochemists to expect that the protein-protein contacts in virus structures would always be specific in the same "lock and key" sense. The evidence on phage assembly suggests rather strongly that this need not always be the case. The detailed nature of the contacts between proteins that touch each other across the symmetry mismatch boundaries are not yet known, nor is the biological significance of this phenomenon, but it seems likely that it is an important part of the assembly strategy.

Determination of shell shape appears to be fundamentally a problem of three dimensions, but determination of tail length can be considered to be a one-dimensional process. Thus, it now seems firmly established that there is a tape-measure protein that spans the length of the tail and indicates when shaft polymerization should stop. The mechanistic details of how it does so are still unknown but should be reasonably susceptible to experimental elucidation.

It is probably fair to say that we now know more about assembly and shape determination for the tailed phages than for any other biological structure of comparable complexity. Nevertheless, it is also true that the results we can expect in the next few years will be even more informative about principles of biological assembly than the results to date. Now that the general outlines of the assembly pathways are mostly worked out, interest is shifting to the mechanistic details of how assembly steps are regulated. In particular, more detailed structural information about the individual proteins and the complexes they make is needed for any real understanding of how they influence assembly behavior of one another. The high resolution studies on TMV and the small isometric plant viruses point toward the quality of information we can expect. Given the greater complexity of the structures of the tailed phages and of their assembly pathways, the amount of useful detail we will see, and the level of insight into how proteins interact to accomplish biological form determination, should be correspondingly large. On another front, studies of *in vitro* assembly using purified components are well underway, and it is not likely to be long before an entire assembly pathway can be carried out using only purified, assembly-naive components. It is also certain that more detailed investigations will reveal control mechanisms that have not yet surfaced. As an obvious example, it would not be surprising if part of the control of shell or scaffold polymerization involves conformational switching of the protein subunits.

It is already abundantly clear that the components of phages are not simply little bricks that quietly slide into place as the structure is assembled. Rather, they are fascinating little machines that bend and stretch, communicate and interact with each other, and in the process, build a structure that is an elegant example of the efficient integration of structure and function, all this to a precise set of rules that we have just begun to learn.

ACKNOWLEDGEMENTS

I thank my colleagues in this field for the sorts of discussions and interactions that make phage assembly not only intellectually stimulating but also fun. I especially thank Sherwood Casjens who has made major contributions, both intellectual and editorial, to this chapter. Work in the author's laboratory was supported by grants from the National Institutes of Health (AI12227) and the National Science Foundation (PCM-7910461).

REFERENCES

Aebi. U., Bijlenga, R., v.d. Broek, J., v.d. Broek, R., Eiserling, F., Kellenberger, C., Kellenberger, E., Mesyanzhinov, V., Muller, L., Showe,M., Smith, R., and Steven, A. 1974. *J. Supramol. Struct.*, 2, 253.

Anderson, D.L., Hickman, D.D., and Reilly, B.E. 1966. *J. Bacteriol.*, 91, 2081.

Barrett, K.J., Marsh, M.L., and Calendar, R. 1976. *J. Mol. Biol.*, 106, 683.

Berget, P.B. and Warner, H.R. 1975. *J. Virol.*, 16, 1669.

Black, L.W. and Showe, M.K. 1983. 219. In *Bacteriophage T4*, C. Matthews et al., (eds). American Society for Microbiology.

Bloomer, A.C., Champness, J.N., Bricogne, G., Staden, R., and Klug, A. 1978. *Nature*, 276, 362.

Branton, D. and Klug, A. 1975. *J. Mol. Biol.*, 92, 559.

Carrascosa, J.L., Mendez, E., Corral, J., Rubio, V., Ramirez, G., Salas, M., and Vinuela, E. 1981. *Virology*, 111, 401.

Carrascosa, J.L., Vinuela, E., Garcia, N., and Santisteban, A. 1982. *J. Mol. Biol.*, 154, 311.

Carrascosa, J.L., Garcia, J.A., and Salas, M. 1982. *J. Mol. Biol.*, 158, 731.

Casjens, S.R. 1979. *J. Mol. Biol.*, 131, 1.

Casjens, S.R. and Hendrix, R.W. 1974. *J. Mol. Biol.*, 88, 535.

Casjens, S.R. and King, J. 1974. *J. Supramolec. Struct.*, 2, 202.

Caspar, D.L.D. and Klug, A. 1962. *Cold Spring Harbor Symp. Quant. Biol.*, 27, 1.

Chou, P.Y. and Fasman, G.D. 1978. *Ann. Rev. Biochem.*, 47, 251.

Coombs, D.H. and Eiserling, F.A. 1977. *J. Mol. Biol.*, 116, 375.

Cornick, G., Sigler, P.B., and Ginsberg, H.S. 1971. *J. Mol. Biol.*, 57, 397.

Crowther, R.A. and Franklin, R.M. 1972. *J. Mol. Biol.*, 68, 181.

Dhillon, K.S., Dhillon, T.S., Lai, A.N.C. and Linn, S. 1980. *J. Gen.Virol.*, 50, 217.

Dhillon, T.S., Dhillon, E.K.S., and Lai, A.N.C. 1981. *Virology*, 109, 198.

Doherty, D.H. 1982. *J. Virol.*, 43, 655.

Driedonks, R.A., Engel, A., tenHeggler, B., and van Driel, R. 1981. *J. Mol. Biol.*, 152, 641.

Duda, R.L. and Eiserling, F.A. 1982. *J. Virol.*, 43, 714.

Durham, A.C., Hendry, D.A., and VonWechmar, M.B. 1977. *Virology*, 77, 524.

Earnshaw, W.C. and King, J. 1978. *J. Mol. Biol.*, 126, 721.

Earnshaw, W.C., Casjens, S.R., and Harrison, S.C. 1976. *J. Mol. Biol.*, 104, 387.

Earnshaw, W.C., Hendrix, R.W. and King, J. 1979. *J. Mol. Biol.*, 134, 575.

Elliot, J. and Arber, W. 1978. Molec. Gen. Genet., 161, 1.

Engel, A., van Driel, R., and Driedonks, R. 1982. *J. Ultrastr. Res.*, 80, 12.

Fuller, M.T. and King, J. 1982. *J. Mol. Biol.*, 156, 633.

Geisselsoder, J., Sedivy, J.M., Walsh, R.B., and Goldstein, R. 1982. *J. Ultrastr. Res.*, 79, 165.

Georgopoulos, C.P., Hendrix, R.W., Kaiser, A.D., and Wood, W.B. 1972. *Nature New Biol.*, 239, 38.

Georgopoulos, C.P., Hendrix, R.W., Casjens, S.R. and Kaiser, A.D. 1973. *J. Mol. Biol.*, 76, 45.

Gope, R. and Serwer, P. 1983. *J. Virol.*, 47, 96.

Hagen, E.W., Reilly, B.E., Tosi, M.E., and Anderson, D.L. 1976. *J. Virol.*,19, 501.

Hendrix, R.W. 1978. *Proc. Natl. Acad. Sci. USA*, 75, 4779.

Hendrix, R.W. 1979. *J. Mol. Biol.*, 129, 375.

Hendrix, R.W., and Casjens, S.R. 1974. *Proc. Natl. Acad. Sci. USA*,71, 1451.

Hendrix, R.W. and Casjens, S.R. 1974. *Virology*, 61, 156.

Hendrix, R.W. and Casjens, S.R. 1975. *J. Mol. Biol.*, 91, 187.

Hendrix, R.W. and Katsura, I. 1984. *Cell*, 39, 691.

Hohn, B. 1983. *Proc. Natl. Acad. Sci. USA*, 80, 7456.

Hohn, B., Wurtz, M., Klein, B., Lustig, A. and Hohn, T. 1974. *J. Supramol. Struct.*, 2, 302.

Hohn, T., Hohn, B., Engel, A., Wurtz, M. and Smith, P.R. 1979. *J. Mol. Biol.*, 129, 359.

Hohn, T., Wurtz, M., and Hohn, B. 1976. *Phil. Trans. R. Soc. Lond. B*, 276, 51.

Hsiao, C.L. and Black, L.W. 1977. *Proc. Natl. Acad. Sci. USA*, 74, 3652.

Hsiao, C.L. and Black, L.W. 1978. *Virology*, 91, 15.

Imber, R., Tsugita, A., Wurtz, M., and Hohn, T. 1980. *J. Mol. Biol.*, 139, 277.

Ishii, T. and Yanagida, M. 1975. *J. Mol. Biol.*, 97, 655.

Ishii, T. and Yanagida, M. 1977. *J. Mol. Biol.*, 109, 487.

Joshi, A., Siddiqi, J.Z., Rao, G.R.K., and Chakravorty, M. 1982. *J. Virol.*, 41, 1038.

Kar, S. 1983. Ph.D. Thesis, University of Pittsburgh.

Katsura, I. 1983. *J. Mol. Biol.*, 171, 297.

Katsura, I. and Kuhl, P.W. 1974. *J. Supramol. Struct.*, 2, 239.

Katsura, I. and Kuhl, P.W. 1975. *Virology*, 63, 238.

Katsura, I. and Kuhl, P.W. 1975. *J. Mol. Biol.*, 91, 257.

Kellenberger, E. 1976. *Phil. Trans. R. Soc. Lond. B*, 276, 27.

King, J. 1971. *J. Mol. Biol.*, 58, 693.

King, J. and Casjens, S.R. 1974. *Nature*, 251, 112.

Kistler, J., Aebi, U., Onorato, L., tenHeggeler, B., and Showe, M.K. 1978. *J. Mol. Biol.*, 126, 571.

Kochan, J., Carrascosa, J.L., and Murialdo, H. 1984. *J. Mol. Biol.*, in press.

Kunzler, P. and Berger, H. 1981. *J. Mol. Biol.*, 153, 961.

Laemmli, U.K., Molbert, E., Showe, M.K., and Kellenberger, E. 1970. *J. Mol. Biol.*, 49, 99.

Laemmli, U.K. and Eiserling, F.A. 1968. *Mol. Gen. Genet.*, 101 333.

Laemmli, U.K., Beguin, F. and Gujer. 1970. *J. Mol. Biol.*, 47, 69.

Leonard, K.R., Kleinschmidt, A.K., and Lake, J.A. 1973. *J. Mol. Biol.*,81, 349.

McNicol, L.A., Simon, L.D., and Black, L.W. 1977. *J. Mol. Biol.*, 116, 261.

Moody, M.F. 1965. *Virology*, 26, 567.

Moody, M.F. and Makowski, L. 1981. *J. Mol. Biol.*, 150, 217.

Murialdo, H. 1979. *Virology*, 96, 341.

Murialdo, H. and Becker, A. 1977. *Proc. Nat. Acad. Sci.*, 74, 906.

Murialdo, H. and Becker, A. 1978. *Microbiol. Rev.*, 42, 529.

Murialdo, H. and Becker, A. 1978. *J. Mol. Biol.*, 125, 57.

Murialdo, H. and Ray, P.N. 1975. *Nature*, 257, 815.

Onorato, L., Stirmer, B., and Showe, M.K. 1978. *J. Virol.*, 27, 409.

Papadopoulos, S. and Smith, P.R. 1982. *J. Ultrastr. Res.*, 80, 62.

Parker, M.L. and Eiserling, F.A. 1983. *J. Virol.*, 46, 239.

Paulson, J.R. and Laemmli, U.K. 1977. *J. Mol. Biol.*, 111, 459.

Ray, P. and Murialdo, H. 1975. *Virology*, 64, 247.

Reilly, B.E., Nelson, R.A., and Anderson, D.L. 1977. *J. Virol.*, 24, 363.

Revel, H.R., Stitt, B.L., Lielausis, I., and Wood, W.B. 1980. *J. Virol.*, 33, 366.

Robinson, I.K. and Harrison, S.C. 1982. *Nature*, 297, 563.

Roessner, C.A. and Ihler, G.M. 1984. *J. Bacteriol.*, 157, 165.

Sauer, B., Ow, D., Ling, L., and Calendar, R. 1981. *J. Mol. Biol.*, 145, 29.

Scandella, D. and Arber, W. 1976. *Virology*, 69, 206.

Schwartz, M. 1976. *J. Mol. Biol.*, 103, 521.

Serwer, P. 1980. *J. Mol. Biol.*, 138, 65.

Shaw, J.E. and Murialdo, H. 1980. *Nature*, 283, 30.

Shore, D., Deho, G., Tsipis, J., and Goldstein, R. 1978. *Proc. Natl. Acad. Sci. USA*, 75, 400.

Showe, M.K. and Onorato, L. 1978. *Proc. Natl. Acad. Sci. USA*, 75, 4165.

Sternberg, N. and Weisberg, R. 1977. *J. Mol. Biol.*, 117, 733.

Steven, A.C., Couture, E., Aebi, U., and Showe, M.K. 1976. *J. Mol. Biol.*, 106, 187.

Stubbs, G., Warren, S., and Holmes, K. 1977. *Nature*, 267, 216.

Traub, F. and Maeder, M. 1984. *J. Virol.*, 49, 892.

Tsui, L.. and Hendrix, R.W. 1980. *J. Mol. Biol.*, 142, 419.

Tsui, L.. and Hendrix, R.W. 1983. *Virology*, 125, 257.

van Driel, R. 1980. *J. Mol. Biol.*, 138, 27.

van Driel, R. and Couture, E. 1978. *J. Mol. Biol.*, 123, 713.

Vinuela, E., Camacho, A., Jiminez, F., Carrascosa, J.L., Ramirez, G., andSalas, M. 1976. *Phil. Trans. Roy. Soc. Lond. B*, 276, 29.

Volker, T.A., Kuhn, A., Showe, M.K., and Bickle, T.A. 1982. *J. Mol. Biol.*, 161, 491.

Wagenknecht, T. and Bloomfield, V.A. 1975. *Biopolymers*, 14, 2297.

Walker, J.E., Auffret, A.D., Carne, A., Gurnett, A., Hansch, P., Hill, D., and Saraste, M. 1982. *Eur. J. Biochem.*, 123, 253.

Wood, W.B. and King, J. 1979. 581. In *Comprehensive Virology*, vol. 13. Fraenkel. and Wagner, R. (eds.). Plenum.

Wurtz, M., Kistler, J. and Hohn, T. 1976. *J. Mol. Biol.*, 101, 39.

Yanagida, M., Boy de la Tour, E., Alff-Steinberger, C., and Kellenberger, E. 1970. *J. Mol. Biol.*, 50, 35.

Youderian, P. 1978. Ph.D. Thesis, Massachusetts Institute of Technology.

Zweig, M. and Cummings, D. 1973. *J. Mol. Biol.*, 80, 505.

6
Assembly of Enveloped Animal Viruses

Ellen G. Strauss
and James H. Strauss
California Institute of Technology

INTRODUCTION

Enveloped viruses have the genomic nucleic acid, DNA or RNA, complexed with one or more species of basic protein to form a nucleocapsid, of either helical or icosahedral symmetry, which is in turn surrounded by a lipid bilayer containing virus-encoded polypeptides. For most viruses the nucleocapsid assembles as a fairly well-defined structure in the host cell cytoplasm (or nucleoplasm in the case of herpesviruses), and the bilayer is acquired by budding through virus-modified regions of a pre-existing host membrane. Most RNA viruses bud from the cell plasmalemma, thereby accomplishing in one event both the acquisition of the envelope and release into the extracellular space. Some RNA viruses mature in association with intracellular membranes such as the rough endoplasmic reticulum (coronaviruses), or the smooth endoplasmic reticulum and the Golgi complex (bunyaviruses and flaviviruses), and accumulate in membrane-bound vacuoles. They are released from these vacuoles either by discharge of the vacuoles at the cell surface or are not released until the eventual disintegration of the host cell. The poxviruses, which contain DNA, also mature in association with internal membranes, but in this case the assembly process is complex and poorly understood and seems to involve formation of new membranes. Finally, the herpesviruses, which also contain DNA, acquire their envelope from the nuclear membrane.

Thus, the process of the assembly of enveloped viruses can be divided into three processes, all of which are coordinated to produce the final virion particle: (1) assembly of the nucleocapsid, (2) synthesis, modification, and transport of the viral membrane proteins to the site of budding, and (3) acquisition of the envelope by an interaction between the nucleocapsid and the virus-modified membrane, resulting in closure of the membrane around the core with concomitant exclusion of host integral membrane proteins. Construction of the nucleocapsid may be considered to be a special case of the self-assembly of nucleic acids with proteins, exemplified by the simple plant viruses and the nonenveloped virus groups. The synthesis and transport of viral membrane proteins, especially the glycoproteins (both peripheral and integral), appears to follow the same general pathways as those employed in the synthesis of secretory proteins and the host membrane polypeptides. Indeed, the virus systems have served as models for membrane biogenesis because of their relatively simple protein composition. In addition, in those systems in which virus proteins are addressed to specific regions of a host plasmalemma and are vectorially transported to a particular membrane, the virus-host system has been used to investigate the mechanisms by which cellular proteins are sorted and directed to particular membranous compartments or organelles.

The third process, recognition by the nucleoprotein of regions of virus-modified membrane and specific interaction between the membrane and the capsid, has no direct correlate in the uninfected cell. This process can be further subdivided into those in which there are direct interactions between capsids and transmembranous glycoproteins (which occurs with alphaviruses, bunyaviruses, and probably flaviviruses) and those in which the glycoproteins are more loosely organized and interact only indirectly with the nucleocapsid through a hydrophobic matrix protein. In these cases the matrix protein, which is virus-encoded, coats the cytoplasmic face of the bilayer and interacts with both the glycoproteins and with the nucleoprotein. This latter organization appears to be shared by most of the negative-strand viruses. For two other groups the interaction between capsid and envelope is less well defined. Coronavirus nucleocapsids apparently interact with the E1 glycoprotein, which has characteristics intermediate between those of matrix proteins and those of integral membrane glycoproteins. On the other hand, retrovirus maturation is a complex phenomenon that involves a number of protein-nucleic acid, protein-lipid, and protein-protein interactions, and in this case the nucleocapsid is not preassembled.

Table 1 gives an outline of the various groups of enveloped viruses to be considered. Included in the table are the geometry of the virus nucleocapsids and a listing of the structural polypeptides, both envelope and capsid. With the exception of the herpesviruses, which will be considered separately, all of those shown are RNA-containing viruses. The Poxviridae have been excluded from consideration because their morphogenesis is very complex and poorly understood and appears to be different in character from that of other enveloped viruses.

ASSEMBLY OF NUCLEOCAPSIDS OF ENVELOPED RNA VIRUSES

There are two architectural prototypes of nucleocapsids among the enveloped viruses considered in Table 1, and both fulfill the geometric criteria for assembly of a regular solid from identical subunits. These two models are the helix and the icosahedron (Caspar and Klug, 1962; Chapter 1, this volume). Most viral nucleocapsids are composed primarily of one species of nucleocapsid protein complexed with the genome nucleic acid (RNA for all cases except the herpesviruses, which are considered separately). Negative-strand viruses also contain small numbers of polymerase molecules in their nucleocapsids, owing to the necessity to introduce into the host cell at infection all the components required for primary transcription. Negative-strand viruses all contain helical nucleocapsids that also serve as the templates for replication and transcription.

In most cases it has been assumed that formation of these regular structures, whether helical or icosahedral, proceeds via a self-assembly pathway that includes sequential addition of monomeric capsid protein units (beginning with the interaction of the first core protein) to a single encapsidation initiation site on the viral genome. For these viruses, a specific protein-RNA interaction appears to be essential for encapsidation, since heterologous RNAs are rarely packaged and ordered protein-protein aggregates or empty protein shells are not observed. In general, once initiated, encapsidation is rapid, and few incomplete capsids or intermediates in capsid assembly have been

Table 1 Structural features of enveloped animal viruses

Virus	Symmetry of nucleo-capsid*	Structural proteins			
		Major capsid	Minor capsid	Matrix	Integral membrane
(-)-strand RNA viruses					
Arenavirus	H	N	?	?	1 or 2**
Bunyavirus	H	N	L	None	G1,G2
Orthomyxovirus	H	NP	PA,PB1, PB2	M	HA $HA_1 + HA_2$ NA
Paramyxovirus	H	NP	C,P,L	M	F $F_1 + F_2$ HN
Rhabdovirus	H	N	NS,L	M	G
(+)-strand RNA viruses					
Alphavirus	I	C	None	None	E1,E2,(E3)
Flavivirus	I	C	None	None	M,E
Coronavirus	H	N		?	E2,E1▽
Retrovirus▼	C	p12 p33	p15,p10 (p22)	--	gp85,gp37
DNA viruses					
Herpesvirus	I	40K	Several	?	gB,gC,gD,gE

* H = helical; I = icosahedral; C = complex icosahedral.
** Depending on species.
▽ Coronavirus "E1" protein is an unusual integral membrane glycoprotein with some characteristics suggestive of a matrix protein.
▼ Avian retrovirus nomenclature.

detected in virus-infected cells. The assembly of nucleocapsids of particular virus groups will be discussed in more detail below.

Alphaviruses are the most intensively studied of the groups of enveloped viruses possessing icosahedral nucleocapsids. The RNA genome (ca. 11,700 nucleotides) is complexed with multiple copies of a single species of capsid protein (C) of molecular weight ca. 30,000. The number of C proteins per genome have been variously estimated from composition studies as 180 (which would result in an icosahedron of $T = 3$ symmetry) and 240 (for $T = 4$), and no final answer has been reached (reviewed in Simons and Warren, 1983). The amino acid sequences of the capsid proteins of Semliki Forest virus (Garoff et al., 1980), Sindbis virus (Rice and Strauss, 1981), and Ross River virus (Dalgarno et al., 1983) have been deduced from the nucleotide sequences of their subgenomic mRNA as well as by conventional methods of protein sequencing (Boege et al., 1981). In each case the N-terminal half of the protein contains clusters of basic amino acids and large numbers of proline and glutamine residues, but the amino acid sequence is not conserved. This region of the molecule is very basic in character and is thought to interact nonspecifically with the genomic RNA, perhaps penetrating into the interior of the capsid, where it could serve to neutralize minus charges of the RNA. Indeed, *in vitro* reconstitution experiments (Wengler et al., 1982) have shown that Sindbis capsid protein will assemble into morphologically similar structures with either viral RNA or a selection of heterologous nucleic acids. (However, *in vivo* the interaction is quite specific, and the subgenomic RNA, minus strands, or host RNAs are not packaged; this specificity is believed to arise from a specific initiation event. See Chapter 3, this volume.) The C-terminal half of the capsid protein is highly conserved among the alphaviruses and undergoes specific protein-protein interactions both to build the icosahedron and to recognize the membrane glycoproteins during budding (see below). Capsid protein is not found in soluble form in infected cells and is incorporated while nascent or soon thereafter into ribonucleoprotein complexes (Söderlund, 1973). Alphavirus capsids are fairly stable and can be isolated by detergent treatment from infected cells or from virions. However, RNA in intact capsids is susceptible to RNase digestion, indicating that either capsids are permeable or that some RNA is exposed upon their surface (Söderlund et al., 1979).

Much less information is available about the assembly and physicochemical characteristics of the icosahedral nucleocapsids of the flaviviruses. The genomic RNA remains complexed with capsid protein C (formerly called V2, Westwood et al., 1980) after detergent treatment of intact virions and electron microscopy of virion-derived cores shows that they are icosahedrons. However, the capsid structures appear to be much less stable than those of alphaviruses and it has not been possible to isolate nucleocapsids from the cytoplasm of infected cells. Such capsids have not been unambiguously identified in electron micrographs of infected cells, and it has been postulated that capsid assembly is concomitant with and perhaps dependent on the budding event producing mature enveloped particles (Westaway, 1980).

Assembly of retrovirus nucleocapsids is a complex phenomenon and also appears to occur concomitantly with budding; i.e., RNA, core polypeptides, and envelope proteins coalesce at a membrane site, forming the C-type structures visible in electron micrographs (Bolognesi et al., 1978). For avian viruses the major precursor to the core polypeptides ($pr76^{gag}$) is cleaved to five polypeptides (p19, p10, p27, p12, p15), which together account for 80% of the virion protein (Schwartz et al., 1983). The genomic RNA and p12 are complexed into a ribonucleoprotein strand that is in turn protected by an icosahedral shell composed primarily of p27. The location of p15 is unknown, but it may be a constituent of the capsomers visible on the surface of the core, as well as containing the viral protease activity to process $pr76^{gag}$. P10 is within the space between the membrane and the core (Pepinsky and Vogt, 1983), and its function is obscure. P19 is probably bifunctional, since it is located immediately inside the bilayer and is found associated with lipids (Pepinsky and Vogt, 1979, 1984), as well as binding specifically to a number of sites on the genomic RNA (Darlix and Spahr, 1982). In this dual role the p19 region of $pr76^{gag}$ may form the contact between the genome RNA and the membrane (which initiates virion assembly). Recently the sites on p19 involved in protein-protein interactions have been mapped to the carboxy terminus of p19, and protein-lipid interaction sites were found near the amino terminus (Pepinsky and Vogt, 1984).

Coronaviruses are the only plus-stranded enveloped viruses with a helical nucleocapsid, and little is known of its mode of synthesis. The genomic RNA is complexed with a large number of molecules of a single species of capsid protein, N, of molecular weight 50,000, but the quantitation of the relative amounts of protein and RNA is insufficient for further characterization the nucleocapsid structure (Siddell et al., 1982).

The remaining five groups of RNA viruses (rhabdoviruses, orthomyxoviruses, paramyxoviruses, bunyaviruses, and arenaviruses) all contain negative-strand RNA, and the first four groups have helical nucleocapsids with related properties. The similarities between the nucleocapsids of rhabdovirus, orthomyxovirus, and paramyxovirus are probably due to the fact that these groups were derived evolutionarily from a common archetypical negative-strand virus (discussed in Strauss and Strauss, 1983). Similarities in both morphology and replication strategy suggest that the Bunyaviridae also diverged from the same ancestor (Patterson et al., 1984); too little information is available on arenaviruses to decide whether they too are related or whether they acquired a similar structure by convergent evolution. Negative-strand viruses have a nucleocapsid that fulfills the dual role of protecting the RNA on the one hand and serving as a replication/transcription template.

The amino acid sequence of the nucleocapsid proteins of the rhabdovirus vesicular stomatitis virus, VSV (Gallione et al., 1981), the orthomyxovirus influenza A, (Winter and Fields, 1981), the coronavirus mouse hepatitis virus, (Skinner and Siddell, 1983), and two members of the bunyavirus family (Akashi and Bishop, 1983), as well as the N protein of the arenavirus Pichinde (Auperin et al., 1984), and the NP gene of the paramyxovirus Sendai (Shioda et al., 1983; Morgan et al., 1984), have been deduced from the RNA sequence of the region encoding them, and although they possess a net positive charge, they each lack the clusters of basic amino acids seen in alphaviruses. This difference may indicate that the capsid proteins interact with the RNA at a number of locations along their length. Assembly into the helical structures requires the presence of RNA; for example, it has been shown that when RNA is present, VSV N protein that has been synthesized *in vitro* forms nucleocapsids that are indistinguishable from virion nucleocapsids (Patton et al., 1983). Helical nucleocapsids of rhabdoviruses and paramyxoviruses are single helices that respond reversibly to changes in ionic environment (Heggeness et al., 1980). In contrast, influenza nucleocapsids appear to be double helices with the RNA looped back upon itself, which unwind in low salt (Heggeness et al., 1982). Arenavirus nuclocapsids are somewhat different in structure from all three of these groups. The ribonucleoprotein is a linear array of subunits (similar to nucleosomes) that folds into higher-order helical structures in the form of closed circles (Young and Howard, 1983). Approximate calculations from the sizes of the genomes and the number of nucleoprotein molecules per virion indicate ratios of 7 to 20 nucleotides per polypeptide for the ribonucleoproteins of the negative-strand viruses. This calculation illustrates that these flexible helices differ markedly in character from helical assemblies such as TMV rods, for which it has been determined that each protein subunit interacts with exactly three nucleotides and that the isolated protein assumes a variety of multimeric forms in the absence of RNA (Klug, 1979).

SYNTHESIS, MODIFICATION, AND TRANSPORT OF VIRAL MEMBRANE GLYCOPROTEINS

Overview of Glycoprotein Synthesis and Transport

The pathways by which viral envelope glycoproteins are synthesized, modified, and transported appear to be similar to those for the synthesis of a large number of host-cell membrane proteins and secreted polypeptides. These proteins are synthesized on polyribosomes bound to the endoplasmic reticulum and cotranslationally inserted into the lumen by a hydrophobic stretch of amino acids called the signal sequence or the signal peptide. This signal sequence may be N-terminal or internal, and may or may not be subsequently removed from the growing polypeptide by signalase (reviewed in Sabatini et al., 1982; Strauss and Strauss, 1983).

Cotranslational transfer of the growing polypeptide chain through the endoplasmic reticulum membrane continues until a "stop transfer" signal is reached, or for secreted proteins, until the completed polypeptide is released into the lumen of the endoplasmic reticulum. The stop-transfer signal for integral membrane proteins resides in a second highly hydrophobic region, usually near the carboxyl terminus, that forms the membrane-associated anchor of the protein. For virus integral proteins such as the G protein of VSV, the envelope glycoproteins E1 and E2 of alphaviruses, the hemagglutinin (HA) of influenza virus, and the fusion protein F of paramyxoviruses, a variable number of amino acid residues that are usually highly hydrophilic in character remain on the cytoplasmic side of the membrane. This number of amino acid residues varies from just two arginines in the case of alphavirus E1 (Rice et al., 1982), to 11 amino acids in HA2 of influenza (Min Jou et al., 1980), to 44 residues in the G protein of rabies virus (Anilionis et al., 1981). There are two well-documented cases of N-terminal anchors. For influenza neuraminidase, the N-terminal signal sequence appears to remain membrane-associated throughout synthesis and serves as the membrane anchor, leaving the carboxyl end of the molecule free in the lumen (Blok et al., 1982). A similar structure has been found for the hemagglutinin-neuraminidase (HN) protein of the paramyxovirus SV5 (Hiebert et al., 1985). Thus, the orthomyxoviruses and the paramyxoviruses each appear to have one glycoprotein with an N-terminal anchor, HN and NA, respectively, and a second glycoprotein with a C-terminal anchor whose function is activated by proteolytic cleavage of the external N-terminal portion, HA and F, respectively.

In the endoplasmic reticulum the first modification of the proteins occurs: the addition of the simple carbohydrate chains composed of glucosamine and mannose residues. These chains are transferred en bloc from a dolichol intermediate and are attached by N-glycosidic linkages to asparagines present in the sequences Asn-X-Thr/Ser, in which X can be any amino acid (reviewed in Hubbard and Ivatt, 1981). Most, but not all, of these sequences identified in viral membrane protein sequences are glycosylated.

From the rough endoplasmic reticulum most viral glycoproteins migrate via smooth membranes to the Golgi apparatus where a complex series of modifications occurs. These include trimming of some of the mannose residues from the carbohydrate chains; addition of galactose, fucose, and sialic acid to produce the characteristic complex oligosaccharides found in virions; addition of fatty acid residues to some proteins; and a number of proteolytic cleavage events (J. Green et al., 1981; Griffiths et al., 1982). The Golgi apparatus, a multilamellar "stack" of membranes located in the perinuclear region, appears to be composed of at least two functional compartments: the *cis* Golgi and the *trans* Golgi. Proteins appear to enter the stack at the *cis* Golgi and to be transported vectorially across the stack while being modified, so only fully mature forms reach the *trans* Golgi (Rothman, 1981). Addition of lipid occurs quite early during this transit, while complex oligosaccharide formation, which is often assayed as resistance to the enzyme endoglycosidase H, occurs late (reviewed in Tarkatoff, 1983). The mature glycoproteins are then transported to the sites of viral assembly, which may be variously smooth membranes or the plasmalemma of the cell (reviewed in Simons and Garoff, 1980). In at least one instance, the maturation of VSV G protein, the transport from endoplasmic reticulum to the Golgi and from the Golgi to the plasmalemma was reported to be mediated by clathrin-coated vesicles (Rothman and Fine, 1980); however, other authors have questioned the validity of this result (Wehland et al., 1982).

In order to understand the sequence of modifications and their role in proper transport of glycoproteins to viral assembly sites, investigators have perturbed these virus-host systems in a number of ways. These include prevention of N-linked glycosylation with the drug tunicamycin, disruption of Golgi function with the ionophore monensin, and the study of conditional lethal mutants (*ts* mutants) whose proteins are not transported at the nonpermissive temperature. More recently, transport has been studied by manipulating cloned copies of glycoprotein genes to produce transient expression of molecules lacking selected regions, such as the cytoplasmic-COOH terminus or the anchor region. Although different virus systems (discussed in more detail below) have differing sensitivities to these perturbations, the most general conclusion from these studies is that the three-dimensional conformation of the protein is crucial for proper transport and that no particular contiguous sequence of amino acids forms a "transport" address analogous to the signal peptide for initial transmembrane insertion. Complicating this question is the fact that for many systems the measure of transport has been assay of release of infectious virus. Many of these incompletely modified proteins are found in the "correct" membranes and are even assembled into particles but in a form such that the resulting virion is noninfectious.

Role of Glycosylation in Transport

In an effort to resolve the role of glycosylation in the transport and function of viral glycoproteins many systems have been studied in the presence of tunicamycin, a drug that inhibits glycosylation with N-linked carbohydrates (Takatsuki et al., 1971). For alphaviruses the nonglycosylated forms of the envelope proteins E2 and E1 are inserted into intracellular membranes normally (Garoff and Schwarz, 1978), but no envelope protein is detected at the cell surface and no budded viruses are produced (Leavitt et al., 1977a). It has been suggested that these proteins aggregate in tunicamycin-treated cells and that the effects on transport are not a result of the lack of carbohydrate but of the altered solubility of the nonglycosylated forms (Leavitt et al., 1977b).

For the coronaviruses, treatment of infected cells with tunicamycin has revealed two different forms of carbohydrate attachment. The E2 protein, which forms the projections of "peplomers" on the exterior of the virion, is not glycosylated in tunicamycin and indeed is difficult to find in treated cells, which suggests that the nonglycosylated form is more easily degraded (Holmes et al., 1981). The second glycoprotein of coronaviruses (E1) has more of the character of a matrix protein: only 10% of the molecule is external to the lipid bilayer, and this contains the carbohydrates. The remainder of the molecule is membrane-associated and forms specific interactions with the nucleocapsid (Siddell, 1982; Rottier et al., 1984). The glycosylation of the E1 moiety is unaffected by tunicamycin, indicating that it contains O-linked oligosaccharides (Holmes et al., 1981).

Bunyaviruses, which have been studied less completely, have two external glycoproteins, G1 and G2. G1 of Uukuniemi virus contains exclusively complex carbohydrates resistant to endoglycosidase H, while G2 contains a heterogeneous mixture of carbohydrates that are primarily endo H-sensitive (Kuismanen, 1984). Glycosylation of their carbohydrates is inhibited by tunicamycin (Cash et al., 1980), but the transport of viral glycoproteins to the cell surface is very slow (Madoff and Lenard, 1982), even during normal infection, and the effect of glycosylation on transport is difficult to assess.

The glycopeptides of Tacaribe virus, an arenavirus, have also been examined and appear to contain both simple and complex oligosaccharides (Boersma et al., 1982), but the role of these in transport is unknown.

For the remaining group (rhabdoviruses, orthomyxoviruses, and paramyxoviruses), all of which mature at the cell surface, the role of carbohydrate in transport, budding, and infectivity varies widely. Compans and Klenk (1979) reported that some influenza strains produce particles, albeit in reduced yield and with lowered specific infectivity, in the presence of tunicamycin, suggesting that carbohydrate is not absolutely required for transport to the cell plasmalemma. For paramyxoviruses the nonglycosylated forms of both F (the fusion protein) and HN (the hemagglutinin neuraminidase) are properly inserted into the endoplasmic reticulum and migrate to the smooth membranes. Some authors have found nonglycosylated F, but not HN, inserted into plasma membranes, producing normal yields of particles, albeit noninfectious (Morrison et al., 1981). However, others have been unable to detect either polypeptide at the cell surface and found neither infectious virus nor any budding figures in tunicamycin-treated cells (Nakamura et al., 1982). For rhabdoviruses the situation is more complex, and different virus strains differ in their sensitivity to tunicamycin (Gibson et al., 1980). Furthermore, point mutations in the VSV G protein can alter the requirement for glycosylation for transport (Chatis and Morrison, 1981).

From all of these apparently contradictory results emerges a general picture of the role of carbohydrate chains in transport of glycoproteins to sites of virus maturation. First, the "address" for transport does not appear to reside in the oligosaccharides, since glycosylation itself is not required for transport. Second, the conformation of the protein is critically important for transport, and the effect of oligosaccharides in maintaining conformation differs markedly in various systems. Glycosylation appears necessary for transport in systems like alphaviruses in which nonglycosylated membrane proteins have been shown to have different properties in solution from their glycosylated counterparts. On the other hand, the example cited above illustrates that the conformation of the G protein of VSV required for transport can be rendered independent of glycosylation by the introduction of point mutations in the polypeptide.

Role of Conformation in Transport

The role of conformation in transport has also been demonstrated by studies of mutants of alphaviruses that are temperature-sensitive for transport of the glycoproteins to the plasmalemma. Immunofluorescence studies (Saraste et al., 1980) showed that for two E1 mutants, glycoproteins moieties did not reach the plasmalemma at the nonpermissive temperature. Instead, a bright fluorescence was detected in the perinuclear region (Golgi apparatus). However, these preformed glycoproteins were fully functional and, upon shift down to the permissive temperature, were transported to the cell surface and incorporated into virions. To examine this phenomenon more closely Arias et al. (1983) determined the nucleotide sequence in the region encoding both E1 and E2 for two mutants defective in hemagglutination and transport and for their revertants. Three point mutations were located in E1: one was a double mutant, the other a single mutant. Revertants were same-site revertants to the same or a functionally equivalent amino acid. Most interesting, however, was the fact that in a molecule of 422 amino acids, the mutations altered residues 106, 176, and 267, widely separated in the linear sequence, indicating that the three-dimensional conformation, which the mutant polypeptide could maintain at the permissive but not the nonpermissive temperature, was crucial for proper transport to the plasmalemma. The location of these three mutations in the glycoprotein is diagrammed schematically in Figure 1; several landmark features of the glycoprotein are also indicated. Note that all three changes that confer temperature sensitivity for transport are located in the extracellular region of E1.

Maturation-defective mutants with phenotypes similar to those of *ts10* and *ts23* of Sindbis virus have also been isolated for the rhabdovirus VSV. Two such G-protein mutants, which are arrested at different points in their modification and transport at the nonpermissive temperature, have been described. One mutant is blocked at an early pre-Golgi step such that the G protein possesses only simple oligosaccharide chains; a second mutant is blocked later and has complex chains that are deficient in fucose residues. Neither mutant has fatty acids added at the nonpermissive temperature (see below). Both mutants are reversible upon shift to the permissive temperature, but the amino acid substitutions responsible for these phenotypes have not been determined (Lodish and Kong, 1983).

The effect of glycosylation on the conformation of the protein can also show up as changes in enzymatic activity. In a recent study influenza neuraminidase synthesized during glucose starvation was found to be underglycosylated and enzymatically inactive (Griffin et al., 1983).

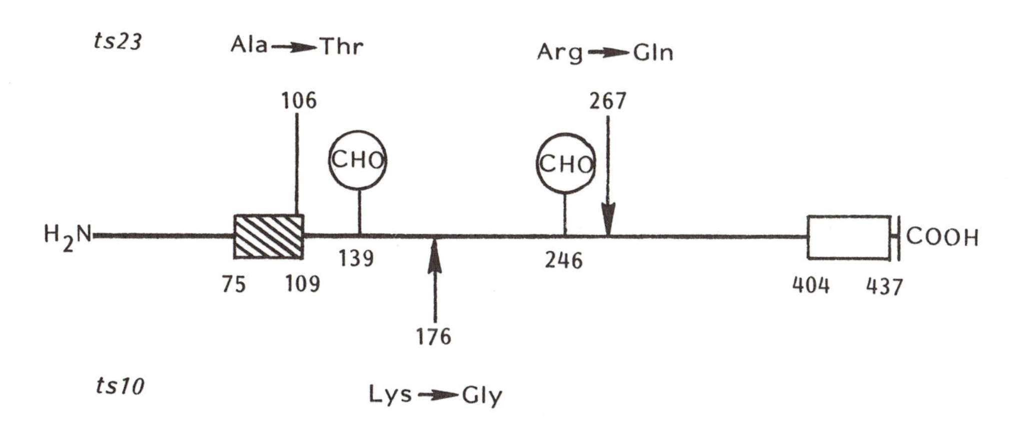

FIGURE 1 Schematic representation of glycoprotein E1 of Sindbis virus. The changes in *ts10* and *ts23* responsible for their temperature-sensitive phenotype are shown in this figure, which is drawn to scale. Several landmark features of E1 are indicated, including attachment sites of the carbohydrate chains as circles, the membrane spanning domain of the protein (residues 404 to 437) as an open box, and a hydrophobic, highly conserved segment from residues 75 to 109 which may be involved in membrane fusion upon infection, as a hatched box. (From Arias et al. (1983), with permission.)

Role of Fatty Acid Acylation in Transport

Another glycoprotein modification whose role in transport and assembly is not well understood is fatty acid acylation (Schmidt et al., 1979; Schmidt and Schlesinger, 1979; reviewed in Schmidt, 1983) of residues in the hydrophobic membrane spanning regions of the proteins (Rice et al., 1982; Capone et al., 1982). Although it was originally postulated that palmitate was attached by ester linkages to serine residues, it has now been shown conclusively that for VSV acylation occurs by means of thiol linkages to cysteine (Rose et al., 1984). When the codon specifying the cysteine used was changed to one specifying serine, G protein was synthesized that lacked fatty acids. Lack of acylation did not, in this case, affect glycosylation or transport.

Lipids have been found attached to the G protein of rhabdoviruses, both glycoproteins E1 and E2 of alphaviruses, the F protein of paramyxoviruses, the G1 protein of the bunyavirus La Crosse virus, the peplomer protein of coronaviruses, and the HA2 polypeptide of influenza. Other membrane glycoproteins, such as the HN protein of paramyxoviruses, the E1 glycoprotein of coronaviruses, and neuraminidase of influenza, lack attached lipids (Schmidt, 1982). The role of these covalently attached lipids in transport of membrane proteins and subsequent morphogenesis is unclear at present, but studies with cerulenin, an antibiotic that inhibits fatty acid synthesis (Omura, 1976) have shown that nonacylated Sindbis E1 and E2 and nonacylated VSV G protein were transported to the cell surface. However, particle assembly was inhibited (Schlesinger and Malfer, 1982), perhaps by altering the hydrophobicity of the membrane anchor and its conformation in the bilayer during budding. From studies with G protein mutants of VSV that are not acylated, Lodish and Kong (1983) have suggested that addition of fatty acid may be required either for specific translocation of G from the Golgi to the plasmalemma or for the budding event itself. The studies of Rose et al., (1984) described above suggest that acylation of the VSV G protein is not required for transport, but these studies do not reveal whether the nonacylated protein can be incorporated into virions.

Effect of Monensin

Another approach to dissecting the effects of modification of membrane glycoproteins has been to examine virus production in the presence of the ionophore monensin. This ionophore alters the ionic environment within the Golgi complex and blocks transport of molecules through the Golgi stack (Tarkatoff, 1983); this leads to the intracellular accumulation of polypeptides lacking the Golgi-produced modifications, such as complex oligosaccharides and covalently linked fatty acids. Monensin has been shown to block transport of alphavirus and rhabdovirus proteins to the cell plasmalemma (Kääriäinen et al., 1980; Pesonen and Kääriäinen, 1982; Johnson and Schlesinger, 1980), and virions are seen budding into cytoplasmic vacuoles, presumably at the sites in which glycoproteins accumulate when further transport is blocked. In cells that exhibit a polarity for virus maturation (see below) such as MDBK, monensin completely inhibits VSV from maturing at the basolateral membranes but has no effect on influenza virus maturation at the apical surfaces (Alonso and Compans, 1981).

Transport of the major retrovirus glycoprotein (gp70) to the cell surface is also blocked by monensin, and assembly and release of infectious virus is prevented (Srinivas et al., 1982; Chatterjee et al., 1982). Particles are produced, both from the plasmalemma and from intracellular vacuoles, but the particles lack gp70 and the second glycoprotein (gp35 or p15(E) depending on the system), incorporating instead the uncleaved *env* precursor (Chatterjee et al., 1982).

Experiments with Truncated or Chimeric Proteins

In separate experiments it has been shown that the hydrophobic membrane-spanning regions may not be essential for proper transport of integral membrane glycoproteins but are required to retain the protein in the bilayer. (This is directly analogous to the situation found for cell-associated and secreted forms of immunoglobulins. Membrane-associated IgM chains contain a hydrophobic root of 41 amino acids that is replaced in the secreted form of IgM molecules by a hydrophilic 21-amino acid carboxyl terminus containing carbohydrate (Kehry et al., 1980).) The cloned genes for both the VSV G protein and the HA protein of influenza have been inserted into SV40 vectors that contain the SV40 promoter and that express the inserted gene in a transient fashion. The cloned gene can then be chemically manipulated to remove sequences at will, creating specific deletions, or to combine coding sequences to produce chimeric proteins. Such studies have shown that a truncated G protein lacking 79 amino acids at the carboxyl terminus is glycosylated and transported normally; however, it is secreted rather than accumulating at the cell surface (Rose and Bergmann, 1982, 1983). Influenza HA deleted in the carboxyl region was also glycosylated and secreted (Sveda et al., 1982). Similarly, alphavirus E2 is properly glycosylated and transported if its cytoplasmic domain is removed (Garoff et al., 1983), but after deletion of the hydrophobic anchor region, E2 could not be localized in cells and was presumably secreted. Substitution of the E2 membrane anchor region with that of influenza HA or the VSV G protein resulted in a chimeric E2 that was stably integrated into the cell surface (Riedel, 1985). Thus, it appears that the membrane anchor and cytoplasmic domains of viral glycoproteins are not absolutely required for transport and modification of these proteins, but that a hydrophobic transmembrane domain, even if from a heterologous virus, is essential for stable integration to cell surface membranes.

In related experiments Rose and Bergmann (1983) found that deletions in the cytoplasmic domain of VSV G protein that left the hydrophobic anchor intact resulted in a protein that was inserted into the endoplasmic reticulum but that was not transported properly. Some of these altered proteins never left the endoplasmic reticulum; others were transported eventually to the cell surface but at greatly reduced rates. Thus, altering the cytoplasmic domain of this protein does affect transport, but in an unpredictable fashion, and the results suggest once again that the overall conformation of the protein is important for proper transport.

In another set of experiments to probe the effect of the transmembrane and cytoplasmic domains on transport, Guan and Rose (1984) expressed a chimeric protein in which rat growth hormone (a polypeptide that is normally secreted) was fused to the transmembrane and cytoplasmic domains of the VSV G protein. This chimeric protein was transported to the Golgi apparatus and esterified with palmitic acid, but was not further transported to the cell surface.

Other experiments have probed the function of the signal or leader sequence in transport. When deletions were made in the signal peptide of influenza HA, the polypeptide was not glycosylated and was not associated with membranes, either intracellularly or at the surface (Gething and Sambrook, 1982; Sekikawa and Lai, 1983); the amount of HA that accumulated in the cytoplasm depended upon the exact site of the deletion. When signal sequences from VSV G protein (McQueen et al., 1984) or influenza NA (Bos et al., 1984) were substituted for the normal HA signal peptide, HA was translocated across the membranes of the ER but did not reach the Golgi apparatus or the cell surface. These chimeric proteins presumably assume an altered configuration, indicating that not only is a signal sequence required for insertion into the membranes, as expected, but that a specific signal sequence may also be necessary for proper transport.

Thus, in general, virus-specific glycoproteins follow a unidirectional and well-defined pathway from rough ER to smooth ER to *cis* Golgi to *trans* Golgi, and from there in most cases to the plasmalemma. In contrast to the signal peptide necessary for original insertion into the rough ER, there does not appear to be a simple linear domain to serve as the address for transport; instead, the overall conformation of the protein seems to be essential. Concomitantly, as the polypeptides pass from one organelle compartment to the next, a regular sequence of post-translational modifications occurs. No one of these modifications contains a specific address either, but at least some may serve to stabilize the configuration of a protein and facilitate its movement along the transport pathway.

Vectorial Transport of Proteins

Most of the studies of viral glycoprotein transport and secretion in monolayer cells in culture (using viruses that normally bud from the plasmalemma) have shown the viral polypeptides to be distributed evenly over the exterior of the cells, both at the surface adherent to the substrate and at the surface exposed to the culture fluid. However, certain cell types maintain a morphological and biochemical asymmetry in culture; the surface in the medium corresponds to the apical region, and the surface next to the plate corresponding to the basolateral surface, as defined in the organ from which the cells were derived. These two compartments are kept separate by the formation of tight junctions at the lateral margins of the cells making up the monolayer. Electron micrographic studies have shown that orthomyxoviruses and paramyxoviruses bud preferentially from the apical surface, whereas rhabdoviruses and retroviruses (and probably alphaviruses) bud predominantly from the basolateral surface (Table 2) (Rodriguez-Boulan and Sabatini, 1978; Roth et al., 1983b). As expected, it has been possible to show that the respective viral glycoproteins distribute on the surface of the cell in the same polarized pattern (Rodriguez-Boulan and Pendergast, 1980).

In agreement with the earlier results on the effects of carbohydrate addition on the fate of viral glycoproteins, it was shown that for this polarized distribution, glycosylation did not affect the sorting of the glycoproteins. In the presence of tunicamycin VSV glycoproteins were still found associated with basolateral membranes, whereas influenza proteins were found on the apical surface (Roth et al., 1979; R. Green et al., 1981). Furthermore, this distribution is a function of the particular glycoprotein moiety alone and is not dependent on the presence of other viral components, such as the matrix protein or the nucleocapsid. Expression of a cloned gene for the hemagglutinin of influenza resulted in accumulation of the gene product in the appropriate membrane compartment in the absence of other viral components (Roth et al., 1983a).

BUDDING AND MATURATION

In three groups of viruses (alphaviruses, coronaviruses, and bunyaviruses) budding of mature virions appears to take place by

Table 2 Sites of maturation

Virus group	Membrane	Location in polarized cells	References
Retrovirus	Plasma membrane	Basolateral	2
Rhabdovirus	Plasma membrane	Basolateral	1
Alphavirus	Plasma membrane	Basolateral?	3
Orthomyxovirus	Plasma membrane	Apical	1
Paramyxovirus	Plasma membrane	Apical	1
Bunyavirus	Smooth ER		
Coronavirus	Rough ER		
Flavivirus	Smooth ER, vacuoles		

References: (1) Rodriguez-Boulan and Pendergast, 1980; (2) Roth et al., 1983b; (3) Simons and Warren, 1983.

direct recognition by the nucleocapsid of the cytoplasmic domains of integral membrane proteins, but the details of the interaction and the cellular site of maturation are different for each case.

Budding of alphaviruses has been the most extensively documented (for reviews, see Strauss and Strauss, 1977, 1983; Simons and Garoff, 1980; Simons and Warren, 1983). All three virus proteins (C, E1, and E2) are present in equimolar ratios (as determined by chemical analysis of purified virions), and the glycoprotein spikes external to the bilayer (in addition to the nucleocapsid subunits) are present in an icosahedral array. This observation has led to the hypothesis that each capsid protein interacts with a single E2-E1 heterodimer (probably with the cytoplasmic domain of E2). Interaction with the capsid fixes the hitherto freely diffusing glycoprotein complex, and successive interactions sterically exclude host polypeptides from the vicinity of the budding virus in the plasmalemma (Strauss, 1978). Specific interaction of the capsid and the glycoproteins has been demonstrated by crosslinking studies with bifunctional reagents and by the fact that mature virions can be stripped of lipids by mild detergent treatment, leaving the glycoproteins still in association with the ribonucleoprotein core (reviewed in Simons and Warren, 1983). This direct association also explains the asymmetric phenotypic mixing seen between alphaviruses and VSV; that is, VSV cores, which recognize their matrix protein on the inner surface of the plasmalemma (see below), will bud through membranes containing Sindbis virus glycoproteins, but Sindbis cores cannot bud into membranes containing VSV-specific protein (Zavadova et al., 1977). This type of alphavirus phenotypic mixing has only been seen between closely related alphaviruses, such as Sindbis and Western equine encephalitis (Burge and Pfefferkorn, 1966; Strauss et al., 1983).

Coronaviruses bud into internal membranes, notably the rough endoplasmic reticulum, and the specific interactions that are responsible are not known. However, assembly probably involves the cytoplasmic domain of the E1 polypeptide, a curious transmembrane protein that has an external glycosylated domain comprising about 10% of the mass, a small domain on the cytoplasmic face, and the bulk of the protein buried within the membrane (Rottier et al., 1984). Glycoprotein E1 is apparently intimately involved in maturation of coronaviruses, since it is localized almost exclusively in the perinuclear region of host cells (Dubois-Dalq et al., 1982). Furthermore, gentle disruption of coronavirus virions with nonionic detergents under certain conditions results in quantitative binding of E1 to the nucleocapsids (reviewed in Sturman and Holmes, 1983). Thus, this polypeptide is more akin to the matrix proteins discussed below. Association of the helical capsid with the more peripheral peplomer protein has not been demonstrated (Siddell et al., 1982). The peplomer protein is found throughout coronavirus-infected cells, and thus it has been suggested that the localization of the E1 polypeptide determines the site of virion maturation (Holmes et al., 1984).

Finally, two groups of viruses bud into intracellular vacuoles, the flaviviruses and the bunyaviruses. Formerly grouped with alphaviruses, pestiviruses, and rubiviruses in the family Togaviridae, flaviviruses have now been reclassified as a separate family,

Flaviviridae, on the basis of their unique genome organization and replication strategy. They are morphologically similar to alphaviruses, but their mode and site of assembly have more in common with bunyaviruses. Neither preformed capsids nor budding figures have been seen for flaviviruses, but fully enveloped particles are seen within vacuoles (Westaway, 1980). Although it is thought that the capsid interacts with the transmembrane portion of the external glycoprotein (E or V1), the role of a small nonglycosylated envelope protein (M or V3), which is derived by proteolytic cleavage of a glycosylated precursor (Shapiro et al., 1972; C. M. Rice et al., submitted), has not been precisely determined, and the driving mechanism for budding is unknown.

Morphogenetic studies of Bunyaviridae have been conducted with representatives of the phleboviruses or sandfly fever genus (Smith and Pifat, 1982) as well as Uukuviruses (Kuismanen et al., 1982). Although these viruses are negative-strand viruses with helical nucleocapsids, the virions lack a matrix protein and have two integral membrane glycoproteins, which differ in their attached oligosaccharides and may also differ in their roles in maturation (Kuismanen, 1984). Viral antigens in infected cells are seen only in the perinuclear membrane, closely associated with the Golgi, and not in cell surface membranes. Nucleocapsids can only be observed at budding sites and are not found as preformed structures in the cytoplasm, making the assembly of these virions at least superficially similar to both flavivirus and C-type retrovirus assembly. Immunofluorescence studies have shown a correlation between the characteristic vacuolization of the Golgi complex seen in Uukuniemi-infected cells and accumulation of viral nucleoprotein and glycoproteins in these membranes (Kuismanen et al., 1984). Protease-protection studies have shown that one of the two glycoproteins is a transmembrane entity with approximately 12% of its mass on the cytoplasmic face before budding, and it is postulated that budding occurs by direct interaction of this domain with the nucleocapsid (Smith and Pifat, 1982).

Another negative-strand virus group, arenaviruses, also appears to lack a matrix protein, but nothing is known of either their site or mode of budding.

Maturation of retroviruses is a complex process. The primary translation product of the *env* gene is a glycosylated precursor $pr95^{env}$ that is processed to two mature glycoproteins (gp85 and gp35 for avian retroviruses and gp70 and p15E for murine viruses), that are linked by disulfide bonds (Shinnick et al., 1981; Schwartz et al., 1983). Interestingly, the consensus cleavage sequences between the two glycoproteins of Rous sarcoma virus and the murine viruses is Arg-X-Lys-Arg, in which X can vary and in which the first cleavage event occurs after the second Arg. This sequence is the same as the cleavage sequence Arg-Glu-Lys-Arg for the H7 hemagglutinin in the pathogenic influenza strains (Bosch et al., 1981) and very similar to the consensus sequence for the cleavage site between E3 and E2 of alphaviruses (Arg-X-Lys/Arg-Arg) (Garoff et al., 1980b; Rice and Strauss, 1981; Dalgarno et al., 1983). It has been suggested that one host protease is responsible for all of these events, possibly the same activity that cleaves proparathyroid hormone and proalbumin, though in the last two cases the cleavage recognition site appears to consist of merely two adjacent basic residues (Dean and Judah, 1980). An enzyme of similar specificity (Arg-Glu-Arg-Lys↓Ala) has been isolated from brain cells and shown to be responsible for processing of the peptide hormone somatostatin (Gluschankof et al., 1984).

Immunoprecipitation studies have shown that p19 (a *gag* gene product) remains associated with gp85 and gp35, and the hypothesis has been put forward that this association of an inner antigen with the envelope proteins is essential for assembly, because mutants in p19 fail to bud properly (Rohrscheider et al., 1976; reviewed in Bolognesi et al., 1978).

Immunoprecipitation studies have shown that p19 (a *gag* gene product) remains associated with gp85 and gp35, and the hypothesis has been put forward that this association of an inner antigen with the envelope proteins is essential for assembly, because mutants in p19 fail to bud properly (Rohrscheider et al., 1976; reviewed in Bolognesi et al., 1978).

As noted in Table 2 retroviruses mature at the basolateral surface of polarized epithelial cells, and the envelope proteins are only found inserted into the basolateral membranes (Roth et al., 1983b). Whether the gp85 and gp35 (avian) or gp70 and p15E (murine) moieties are separately addressed to this destination or whether the address is a function of the disulfide-linked complex is unknown. A small amount of an uncleaved precursor $pr95^{env}$ is found at the surface of cells infected with Rous sarcoma virus, but this may represent a deadend pathway, since the precursor contains only high-mannose oligosaccharides and is not incorporated into virions. Normal processing of the carbohydrates to the complex chains. found on mature gp85 and gp35, as well as proteolytic cleavage of $pr95^{env}$ are postulated normally to occur in the Golgi (Bosch et al., 1982).

Budding and maturation of the remaining three virus groups, all negative-strand viruses, share many characteristics, probably reflecting a relatively recent evolutionary divergence of the rhabdoviruses, orthomyxoviruses, and paramyxoviruses from a common ancestral type. Other aspects of the molecular details of the replication of these viruses also suggest that these groups are closely related (reviewed in Strauss and Strauss, 1983). In each case there is a nucleocapsid with helical architecture that interacts quite specifically with a matrix protein (M) lining the inner surface of the cell plasmalemma, and in at least one case the matrix protein is present as a regular pseudocrystalline array (Büechi and Bächi, 1982). Little is known about the transport of M proteins to the cell surface. They are not glycosylated and do not pass through the modification pathway described above for surface glycoproteins; indeed they may reach the cell surface by simple diffusion. In most cases the ratio of matrix protein to nucleocapsid protein found in virions for a particular virus is constant. Inserted into the bilayer, and in some cases shown to be truly transmembranous, are the glycoproteins (G of VSV, HA and NA of influenza, and F and HN of paramyxoviruses). The ratio of glycoprotein to matrix is not fixed, implying a much less specific association between these two (Lodish and Porter, 1980). Moreover, many of these viruses show much more latitude in the amount of host polypeptide present in their membranes. VSV has been shown to incorporate H2 antigens into mature virions, and paramyxoviruses can contain significant amounts of a number of host proteins, notably actin (Wang et al., 1976; reviewed in Lodish et al., 1981). Indeed, actin has been shown to complex specifically with the matrix protein of Newcastle disease virus and Sendai virus (Giuffre et al., 1982).

For the best-studied case, the rhabdovirus VSV, proper assembly of nucleocapsid with a bilayer containing the M protein is essential, since M protein mutants do not assemble (Weiss and Bennett, 1980); however, VSV will form phenotypically mixed particles containing glycoproteins encoded by many other groups including paramyxo-, retro-, alpha-, and even herpesviruses (Pringle, 1977). The presence of M protein is essential for stabilization of the glycoproteins in the membrane (Yoshida et al., 1979), and, as noted above, M probably interacts specifically with VSV nucleocapsids as well (Jacobs and Penhoet, 1982).

The role of M protein in the assembly of orthomyxoviruses and paramyxoviruses has not been so thoroughly explored (for a review, see Simons and Garoff, 1980). Influenza M protein associates readily with lipid bilayers by means of two hydrophobic domains (Gregoriades and Frangione, 1981). M protein also interacts spontaneously with preformed liposomes (Bucher et al., 1980), and such M-containing liposomes in turn interact preferentially with the neuraminidase glycoprotein (Davis and Bucher, 1981). In cells in which influenza infection is abortive (those that fail to produce progeny virions), the synthesis of the M polypeptide is specifically inhibited (Lohmeyer et al., 1979). Some direct interaction between matrix proteins and surface glycoproteins must occur, at least in paramyxoviruses, since mutants in paramyxovirus M protein results in virions of reduced infectivity containing lesser amounts of F protein (Peeples and Bratt, 1984). Thus, M protein appears to be essential for proper organization of the virus-modified membrane, interaction with the nucleocapsid, and successful virion budding.

All of the enveloped viruses acquire their envelopes by budding through preexisting cellular membranes, and the arrangement of the viral lipids reflects those of the host membrane. Although particular virus envelope proteins may have a slight influence on the lipid composition of their envelopes, extensive analysis has shown that the lipid composition of viruses primarily reflects that of the host. The same virus, grown in hosts with different membrane lipids, contains identical virus-specific proteins embedded in a bilayer that matches the lipids of the membrane through which they budded. Similarly, lipids present in virions in most cases studied have been found to have been synthesized prior to infection (reviewed in Compans and Klenk, 1979).

MORPHOGENESIS AND ASSEMBLY OF HERPESVIRUS

The pathway of assembly of herpesviruses shows so few parallels with the other virus groups above that it will be considered separately. The herpesvirion consists of four major components, the DNA-containing core and three outer protective layers, called in order of increasing radius: the capsid, the tegument, and the envelope (reviewed in Roizman, 1977). The structure of the inner core has not been completely determined but has been described as "a fibrous spool on which the DNA is wound" (Roizman, 1982). The capsid appears to be a regular icosahedral shell composed of 162 capsomeres, but the virion proteins that comprise the pentameric and hexameric structures are unclear. The tegument is an amorphous layer separating the capsid and the envelope, and the amount of tegument is variable for individual virions in a population and variable among viruses.

Unlike the previous examples, assembly of herpesviruses occurs in or near the nucleus of the cell. There are three species of major capsid proteins, one of which also exists in a precursor form. Studies with temperature-sensitive mutants have shown that capsid mutants in several different complementation groups do not process the precursor to the 35K major capsid protein and that none of the capsid proteins accumulate in the nucleus of cells infected with capsid mutants (Ladin et al., 1982). This indicates that the presence of all three polypeptides in functional form and continued capsid assembly is necessary for the transport of capsid proteins from their site of synthesis in the cytoplasm to the capsid assembly site in the nucleus. Subsequently these capsids are filled with DNA in an event that occurs simultaneously with cleavage of genome-sized DNAs from a concatameric precursor (Vlazny et al., 1982). These processes seem to involve the nuclear matrix of the cell (Bibor-Hardy et al., 1982).

Herpesviruses have at least four glycoproteins in their envelopes, gB, gC, gD, and gE, which contain unique antigenic sites (Norrild, 1980). Another glycoprotein, gA, has been found to be a form of gB (Spear, 1984). Owing to the complexity of their electrophoretically separable glycosylated forms the enumeration of the herpesvirus glycoproteins and clarification of the relationships between glycoproteins of different herpesviruses will probably be best accomplished by mapping and sequence analysis of their genes (Spear, 1984). For example, this approach has shown that the gF gene of HSV 2 is the homolog of the gC gene of HSV 1 (Dowbenko and Lasky, 1984). These glycoproteins are found in the plasma membrane of infected cells as well as in virions (Roizman, 1977). In the presence of tunicamycin, gA, gB, and gD are produced as underglycosylated polypeptides (Norrild and Pederson, 1982), indicating that their carbohydrate moieties are attached to asparagine by N-glycosidic bonds. In contrast, glycosylation of gC occurs in tunicamycin-treated cells, indicating that some of the glucosamine and galactose residues associated with this polypeptide are attached by O-glycosidic bonds (Wenske and Courtney, 1983). Glycosylation appears to be necessary for the transport of herpes glycoproteins to the surface of infected cells (Norrild and Pederson, 1982).

The presence of herpesvirus glycoproteins in the host plasma membrane remains somewhat of a conundrum, since in contrast to the other enveloped viruses, infectious virions appear to bud primarily from the inner lamella of the nuclear membrane in sites morphologically altered by the presence of the integral membrane glycoproteins, as well as by components of the tegument (reviewed in Roizman, 1977). How the exit of virions from this location is accomplished is unknown, but it has been postulated that either the perinuclear space may be contiguous with other membranous organelles of the cell or that the virions are exported by vacuoles to the cell surface. Recent experiments have implicated the Golgi membranes, since the ionophore monensin completely inhibits the release of herpes virions under conditions in which infectious virus particles still accumulate within the cell (Johnson and Spear, 1982). One intriguing possibility is that virions that contain immature forms of the glycoproteins may be assembled at the nuclear membrane. As these virus particles are transported to the cell surface, they pass through various membrane-bound organelle compartments, and it has been suggested that the viral glycoproteins are further modified by host cell enzymes after assembly (Spear, 1984). This implies that only the N-linked high-mannose oligosaccharides that are attached in the rough endoplasmic reticulum are required for assembly of particles with a morphology and infectivity very similar to that of mature virus.

CONCLUDING REMARKS

This review was initially completed in June, 1983, and limited revisions were made in March, 1985. It was interesting to note the uneven progress in various aspects of the field of virus assembly within the last two years. In those areas to which the powerful tools of recombinant DNA techniques have been applied, for example, site-specific mutagenesis, expression of cloned copies of viral genes, and nucleotide-sequence determination, advances have been quite rapid. On the other hand, progress in unraveling the steps of complex morphogenetic pathways, such as those employed by the retroviruses and the herpesviruses, has been more modest. However, there are many examples in which enveloped animal viruses are fulfilling their potential as model systems for membrane biogenesis. The complete nucleotide sequences of the genomes of orthomyxoviruses (reviewed in Lamb and Choppin, 1983), murine retroviruses (Shinnick et al., 1981), avian retroviruses (Schwartz et al., 1983), alphaviruses (Strauss et al., 1984), and flaviviruses (C. M. Rice, submitted) have been determined. In addition, the nucleotide sequences encoding the virus glycoproteins are known for the rhabdoviruses VSV (Rose and Gallione, 1981) and rabies (Anilionis et al., 1981), as well as the HN (Hiebert et al., 1985) and F (Collins et al., 1984; Paterson et al., 1984) proteins of paramyxoviruses and the G1 and G2 proteins of bunyaviruses (Eshita and Bishop, 1984). Furthermore, owing in part to the potential of these glycoproteins as vaccinating antigens, many of these glycoproteins have been cloned and expressed either transiently in a number of systems (see section above, *Experiments with Truncated or Chimeric Proteins*) or inserted into recombinant viruses, notably vaccinia (Paoletti et al., 1984; Mackett et al., 1984). These studies, as discussed above, have illustrated that most, if not all of the information required for modification and transport of a glycoprotein is encoded into its polypeptide moiety and that transport signals targeting a protein to the cell surface, even to the proper cell surface in a polarized host cell, function in the absence of other viral components. Certain features of integral membrane proteins can now be recognized in the deduced amino acid sequence from the nucleotides encoding them. As the data base of comparative glycoprotein sequences expands, we hope that better algorithms will be developed for predicting three-dimensional conformation of the proteins from the linear sequence and for recognizing and locating functional domains. Furthermore, it is now possible to manipulate cloned viral genomes to probe particular questions of interest. The ability to alter individual amino acids at will by site-directed mutagenesis and study the effects of these mutations on the encoded protein (either expressed alone or in the presence of other viral components) is opening up fascinating avenues for future research. In this way we hope to be able to understand the interactions and interdependence between capsid and matrix protein, or between matrix and glycoproteins. With their limited coding capacity, animal viruses depend on cellular mechanisms for synthesis, transport, and modification of proteins. A knowledge, on the molecular level, of how nucleic acids, proteins, lipids, and carbohydrates made at various intracellular locations can come together to form a highly ordered structure such as an infectious virus should be a valuable model for the assembly of higher-order aggregations such as organelles, cells, organs, and ultimately entire organisms.

ACKNOWLEDGEMENTS

Work of the authors is supported by grants AI10793 and AI20612 from the National Institutes of Health and by a grant PCM8316856 from the National Science Foundation.

REFERENCES

Akashi, H. and D. H. L. Bishop. 1983. *J. Virol.*, 45, 1155.

Alonso, F. V. and R. W. Compans, 1981. *J. Cell Biol.*, 89, 700.

Anilionis, A., W. H. Wunner, and P. J. Curtis. 1981. *Nature*, 294, 275.

Arias, C., J. R. Bell, E. M. Lenches, E. G. Strauss, and J. H. Strauss. 1983. *J. Mol. Biol.*, 168, 87.

Auperin, D. D., M. Galinski, and D. H. L. Bishop. 1984. *Virology*, 134, 208.

Bibor-Hardy, V., M. Pouchelet, E. St.-Pierre, M. Herzberg, and R. Simard. 1982. *Virology*, 121, 296.

Blok, J., G. M. Air, W. G. Laver, C. W. Ward, G. G. Lilley, E. F. Woods, C. M. Roxburgh, and A. S. Inglis. 1982. *Virology*, 119, 109.

Boege, U., G. Wengler, G. Wengler, and B. Wittmann-Liebold. 1981.

Virology, 113, 293.

Boersma, D. P., Saleh, K. Nakamura, and R. W. Compans. 1982. *Virology*, 123, 452.

Bolognesi, D. P., R. C. Montelaro, H. Frank, and W. Schafer. 1978. *Science*, 199, 183.

Bos, T. J., A. R. Davis, and D. P. Nayak. 1984. *Proc. Nat. Acad. Sci. USA*, 81, 2327.

Bosch, F. X., W. Garten, H.-D. Klenk, and R. Rott. 1981. *Virology*, 113, 725.

Bosch, J. V., R. T. Schwarz, A. Ziemiecki, and R. R. Friis. 1982. *Virology*, 119, 122.

Bucher, D. J., I. G. Kharitonenkov, J. A. Zakomirdin, V. B. Grigoriev, S. M. Klimenko, and J. F. Davis. 1980. *J. Virol.*, 36, 586.

Büechi, M. and Th. Bächi. 1982. *Virology*, 120, 349.

Burge, B. W. and E. R. Pfefferkorn. 1966. *Nature*, 210, 1397.

Capone, J., F. Toneguzzo, and H. P. Ghosh. 1982. *J. Biol. Chem.*, 257, 16.

Cash, P., L. Hendershot, and D. H. L. Bishop. 1980. *Virology*, 103, 235.

Caspar, D. L. D. and A. Klug. 1962. *Cold Spring Harbor Symp. Quant. Biol.*, 27, 1.

Chatis, P. A. and T. G. Morrison. 1981. *J. Virol.*, 37, 307.

Chatterjee, S., J. A. Bradac, and E. Hunter. 1982. *J. Virol.*, 44, 1003.

Collins, P. L., Y. T. Huang, and G. W. Wertz. 1984. *Proc. Nat. Acad. Sci. USA*, 81, 7683.

Compans, R. W. and H.-D. Klenk. 1979. In *Comprehensive Virology*,, Vol. 13. H. Fraenkel-Conrat and R. Wagner, eds., p. 293. Plenum.

Dalgarno, L., C. M. Rice, and J. H. Strauss. 1983. *Virology*, 129, 170.

Darlix, J.-L. and P. F. Spahr. 1982. *J. Mol. Biol.*, 160, 147.

Davis, J. F. and D. J. Bucher. 1981. In *The Replication of Negative-*

Strand Viruses, D. H. L. Bishop and R. W. Compans, eds., p. 203. Elsevier.

Dean R. T. and J. D. Judah. 1980. In *Comprehensive Biochemistry*. Vol. 19B, M. Florkin and A. Neuburger, eds., p. 233. Elsevier.

Dowbenko, D. J. and L. A. Lasky. 1984. *J. Virol.*, 52, 154.

Dubois-Dalcq, M. E., E. W. Doller, M. V. Haspel, and K. V. Holmes. 1982. *Virology*, 119, 317.

Edbauer, C. A. and R. B. Naso. 1984 *Virology*, 134, 389.

Eshita, T. and D. H. L. Bishop. 1984. *Virology*, 137, 227.

Gallione, C. J., J. R. Greene, L. E. Iverson, and J. K. Rose. 1981. *J. Virol.*, 39, 529.

Garoff, H., A.-M. Frischauf, K. Simons, H. Lehrach, and H. Delius. 1980a. *Proc. Nat. Acad. Sci. USA*, 77, 6376.

Garoff, H., A.-M. Frischauf, K. Simons, H. Lehrach, and H. Delius. 1980b. *Nature*, 288, 236.

Garoff, H. C., C. Kondor-Koch, R. Petterson, and B. Burke. 1983. *J. Cell Biol.*, 97, 652.

Garoff, H. and R. T. Schwarz. 1978. *Nature*, 274, 487.

Gething, M.-J. and J. Sambrook. 1982. *Nature*, 300, 598.

Gibson, R., S. Kornfield, and S. Schlesinger. 1980. *Trend Biochem. Sci.*, November, 290.

Giuffre, R. M., D. R. Tovell, C. M. Kay, and D. L. J. Tyrrell. 1982. *J. Virol.*, 42, 963.

Gluschankof, P., A. Morel, S. Gomez, P. Nicolas, C. Fahy, and P. Cohen. 1984. *Proc. Nat. Acad. Sci. USA*,. 81, 6662.

Gregoriades, A. and B. Frangione. 1981. *J. Virol.*, 40, 323.

Green, J., G. Griffiths, D. Louvard, P. Quinn, and G. Warren. 1981. *J. Mol. Biol.*, 152, 663.

Green, R. F., H. K. Meiss, and E. Rodriguez-Boulan. 1981. *J. Cell Biol.*, 89, 230.

Griffin, J. A., S. Basak, and R. W. Compans. 1983. *Virology*, 125, 324.

Griffiths, G., R. Brands, B. Burke, D. Louvard, and G. Warren. 1982. *J. Cell Biol.*, 95, 781.

Guan, J.-L. and J. K. Rose. 1984. *Cell*, 37, 779.

Heggeness, M. H., A. Scheid, and P. W. Choppin. 1980. *Proc. Nat. Acad. Sci. USA*, 77, 2631.

Heggeness, M. H., P. R. Smith, I. Ulmanen, R. M. Krug, and P. W. Choppin. 1982. *Virology*, 118, 466.

Hiebert, S. W., R. G. Paterson, and R. A. Lamb. 1985. *J. Virol.*, 54, 1.

Holmes, K. V., E. W. Doller, and L. S. Sturman. 1981. *Virology*, 115, 334.

Holmes, K. V., M. F. Frana, S. G. Robbins, and L. S. Sturman. 1984. In *Molecular Biology and Pathogenesis of Corona Viruses*. P. J. M. Rottier, B. A. M. van der Zeijst, W. J. M. Spaan, and M. Horzinek, eds. p. 37. Plenum.

Hubbard, S. C. and R. J. Ivatt. 1981. *Ann. Rev. Biochem.*, 50, 555.

Jacobs, B. L. and E. E. Penhoet. 1982. *J. Virol.*, 44, 1047.

Johnson, D. C. and M. J. Schlesinger. 1980. *Virology*, 103, 407.

Johnson, D. C. and P. G. Spear. 1982. *J. Virol.*, 43, 1102.

Kääriäinen, L., K. Hashimoto, J. Saraste, I. Virtanen, and K. Penttinen. 1980. *J. Cell Biol.*, 87, 783.

Kehry, M., S. Ewald, R. Douglas, C. Sibley, W. Raschke, D. Fambrough, and L. Hood. 1980. *Cell*, 21, 393.

Klug, A. 1979. *Harvey Lectures*, 74, 141.

Kuismanen. E. 1984. *J. Virol.*, 51, 806.

Kuismanen, E. B. Bang, M. Hurme, and R. F. Petterson. 1984. *J. Virol.*, 51, 137.

Kuismanen, E., K. Hedman, J. Saraste, and R. F. Petterson. 1982. *Molec. Cell Biol.*, 2, 1444.

Ladin, B. F., S. Ihara, H. Hampl, and T. Ben-Porat. 1982. *Virology*, 116, 544.

Lamb, R. A. and P. W. Choppin. 1983. *Ann. Rev. Biochem.*, 52, 467.

Leavitt, R., S. Schlesinger, and S. Kornfeld. 1977a. *J. Virol.*, 21, 375.

Leavitt, R., S. Schlesinger, and S. Kornfeld, 1977b. *J. Biol. Chem.*, 252, 9018.

Lodish, H. F., W. A. Braell, A. L. Schwartz, G. J. A. M. Strous, and A. Zilberstein. 1981. *Int. Rev. Cytol.*, Suppl. 12, 247.

Lodish, H. F., and N. Kong. 1983. *Virology*, 125, 335.

Lodish, H. F. and M. Porter. 1980. *J. Virol.*, 33, 52.

Lohmeyer, J., L. T. Talens, and H. D. Klenk. 1979. *J. Gen. Virol.*, 42, 73.

Mackett, M., G. L. Smith, and B. Moss. 1984. *J. Virol.*, 49, 857.

Madoff, D. H. and J. Lenard. 1982. *Cell*, 28, 821.

McQueen, N. L., D. P. Nayak, L. V. Jones, and R. W. Compans. 1984. *Proc. Nat. Acad. Sci. USA*,. 81, 395.

Min Jou, W., M. Verhoeyen, R. Devos, E. Saman, R. Fang, D. Huylebroeck, and W. Fiers. 1980. *Cell*, 19, 683.

Morgan, E. M., G. G. Re, and D. W. Kingsbury. *Virology*, 135, 279.

Morrison, T. G., P. A. Chatis, and D. Simpson. 1981. In *The Replication of Negative-Strand Viruses*. D. H. L. Bishop and R. W. Compans, eds., p. 471. Elsevier.

Nakamura, K., M. Homma, and R. W. Compans. 1982. *Virology*, 119, 474.

Norrild, B. 1980. *Curr. Top. Microbiol. Immunol.*, 90, 67.

Norrild, B. and B. Pedersen. 1982. *J. Virol.*, 43, 395.

Omura, S. 1976. *Bacteriol. Rev.*, 40, 681.

Paoletti, E., B. R. Lipinskas, C. Samsonoff, S. Mercer, and D. Panical. 1984. *Proc. Nat. Acad. Sci. USA*,. 81, 193.

Paterson, R. G., T. J. R. Harris, and R. A. Lamb. 1984. *Proc. Nat. Acad. Sci. USA*,. 81, 6706.

Patterson, J. L., B. Holloway, and D. Kolakofsky. 1984. *J. Virol.*, 52, 215.

Patton, J. T., N. L. Davis, and G. W. Wertz. 1983. *J. Virol.*, 45, 155.

Peeples, M. E. and M. A. Bratt. 1984. *J. Virol.*, 51, 81.

Pepinsky, R. B. and V. M. Vogt. 1979. *J. Mol. Biol.*, 131, 819.

Pepinsky, R. B. and V. M. Vogt. 1983. *J. Virol.*, 45, 648.

Pepinksy, R. B. and V. M. Vogt. 1984. *J. Virol.*, 52, 145.

Pesonen, M. and L. Kääriäinen. 1982. *J. Mol. Biol.*, 158, 213.

Pringle, C. F. 1977. In *Comprehensive Virology,*, Vol. 9, H. Fraenkel-Conrat and R. Wagner, eds., p. 239. Plenum.

Rice, C.M., J. R. Bell, M. W. Hunkapiller, E. G. Strauss, and J. H. Strauss. 1982. *J. Mol. Biol.*, 154, 355.

Rice, C. M. and J. H. Strauss. 1981. *Proc. Nat. Acad. Sci. USA*, 78, 2062.

Riedel, H. 1985. *J. Virol.*, 54, 224.

Ro, J. H.-S. and H. P Ghosh. 1984. *Virology*, 135, 489.

Rodriguez-Boulan, E. and D. B. Sabatini. 1978. *Proc. Nat. Acad. Sci. USA*, 75, 5071.

Rodriguez-Boulan, E. and M. Pendergast. 1980. *Cell*, 20, 45.

Rohrschneider, J. M., H. Diggelmann, H. Ogura, R. P. Friis, and H. Bauer. 1976. *Virology*, 75, 177.

Roizmann, B. 1977. In *The Molecular Biology of Animal Viruses*, D. P. Nayak, ed., p. 769. Dekker.

Roizman, B. 1982. In *The Herpesviruses*, Vol. 1. B. Roizman, ed. p. 1. Plenum.

Rose, J. K., G. A. Adams, and C. J. Gallione. 1984. *Proc. Nat. Acad. Sci. USA*,. 81, 2050.

Rose, J. K. and J. E. Bergmann. 1982. *Cell*, 30, 753.

Rose, J. K. and J. E. Bergmann. 1983. *Cell*, 34, 513.

Rose, J. K., and C. J. Gallione. 1981. *J. Virol.*, 39, 519.

Roth, M. G., R. W. Compans, L. Giusti, A. R. Davis, D. P. Nayk, M.-J. Gething, and J. Sambrooks. 1983a. *Cell*, 33, 435.

Roth, M. G., J. P. Fitzpatrick, and R. W. Compans. 1979. *Proc. Nat. Acad. Sci. USA*, 76, 6430.

Roth, M. G., R. V. Srinivas, and R. W. Compans. 1983b. *J. Virol.*, 45, 1065.

Rothman, J. 1981. *Science,* 213, 1212.

Rothman, J. E. and R. E. Fine. 1980. *Proc. Nat. Acad. Sci. USA,* 77, 780.

Rottier, P., D. Brandenburg, J. Armstrong, B. van der Zeijst, and G. Warren. 1984. *Proc. Nat. Acad. Sci. USA,.* 81, 1421.

Sabatini, D. D., G. Kreibich, T. Morimoto, and M. Adesnik. 1982. *J. Cell Biol.,* 92, 1.

Saraste, J., C.-H. von Bonsdorff, K. Hashimoto, L. Kääriäinen, and S. Keränen. 1980. *Virology,* 100, 229.

Schlesinger, M. J. and C. Malfer. 1981. *J. Biol. Chem.,* 257, 9887.

Schmidt, M. F. G. 1982. *Virology,* 116, 327.

Schmidt, M. F. G. 1983. *Curr. Topics. Microbiol. Immunol,,* 102, 101.

Schmidt, M. F. G., M. Bracha, and M. J. Schlesinger. 1979. *Proc. Nat. Acad. Sci. USA,* 76, 1687.

Schmidt, M. F. G. and M. J. Schlesinger. 1979. *Cell,* 17, 813.

Schwartz, D. E., R. Tizard, and W. Gilbert. 1983. *Cell,* 32, 853.

Sekikawa, K., and C.-J. Lai. 1983. *Proc. Nat. Acad. Sci. USA,.* 80, 3563.

Shapiro, D., W. E. Brandt, and P. K. Russell. 1972. *Virology,* 50, 906.

Shinnick, T. M., R. A. Lerner, and J. G. Sutcliffe. 1981. *Nature,* 293, 543.

Shioda, T., Y. Hidaka, T. Kanda, H. Shibuta, A. Nomoto, and K. Iwasaki. 1983. *Nucl. Acids Res.,* 11, 7317.

Siddell, S., H. Wege, and V. ter Meulen. 1982. *Curr. Topics Microbiol. Immun.,* 100, 131.

Simons, K. and H. Garoff. 1980. *J. Gen. Virol.,* 50, 1.

Simons, K. and G. Warren. 1983. *Adv. Protein Chemistry,* 36, 79.

Skinner, M. A. and S. G. Siddell. 1983. *Nucl. Acids Res.,* 11, 5045.

Smith J. F. and D. Y. Pifat. 1982. *Virology,* 121, 61.

Söderlund, H. 1973. *Intervirology,* 1, 354.

Söderlund, H., C.-H. von Bonsdorff, and I. Ulmanen. 1979. *J. Gen. Virol.*, 45, 15.

Spear, P. 1984. In *The Herpesviruses*, Vol. 3. B. Roizman, ed. p. 315. Plenum.

Srinivas, R. W., L. R. Melsen, and R. W. Compans. 1982. *J. Virol.*, 42, 1067.

Strauss, E. G. 1978. *J. Virol.*, 28, 466.

Strauss, E. G., Rice, C. M., and Strauss, J. H. 1984. *Virology*, 133, 92.

Strauss, E. G. and J. H. Strauss. 1983. *Curr. Topics Micro. Immun.*, 105, 1.

Strauss, E. G., H. Tsukeda, and B. Simizu. 1983. *J. Gen. Virol.*, 64, 1581.

Strauss, J. H. and E. G. Strauss. 1977. In *The Molecular Biology of Animal Viruses*, D. P. Nayak, ed., p. 111. Dekker.

Sturman, L. S. and K. V. Holmes. 1983. *Adv. Virus Res.*, 28, 35.

Sveda, M. M., L. J. Markoff, and C.-J. Lai. 1982. *Cell*, 30, 649.

Takatsuki, A. and G. Tamura. 1971. *J. Antibiot.*, 24, 232.

Tarkatoff, A. M. 1983. *Cell*, 32, 1026.

Vlazny, D. A., A. Kwong, and N. Frenkel. 1982. *Proc. Nat. Acad. Sci. USA*, 79, 1423.

Wang, E., B. A. Wolf, R. A. Lamb, P. W. Choppin, and A. P. Goldberg. 1976. In *Cell Motility*, R. Goldman, T. Pollard, and J. Rosenbaum, eds., p. 589. Cold Spring Harbor Press.

Wehland, J., M. C. Willingham, M. G. Gallo, and I. Pastan. 1982. *Cell*, 28, 831.

Weiss, R. A. ad P. L. P. Bennett. 1980. *Virology*, 100, 252.

Wengler, G., U. Boege, G. Wengler, H. Bischoff, and K. Wahn. 1982. *Virology*, 118, 401.

Wenske, E. A. and R. J. Courtney. 1983. *J. Virol.*, 46, 297.

Westaway, E. G. 1980. In *The Togaviruses*, R. W. Schlesinger, ed., p. 531. Academic Press.

Westaway, E. G., R. W. Schlesinger, J. M. Dalrymple, and D. W. Trent. 1980. *Intervirology*, 14, 114.

Winter, G. and S. Fields. 1981. *Virology*, 114, 423.

Yoshida, T., Y. Nagai, K. Maeno, M. Iinuma, M. Hamaguchi, T. Matsumoto, S. Nagayoshi, and M. Hoshino. 1979. *Virology*, 92, 139.

Young, P. R. and C. R. Howard. 1983. *J. Gen. Virol.*, 64, 833.

Zavadova, Z., J. Zavada, and R. Weiss. 1977. *J. Gen. Virol.*, 37, 557.

7

Structure and Assembly of the Class I Filamentous Bacteriophage

Robert E. Webster and Javier Lopez
Duke University Medical Center

INTRODUCTION

The filamentous bacteriophages comprise a class of viruses that contain a circular single-stranded DNA genome encased in a long helical sheath of protein subunits. Most of the filamentous phages that have been described infect bacteria that harbor conjugative plasmids. The specificity of a particular phage for its host generally has been found to reside in the type of pilus encoded by the plasmid. Thus, *Escherichia coli* cells that contain the F plasmid, and consequently have F sex pili, can be infected by the filamentous phages f1, fd, M13, and ZJ/2. *E. coli* containing the I or N plasmid and possessing the corresponding pili can be infected by filamentous phages If1 or IKe, respectively (Bradley, 1981). Other bacteria can also be infected by specific filamentous phages. For example, *Pseudomonas aeruginosa* strain K is infected by Pf1 and Pf2 (Minamishima et al., 1968), which can adsorb to the polar pili produced by this host (Bradley, 1973). Phage Pf3 infects *Ps. aeruginosa* strain PAO1 containing the RP1 plasmid (Stanisich, 1974), but it is not clear whether the phage requires a pilus for adsorption. The filamentous phage Xf infects *Xanthomonas oryzae*, a plant bacterium (Kuo et al., 1969).

The physical parameters of many of these phage particles have been studied in an attempt to understand their structure (most recently summarized in Day and Wiseman, 1978; Thomas et al., 1983). Based on their x-ray fiber diffraction pattern, these filamentous phage have been classified into two groups (Marvin et al., 1974a; 1974b): class I includes fd, f1, M13, ZJ/2, If1, and IKe, and class II includes Pf1, Pf3, and Xf. Most of the recent structural studies have been done using Pf1, a class II phage, since its x-ray diffraction pattern contains more information than the pattern for the class I phages. However, the genetics, replication, and morphogenesis of the class I phages f1, fd, and M13 have been studied more extensively than any of the class II phages (reviewed in Denhardt et al., 1978; Kornberg, 1980; 1982). These three class I phages are very closely related, since their nucleotide sequences differ at only a few positions (van Wezenbeek et al., 1980; Beck and Zink, 1981; Hill and Petersen, 1982), and the gene products of one phage can complement corresponding mutants of the other phage. Information about the structure or replication of any one of these three phages has been directly applicable to the others. and therefore they are usually considered to be a single group. Since the biology of f1, fd, and M13 is best known, this review will concentrate on this group of filamentous phages.

Assembly of a phage is the process by which the DNA is packaged into a phage particle. To address this aspect of phage development adequately, it is necessary to review briefly the overall replication cycle of bacteriophage f1. The phage is a rod-shaped particle containing a closed, circular single-stranded DNA molecule containing 6407 nucleotides (bacteriophage fd contains one more nucleotide) (see Figure 1). From the DNA sequence (van Wezenbeek et al., 1980; Beck and Zink, 1981; Hill and Petersen, 1982) the existence of 10 genes is predicted. Genetic (Lyons and Zinder, 1972; Simons et al., 1982) and biochemical studies (Model and Zinder, 1974; Smits et al., 1978; Yen and Webster, 1981) confirm this prediction. In addition to these 10 genes, there is a region of 508 nucleotides called the intergenic region, which codes for no protein but contains the origin of replication for synthesis of the viral and complementary strands of the phage DNA (Denhardt et al., 1978). Infection is initiated by specific binding of the phage to the F sex pilus. Binding is an end-to-end interaction between one specific end of the bacteriophage and the tip of the sex pilus (Tzagoloff and Pratt, 1964; Caro and Schnös, 1966). It has been postulated that the attached phage is brought to the cell surface by retraction of the pilus (Marvin and Hohn, 1969; Jacobson, 1972; Burke et al., 1979). Upon interaction with the membrane, the coat proteins depolymerize and become integrated into the host membrane (Trenkner et al., 1967; Smilowitz, 1974; Griffith et al., 1981) while the DNA enters the cytoplasm.

Replication of f1 DNA requires a combination of host- and phage-encoded proteins. The incoming viral single strand is converted to a double-stranded, covalently closed circle called the parental replicative form. This conversion is carried out by pre-existing host enzymes

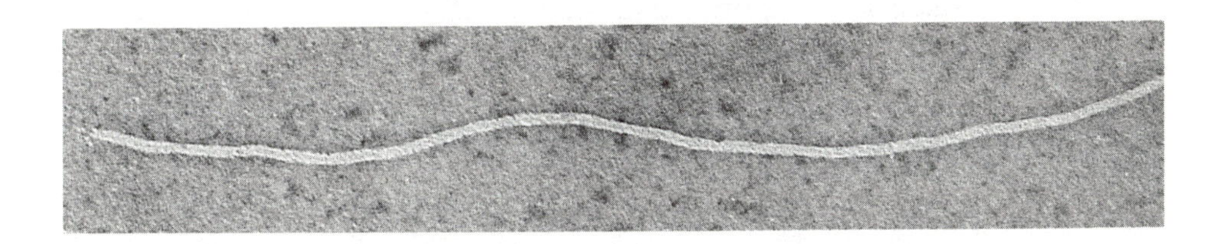

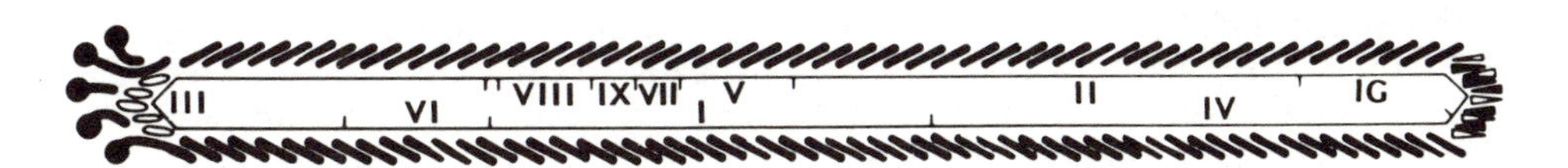

FIGURE 1 The f1 bacteriophage. (Top) Electron micrograph of a negatively-stained f1 bacteriophage. The pIII or A protein is located at the left end of the phage. (Bottom) Schematic representation of the structure of the f1 bacteriophage. The pIII or A protein is the knob-and-stem structure at the left end of the f1 phage (Gray et al., 1981.) The roman numerals mark the positions of the genes in the phage particle. Nucleotide 5537 has been placed at the end of the phage that contains pVII and pIX, a part collectively referred to as the C protein. The figure shows only the relative locations of molecules and does not imply any interactions between proteins or between proteins and DNA. (Figure taken from Webster et al., (1981).)

and does not depend on any phage-specified proteins (Denhardt et al., 1978; Kornberg, 1980; 1982). Parental replicative-form DNA is replicated to form progeny double-stranded DNA molecules which are in turn replicated similarly. The synthesis of the progeny replicative-form molecules is initiated at a specific site located in the intergenic region and requires the action of the phage gene-*II* protein (pII), which nicks the viral strand of the double-stranded replicative-form DNA (Meyer and Geider, 1979a; 1979b). The resulting DNA is replicated semiconservatively by host enzymes, probably by a rolling circle mechanism. The free 3' OH end generated by the pII nick serves as the primer. Replication proceeds around the molecule, displacing the pre-existing viral strand. When the replication fork reaches the origin region, synthesis terminates (Horiuchi, 1980; Dotto and Horiuchi, 1981), generating a double-stranded replicative form and a single-stranded circular viral strand. This viral strand is thought to be converted to a double-stranded replicative form by the host enzymes (Horiuchi et al., 1978; Kornberg, 1982). The gene-*II* protein appears responsible for both the termination event and the circularization of the single-stranded viral DNA (Harth et al., 1981; Geider et al., 1982). As the concentration of the phage gene-*V* protein (a single-stranded DNA-binding protein referred to as pV) increases, synthesis of the complementary strand is inhibited (Mazur and Model, 1973). The result is a shift to viral strand production and its accumulation in the form of a complex termed the pV/f1 ssDNA complex, which is composed of the circular viral DNA and approximately 1300 molecules of the pV (Alberts et al., 1972; Oey and Knippers, 1972; Webster and Cashman, 1973; Pratt et al., 1974; Paradiso and Konigsberg, 1982). The products of phage genes *II* and *V* have been shown to be necessary for the replication of DNA (Pratt et al., 1966; Pratt and Erdahl, 1968). The gene-*X* protein (pX) also appears to be required for DNA synthesis, but its specific function is not known (Fulford and Model, 1984). This protein is produced as a result of an in-phase internal translational restart in gene *II* (Yen and Webster, 1981; Yen and Webster, 1982). Thus, pX has the same amino acid sequence as the carboxy-terminal portion of pII.

Assembly of progeny phage occurs while the DNA contained in the complex is extruded through the cell envelope, picking up the phage capsid proteins and shedding the pV-protein molecules, which are reutilized (Webster and Cashman, 1973; Pratt et al., 1974). Presumably, the pV/f1 ssDNA complex interacts with an "assembly site" on the bacterial membrane, where proteins involved in phage assembly are located. During this assembly and extrusion of the phage particle from the infected bacterium, the host cell does not lyse or stop dividing (Marvin and Hohn, 1969). Rather, there is persistent infection, and phage particles are continuously released at a rate of about 200 per bacterial division (Brown and Dowell, 1968). The phage DNA is inherited by all progeny cells, but it is not clear whether there is spontaneous curing after a number of generations (Merriam, 1977; Lerner and Model, 1981). In any case, at least for several generations, the intracellular phage DNA behaves as a plasmid.

The length of the virion is determined by the size of the DNA. Miniphage containing large deletions of the phage genome are packaged into virions that are shorter than normal (Griffith and Kornberg, 1974). Insertions of DNA into the phage genome result in the production of phage of greater than unit length. The ability of the phage-assembly system to package any length of DNA, together with the ability of the intracellular phage DNA to behave as a plasmid, has made the filamentous phage a valuable vehicle for cloning DNA (for reviews, see Barnes, 1980; Zinder and Boeke, 1982). Foreign DNA can be inserted into the intergenic portion of the replicative form DNA, and after transformation of the host it will lead to the production of chimeric phage particles.

Assembly of the f1 bacteriophage is obviously a very complex reaction. The substrate for this reaction is the pV/f1 ssDNA complex, and the product is the phage particle itself. The assembly site is an integral part of the bacterial cell envelope and involves in some way the products of seven phage genes (*I, III, IV, VI, VII, VIII, IX*), together with bacterial proteins and other membrane components. A prerequisite for determining the mechanism by which viral DNA leaves the pV/f1 ssDNA complex and is assembled into the phage particle is an understanding of the structure of the pV/f1 DNA complex, the structure of the f1 bacteriophage, and the interactions of the phage proteins with the membrane components in the assembly site. Therefore this review will begin by presenting the current view of the structure of both the phage particle and the pV/f1 ssDNA complex. Then, the interaction of phage and bacterial proteins with the membrane will be described, followed by a description of the assembly process (based on somewhat limited knowledge). For convenience, the assembly process will be analyzed as a sequence of three parts: the nucleation step, the extrusion or elongation step, and the termination step.

STRUCTURE OF THE BACTERIOPHAGE

The f1 bacteriophage is a long filamentous particle composed of a circular, single-stranded DNA molecule encased in a flexible tubular protein sheath approximately 890 nm long and 6-7 nm wide (Denhardt et al., 1978) (Figure 1). The sheath is made up of five separate proteins, each encoded by the phage. Along the length of the particle are approximately 2750 molecules of the B protein; this molecule, denoted pVIII (Day and Wiseman, 1978), contains 50 amino acids and is the product of gene *VIII*. One end of the particle contains about five molecules of the 406-amino-acid A protein (pIII), the product of gene *III* (Goldsmith and Konigsberg, 1977; Woolford et al., 1977) and five molecules of the 111-amino-acid D protein (pVI), the product of gene *VI* (Lin et al., 1980; Grant et al., 1981a, Figure 1). Each gene-*III* protein appears as a 50 A knob on the end of a stem 70 A long (Gray et al., 1981; Figures 1 and 2a). The knob consists of the amino-terminal portion of pIII and is involved in the attachment of the phage particle to the tip of the sex pilus, which is the first step in the process of infection (Rossomando, 1970; Gray et al., 1981; Grant et al., 1981a; Nelson et al., 1981: Armstrong et al., 1981). The stem portion presumably interacts with pVI and pVIII to produce a stable end-structure to the phage particle. No distinguishing features can be observed at the other end of the bacteriophage in electron

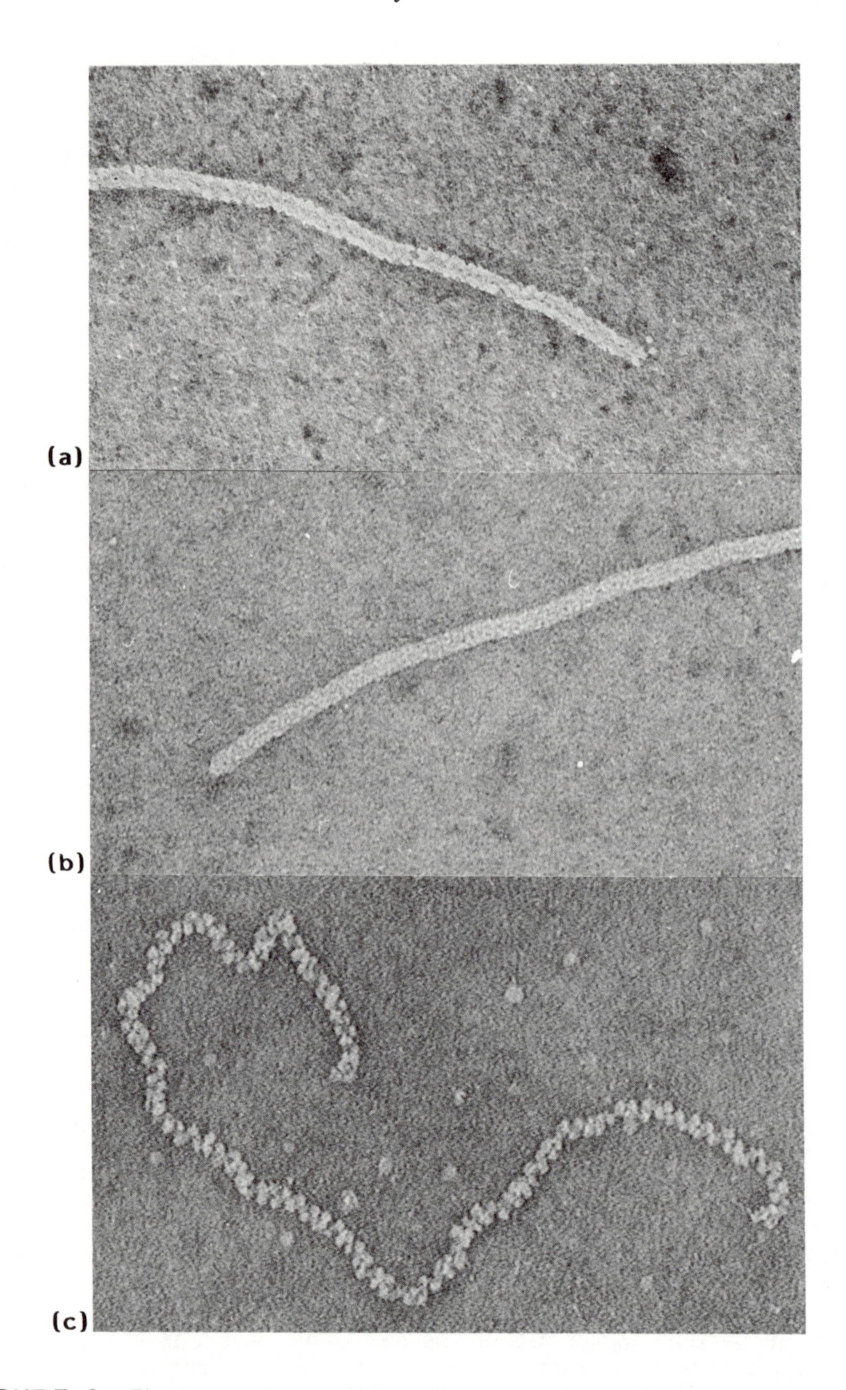

FIGURE 2 Electron micrographs of negatively stained f1 phage and the gV/f1-ssDNA complex. (A) The end of the phage containing pIII and pVI. (B) The end of the phage containing pVII and pIX. (C) The gV/f1-ssDNA complex.

micrographs of negatively-stained virions (Gray et al., 1981, Figure 2b). However, this end of the particle contains approximately 10 molecules of the C protein, a mixture of the 33-amino-acid pVII and the 32-amino-acid pIX proteins, the products of genes *VII* and *IX*, respectively (Grant et al., 1981a; Simons et al., 1981). Presumably, the pVII and pIX proteins must interact with the pVIII to form this end of the phage particle.

The DNA is uniquely oriented within the phage particle, so the intergenic region is near the pVII-pIX end of the virion. The sequence coding for gene *III* is located at the pIII-pVI end (see Figure 1 bottom). Analyses of chimeric phage containing pBR322 DNA inserted into different positions within the intergenic region (Webster et al., 1981) and determination of the DNA content of various defective interfering particles (deletion mutants containing the sequence around the intergenic region, Enea et al., 1977), have indicated that a 375-nucleotide sequence near the junction of gene *IV* and the intergenic region resides at the pVII-pIX end of all phage particles. Contained in these 375 bases is a 78-nucleotide sequence of palindromic symmetry (Schaller, 1978; Beck and Zink, 1981; Hill and Petersen, 1982), which has been shown to form a stable hairpin region, both in solution (Huang and Hearst, 1981) and in the phage particle (Ikoku and Hearst, 1981). Experiments in which the phage are disassembled with organic solvents have led to the proposal that this hairpin region interacts with pVII and/or pIX (Griffith et al., 1981; Lopez and Webster 1982). This interaction is probably electrostatic, since at high salt concentrations this region of the DNA is released from the partially disassembled virion. This hairpin region also is necessary for efficient packaging of DNA into phage particles, as discussed below (Dotto et al., 1981). The gene-*III* and *VI* proteins at the other end of the phage particle do not appear to recognize any specific sequence of the phage genome, since different sequences reside there depending on the length of the viral DNA (Grant et al., 1981b).

Most of the information about the interaction between the pVIII molecules in the phage particle has come from x-ray diffraction analysis of semicrystalline fibers (described in Marvin and Nave, 1982). The majority of these studies have been done on the class II phage Pf1, since its diffraction pattern contains more information. However, the information obtained from the fiber diffraction of fd, together with accurate measurement of other physical parameters such as mass per unit length (Newman et al., 1977; Thomas et al., 1983), have indicated how the pVIII molecules are arranged in class I filamentous phages (Marvin et al., 1974b; Newman et al., 1977; Banner et al., 1981; Makowski and Caspar, 1981; Makowski, 1983). The proteins form a sheath around the DNA with inner and outer diameters of ca. 20 A and 60 A, respectively. The helical domains of the pVIII molecule are aligned within 10 degrees of the long axis of the particle. Recent studies with solid state NMR (Cross et al., 1983), linear dichroism (Fritzsche et al., 1981), and magnetic birefringence (Torbet and Maret, 1981) tend to support the idea that the proteins are aligned approximately parallel to the filament axis of the phage particle. The pVIII molecules overlap each other, with the carboxy-terminal region located on the inner part of the sheath near the DNA and the amino-terminal region exposed on the outer part of the sheath. The symmetry of the coat protein in the virus appears to consist of a fivefold rotational axis combined with a twofold screw axis of pitch 32 A.

A working model of the phage particle that is consistent with these data is shown in Figure 3 (Marzec and Day, 1983; Day and Marzec, personal communication). In accord with the fivefold rotational symmetry, five pVIII subunits are arranged around the axis at the same axial position, though they do not appear to interact with each other. If the phage in Figure 3 is rotated clockwise by 72 degrees, the pVIII helix on the left will occupy the exact position of the one pictured on the right. Similarly, rotation of the phage by 36 degrees followed by an axial translation of 16 A would bring two different pVIII helices into the position of those shown. The overlapping of the pVIII subunits results in the two layers of helices shown in the phage-protein sheath (Figure 3). Since the amino-terminal portion of pVIII, which forms the outer layer, contains more amino acids than the carboxy terminus which forms the inner layer, the cross section of the phage model shown has 15 helices in the outer layer and 10 in the inner layer. It should be pointed out that this is presented as a working model and is not the only one that may be consistent with the data. For example, the structure and angle of the linkage between inner and outer segments of the pVIII is not clear. The angle could be so large that the amino terminal portion may be folded back over the carboxy-terminal region. On the other hand, the angle of the linkage could be so small that each subunit would appear as a gradually curved α-helical segment as suggested by solid-state NMR studies (Cross et al., 1983). This arrangement would be consistent with a proposed model of the phage in which these gently curved pVIII molecules overlap each other like the scales on a fish (Banner et al., 1981; Makowski, 1983).

Much less is understood about the structural relation between the DNA and the protein sheath in class I phages. The antiparallel DNA strands appear to be packaged as a right-handed helix with the phosphodiester backbone on the outside and the bases stacked to some degree on the inside (Casadevall and Day, 1983). Dividing the length of the phage by one half the number of nucleotides in the DNA implies an axial separation between the nucleotides in each of the two strands of approximately 2.7 A. The diffraction patterns of magnetically aligned fd show faint layer lines of a 26.7 A periodicity, which are attributed to the DNA (Banner et al., 1982). These data suggest that the DNA has a pitch of about 27 A with approximately 10 bases per turn of the antiparallel chains. There are 2.3 nucleotides for each pVIII subunit, a non-integral number (Newman et al., 1977); thus, in the phage the symmetry of the DNA appears to be different from that of the protein. NMR spectroscopy indicates that the phosphodiester backbone may not be ordered in any simple way (Cross et al., 1983). In addition, the DNA is not base-paired, and its backbone conformation is not like that of A or B DNA (Thomas et al., 1983). Based on these data, it is difficult to envision any regular pattern of DNA/protein interaction in the particle. However, the data are consistent with the same protein-DNA environment occurring every 12 nucleotides (every 32 A), even though there are 10 nucleotides per

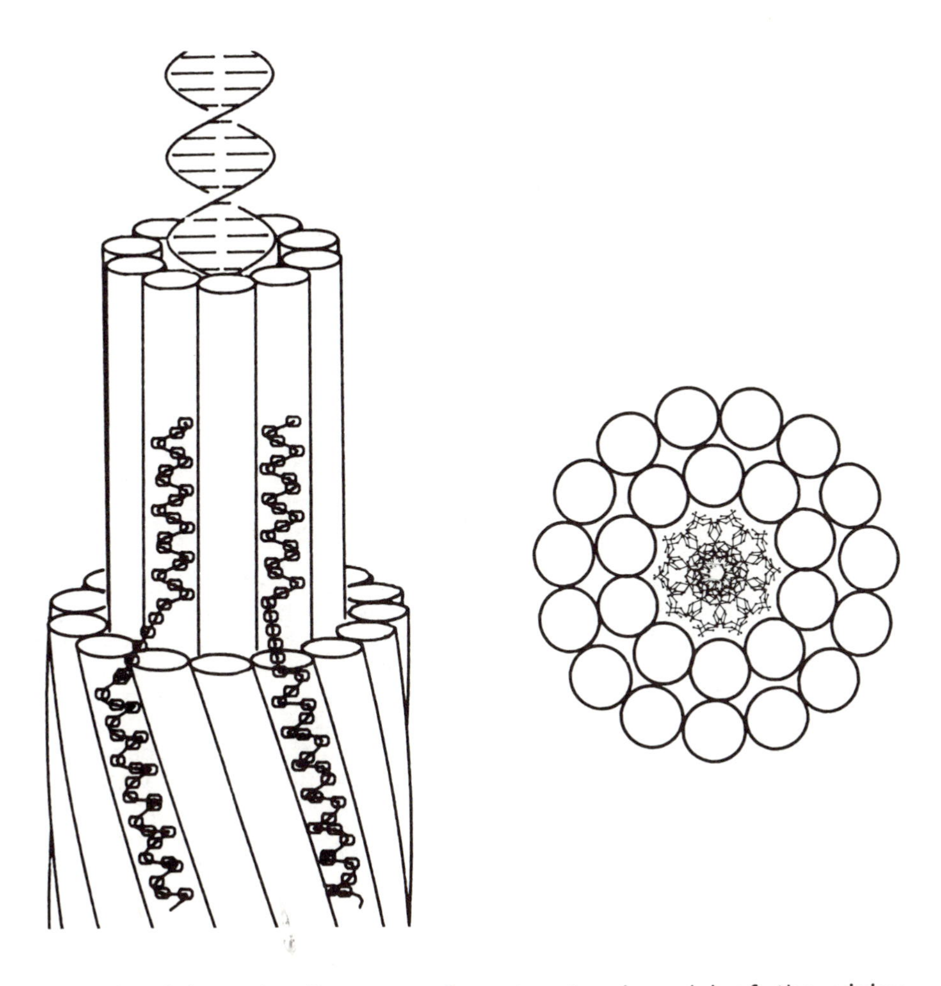

FIGURE 3 Schematic diagrams of a structural model of the virion, as proposed by Marzec and Day (1983; 1984, in preparation). The protein sheath is described in terms of pentamers of identical α-helical subunits, with two translations of 16 Å and rotations of 36° of the pentamers generating an approximate repeat every 32 Å. Two individual subunits of a pentamer are shown, and it is seen how separate α-helical domains of one subunit can contribute to inner and outer layers around the DNA. There are more amino acids in the outer layer than the inner layer. Neighboring pentamers interdigitate in such a way that there can be 10 α-helical tubes in the inner layer and 15 in the outer layer, and close-pack to give the cable-like appearance. How a DNA with ten-fold screw symmetry can fit within such a protein sheath is shown in the end-on view. The pitch connection relationships (Marzec and Day, 1983), coupled with the pentamer translation of 16 Å and a nonintegral ratio of nucleotide to subunit of 2.4, predict a DNA pitch of 26.67 Å and provide an explanation for the 26.7 Å reflection observed by Banner et al. (1981). (Figure courtesy of Loran Day.)

turn of one DNA strand (Banner et al., 1981; Marzec and Day, 1983). A complete understanding of the structural relation of the DNA and protein in the phage particle will be necessary to understand the mechanism of assembly.

STRUCTURE OF THE GENE-V PROTEIN/DNA COMPLEX

The final stage of filamentous phage DNA replication is the synthesis of the single-stranded viral genome. At any one time, there are up to 200 molecules of this covalently closed ssDNA in the cytoplasm of infected bacteria. All of these molecules are found complexed with the phage pV for class I phages (Webster and Cashman, 1973; Pratt et al., 1974). Because the circular DNA in this pV/f1 ssDNA complex is compacted into a linear antiparallel structure whose length is similar to that of the phage particle, the pV/f1 ssDNA complex is presumed to be a true intermediate in phage morphogenesis. A similar complex has been found in bacteria infected with class II phages, but its length is shorter than that of the mature phage particle (Gray et al., 1982b; Kneale and Marvin, 1982; Kneale et al., 1982).

Electron microscope analysis shows the class I pV/f1 ssDNA complex to be a flexible rod about 880 nm long with an apparent diameter of 10 nm (Alberts et al., 1972; Pratt et al., 1974; Gray et al., 1982b, see Figure 2c). Thus, the complex has the same length but is somewhat wider than the phage particle. Morphologically, the complex is a rather open helix with about 97 total helical turns (Gray et al., 1982b). The handedness of the helix for the complex of the class I phages is not known, but the Pf1 complex has been determined to be a right-handed helix (Gray et al., 1982b). Radiochemical tracer (Pratt et al., 1974) and spectral analyses (Pretorius et al., 1975) indicate a stoichiometry of between 4.7 and 5 nucleotides per pV subunit in the complex. Neutron low-angle scattering studies plus electron microscopy have yielded a value for the mass per unit length of 1610 daltons per angstrom, which supports a stoichiometry of five nucleotides per pV molecule in the complex (Torbet et al., 1981). These data indicate that there are 11-13 molecules of pV per helical turn or approximately 1300 molecules of pV in the total complex.

Careful biochemical studies have shown that no phage-specific proteins other than pV are present in the complex isolated from infected bacteria (Paradiso and Konigsberg, 1982; Grant and Webster, 1984a). However, approximately 1-3 molecules of *E. coli* single-stranded-DNA binding protein (SSB) co-purify with the complex and appear to be stably associated with it (Grant and Webster, 1984a). It is not clear whether the SSB associated with the complex is a fortuitous result of its previous involvement in phage DNA synthesis or whether it plays a specific role in complex formation and/or morphogenesis.

The gene-*V* protein has a molecular weight of 9690 and contains 87 amino acids, whose sequence is known (Nakashima et al., 1974). Its structure has been determined by x-ray diffraction analysis to a resolution of 2.3 A (McPherson et al., 1979b; Brayer and McPherson, 1984a,b; see Figure 4). It consists entirely of β structure and short regions of extended chain, with no α helix. The hydrodynamic properties of purified pV indicate that it exists primarily as a dimer in solution (Rasched and Pohl, 1974; Pretorius et al., 1975), a conclusion that is supported by chemical crosslinking experiments with the free protein (Cavalieri et al., 1976). The crystal structure of the free protein reveals a tightly interlocking dimer in which the subunits are related by an intermolecular dyad axis of symmetry (Figure 4) (McPherson et al., 1979b; Brayer and McPherson, 1984a). Formation of this dimer involves the interlocking of extended β loops from each monomer to generate a six-stranded antiparallel β barrel which is stabilized predominantly by hydrophobic interactions.

Purified pV binds strongly and cooperatively to single-stranded DNA (Alberts et al., 1972; Oey and Knippers, 1972; Dunker and Anderson, 1975; Cavalieri et al., 1976) in a sequence-independent manner, though differences in binding affinity have been observed with oligonucleotides of different base compositions. The DNA binds in two shallow, unconstricted channels that lie approximately 30 A apart on either side of the molecular dyad (Brayer and McPherson, 1984a,b; Figure 4). The binding depends on intercalation of the DNA bases with aromatic residues and electrostatic interactions between the phosphate backbone and basic residues in the protein. Chemical-modification studies (Anderson et al., 1975), UV absorbance difference spectra (Day, 1973), and a variety of NMR studies (Anderson et al., 1975; Coleman et al., 1976; Coleman and Armitage, 1978; 1983) show that three tyrosines and one phenylalanine in each pV monomer are involved in stacking interactions between the phosphate backbone of the DNA and specific basic residues in the binding channel (Anderson et al., 1975; Coleman et al., 1976; Alma et al., 1981). All of these studies, together with the refined crystal structure of pV, have shown that this groove can accommodate up to five deoxynucleotides (Brayer and McPherson, 1984a, 1984b). Beginning at the 5' end of the bound pentanucleotide, the sequence of aromatic rings in the binding channel would be: Base 1, Tyr-26, base 2, Phe-73 (from the symmetry-related monomer), base 3, Tyr-34, base 4, Tyr-41, and base 5. The guanidinium group of Arg-80 would be between phosphates 1 and 2, that of Arg-21 would be near phosphate-3, and that of Arg-16 would be between phosphates 3 and 4. Lys-46 would be between phosphates 4 and 5.

The two DNA-binding channels in the pV dimer run identical but antiparallel courses across the face of the molecule (Figure 4). This allows the dimer to bind opposing, antiparallel regions of the DNA and condense them into the ordered structure seen in the electron microscope (Figure 2c). Additional protein-protein interactions, such as those involved in the cooperative binding of pV to DNA, presumably contribute to the final helical structure of the complex.

The position of the DNA and protein relative to the overall structure of the pV/ssDNA complex is not clear. Based on the physical and morphological data about the complex (Torbet et al., 1981; Gray et al., 1982a; 1982b; Kneale and Marvin, 1982; Kneale et al., 1982) and the arrangement of dimers in crystallized complexes with small oligonucleotides (McPherson et al., 1979a; McPherson et al., 1980a,b), the most likely structure is one in which the antiparallel, noncomplementary DNA strands bound to the pV dimers are twisted into a double helix that corresponds to the morphological helix of the complex seen in the electron microscope (see Figure 2c and 5). The

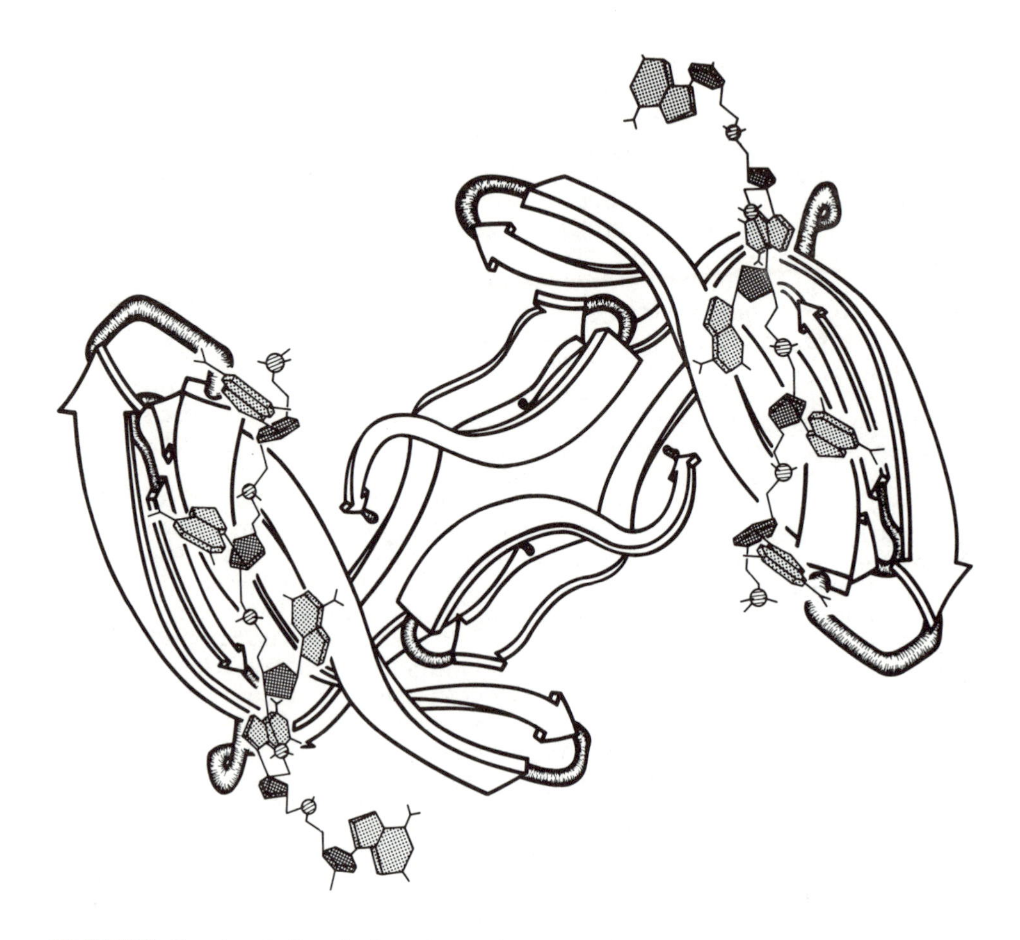

FIGURE 4 A stylized drawing of the gene-*V* protein dimer and the course of bound single-stranded DNA. Features of the bound chains are coded as follows: striped overlay, phosphate groups; cross-hatched overlay, sugar moieties; and dotted overlay, base rings. A total of ten nucleotides are illustrated, five in each of the two bound DNA strands. (Figure courtesy of Alexander McPherson.)

question remains whether the DNA strands are on the outer or inner surface of the complex. Analysis of crystallized complexes of pV with short oligonucleotides led to the suggestion that the DNA-binding sites were on the outside of the pV/f1 ssDNA helical complex (McPherson et al., 1979a; 1980a,b). However, the calculated maximum radius of the DNA backbone in neutron low-angle scattering studies (Day and Wiseman, 1978; Torbet et al., 1981), indicates that the DNA is not on the outer periphery of the helical complex but instead occupies the central region, where it is presumably bound to the inner surface of the pV dimers. In these studies, the DNA is found to have a maximum cross-sectional radius of gyration of 17.6 ± 0.3 A, while that of the protein is about 33.5 ± 0.4 A (Gray et al., 1982a). These results are also supported by calculations of the maximum possible outer radius of the DNA strands in the complex based on geometrical relationships and on the assumptions that both of the coalesced DNA strands are at the same radius and have fully extended sugar-phosphate backbones. Therefore, it appears that the model inferred from crystals of pV complexed with short oligonucleotides is not entirely applicable to the structure of the complex in solution that is formed with long, circular DNA molecules.

Even with the DNA in the interior of the complex, the DNA structure is different from that in the phage and must require several adjustments during maturation of the virion. The radius of the DNA must decrease when it is transferred from the complex to the virion, where it must occupy a core region with a diameter no greater than 20 A. For f1 this implies a 26-degree increase in unit twist, or an increase of 200 turns in the overall helical structure of the DNA relative to that found in the complex (Kneale and Marvin, 1982). The bases of f1 DNA must also change from an array in which they are unstacked and intercalated with aromatic side chains of pV to one in which they are somewhat stacked within the virion (Casadevall and Day, 1983). On the other hand, f1 DNA probably undergoes no change in the relative orientation of the bases and backbone, since in both the virion and the complex the bases are towards the center and the sugar-phosphate backbone is towards the periphery of the core region. This hypothesis follows from the location of the DNA in the interior of the complex (see Figure 5) (Torbet et al., 1981; Gray et al., 1982a) and the location of the DNA backbone near the bottom of the binding channel in each pV molecule (Brayer and McPherson, 1984a,b).

MEMBRANE-ASSOCIATED BACTERIOPHAGE PROTEINS

Mature phage particles have never been detected within infected bacteria. In fact, the only phage-related particle present in the cytoplasm during the replication of the phage is the pV/f1 ssDNA complex. Therefore, it is reasonable to assume that assembly takes place at the membrane itself, presumably at specific assembly sites. Recent evidence from our laboratory suggests that these assembly sites for f1 bacteriophage resemble the adhesion zones (Bayer, 1976) in which the inner and outer membrane meet (Lopez, 1984). Only the products of three phage genes (*II*, *V*, *X*) are necessary for the synthesis of the pV/f1 ssDNA complex (Pratt et al., 1966; Pratt and Erdahl 1968; Fulford and Model, 1984). Five phage genes (*III*, *VI*, *VII*, *VIII*, *IX*) code for the proteins found in the mature phage particle. Infection of bacteria with phage containing mutations in genes *I* and *IV* produce no particles and thus their products are also necessary for assembly of the phage. Since assembly takes place at the membrane, it is logical to expect that the five capsid proteins together with gene-*I* (pI) and gene-*IV* (pIV) proteins would be associated with the membrane at some time following their synthesis and before assembly of proteins into phage. Consistent with this hypothesis is the observation that many of these phage proteins appear to reside in the membrane until assembled into the phage particle.

The phage protein that has been studied most extensively in this respect is pVIII, the major coat protein of the virion. In the f1-infected *E. coli*, large amounts of pVIII are synthesized as precursor molecules containing an additional 23 amino acids at the amino-terminal end (Sugimoto et al., 1977; Chang et al., 1979). This additional sequence of amino acids is analogous to the amino-terminal extension or "signal peptide" found on many exported proteins of *E. coli* (Emr et al., 1980). After insertion of the precursor protein into the cytoplasmic or inner membrane, the amino-terminal portion of the protein is removed by a specific host-encoded protease to yield the mature 50-amino acid pVIII (Chang et al., 1978; Zwizinski and Wickner, 1980). The mature protein then remains as an integral membrane protein until it is assembled into a phage particle (Smilowitz et al., 1972; Webster and Cashman, 1973).

Following synthesis, insertion, and processing, pVIII spans the cytoplasmic membrane asymmetrically with the amino-terminal portion exposed on the outside (Wickner, 1975; Date et al., 1980) and the carboxy terminus on the cytoplasmic surface (Ohkawa and Webster, 1981). Various chemical (Wickner, 1976; Chamberlain et al., 1978) and physical (Hagen et al., 1979; Cross and Opella, 1981; Dettman et al., 1982) studies of pVIII in detergent micelles and phospholipid vesicles have shown that most or all of the hydrophobic residues at positions 21-39 interact with the hydrocarbon bilayer of the lipid membrane (see sequence in Figure 6). The region containing residues 40-50 interacts with the hydrophilic milieu of the cytoplasm and residues 1-20 are exposed to the outside of the cytoplasmic membrane. From data obtained by circular dichroism and Raman spectroscopy (Fodor et al., 1981) the suggestion has been made that pVIII can exist in two different conformations in the membrane: one is composed mostly of β sheet, the other has slightly less than 50% α helix, together with a mixture of β turn and random coil. Other research groups have suggested that only the latter form, which contains 50% α helix, is thermodynamically stable in the membrane (Nozaki et al., 1978). However, it is generally agreed that there is a large difference between the conformation of the membrane-bound and phage-associated

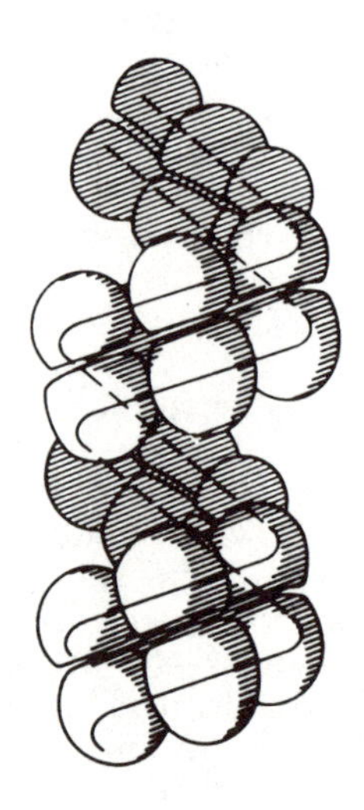

FIGURE 5 A stylized drawing of one helical turn of the gene-*V* protein/f1 ssDNA complex. Each gene-*V* protein monomer is represented by a sphere. The lines represent the paths of the antiparallel DNA strands. This drawing is based on the authors' interpretation of existing data.

```
Gene    Amino acid sequence

          -   - -     +       -               -
VIII    A E G D D P A K A A F D S L Q A S A T E Y I G Y A
                                    +     + +       +
        W A M V V V I V G A T I G I K L F K K F T S K A S

                    -       -   - +         +
III     (336 aa)--Y E F S I D C D K I N L F R G V F A F L
                                          +   +   -
        L Y V A T F M Y V F S T F A N I L R N K E S

                              +
VI      M P V L L G I P L L L R F L G F L L V T L F G Y L
                + +       +
        L T F L K K G F G K I A I A I S L F L A L I I G L
                          -
        N S I L V G Y L S D I S A Q L P S D F V Q G V Q L
                                        +             -
        I L P S N A L P C F Y V I L S V K A A I F I F D V
        +   +           -   - +
        K Q K I V S Y L D W D K

          +       -   -
VII     M E Q V A D F D T I Y Q A M I Q I S V V L C F A L
                      +
        G I I A G G Q R

                                          +
IX      M S V L V Y S F A S F V L G W C L R S G I T Y F T
        +     +
        R L M E T S S
```

FIGURE 6 Amino acid sequences of f1-specific proteins. Charged residues are indicated. Underlined portions represent hydrophobic regions that are or might be inserted into the membrane (see text). The amino acid sequence of the proteins is based on the DNA sequence of the f1 phage (Beck and Zink, 1981; Hill and Petersen, 1982).

pVIII that must be accounted for during the process of assembly. This change in conformational state of pVIII must be easily reversible *in vivo*, since pVIII of an infecting f1 bacteriophage can enter the membrane and be incorporated again into newly synthesized phage particles (Trenkner et al., 1967; Smilowitz, 1974; Armstrong et al., 1983).

The presence of substantial amounts of pVIII in a bacterium unable to produce phage has unusual effects on the membrane. When nonsuppressing bacteria are infected with phage carrying amber mutations in any of a number of genes necessary for morphogenesis, the bacteria stop dividing (Hohn et al., 1971) and accumulate what appears to be excess inner membrane (Schwartz and Zinder, 1968). This phenomenon has been shown to require the accumulation of pVIII in the membrane (Woolford, 1976). The accumulation of pVIII under these conditions appears to affect phospholipid metabolism as well, decreasing the turnover of cardiolipin and the synthesis of phosphatidylethanolamine (Woolford et al., 1974; Chamberlain and Webster, 1976). Based on these and other changes induced by pVIII accumulation it is possible that the excess pVIII shows some type of preferential interaction with negatively charged phospholipids (Chamberlain and Webster, 1976; Pluschke et al., 1978). Phospholipid metabolism appears unaffected late in infection in a bacteria infected with wildtype phage. However, there is a transient alteration of phospholipid metabolism in these bacteria 10-20 minutes after infection (Woolford et al., 1974). During this time period, the pVIII is rapidly being synthesized but is not yet being packaged into mature phage and being released from the bacteria.

The gene-*III* protein, which is part of the adsorption complex of the phage (Figure 1), is also associated with the bacterial membrane. Comparison of its amino acid sequence, as derived from the DNA sequence (Schaller et al., 1978), with the amino-terminal sequence of the mature protein (Goldsmith and Konigsberg, 1977) showed that pIII is synthesized with an 18-amino acid signal peptide at its amino-terminal end. After insertion of the protein into the cytoplasmic membrane the signal peptide is removed and pIII remains as an integral membrane protein until assembled into a phage particle (Webster and Rementer, 1980; Boeke et al., 1982). Recombinant DNA techniques were used to determine the orientation of this protein in the membrane. It was found that a truncated form of the pIII, encoded by a recombinant plasmid and lacking the carboxy terminus, does not remain in the membrane but appears to go through it (Boeke and Model, 1982). Similar investigation using a number of recombinant plasmids containing progressively larger amino-terminal regions of pIII showed that the 23-amino acid hydrophobic region at the carboxy-terminal end of pIII (Figure 6) is necessary for the association of this protein with the membrane (Davis et al., 1985). Thus, pIII appears to span the membrane asymmetrically with the amino-terminal 378 amino acids on the outside and the carboxy-terminal five amino acids on the cytoplasmic side of the inner membrane (Armstrong et al., 1981; Boeke and Model, 1982). As with pVIII, the presence of pIII in bacteria unable to produce phage results in a number of unusual membrane effects. Bacteria containing chimeric plasmids expressing at least the amino-terminal 98 amino acids of pIII exhibit pleiotropic effects on the cell membranes, which include deoxycholate sensitivity, enhanced tolerance to colicins, and increased leakage of β-lactamase, a periplasmic enzyme, into the medium (Boeke et al., 1982).

The product of gene *IV* (pIV) also appears to be tightly associated with the inner membrane. When cytoplasmic membranes are prepared by different techniques from f1-infected bacteria, and the proteins are analyzed by SDS gel-electrophoresis, a band is observed that migrates with the appropriate molecular weight for pIV (Webster and Cashman, 1973; Lin and Pratt, 1974; Webster and Rementer, 1980).

This band is not observed in uninfected bacteria or in bacteria infected with gene-*IV* amber-mutant phage. Examination of the DNA sequence of gene *IV* predicts a protein of 426 amino acids of which the amino-terminal 20 amino acids resemble a signal peptide (van Wezenbeek et al., 1980; Emr et al., 1980; Beck and Zink, 1981; Hill and Petersen, 1982). Consistent with this prediction is the observation that pIV synthesized *in vitro* has a slower mobility on SDS polyacrylamide gels than the pIV from f1-infected bacteria (Russel and Model, personal communication). There are also stretches of hydrophobic amino acids (the longest being 21 amino acids long, between amino acid 315 and 336) that could be inserted into the membrane bilayer. It has not been possible to isolate enough pure pIV to determine more precisely the nature of the interaction of this protein with the membrane.

The other minor proteins of the phage, the products of genes *VI*, *VII*, and *IX* have been relegated to the membrane almost by default, since they have not been detected in the cytoplasm or with the pV/f1 ssDNA complex (Smits et al., 1978; Grant and Webster, 1984a). The amino acid sequences of these proteins, as derived from the DNA sequence show that they have long stretches of hydrophobic amino acids that might interact with the lipid bilayer (Figure 6). The primary translation products for these three proteins, possibly even including the initiating formylmethionine (fMet) are packaged into virions (Simons et al., 1981). It has been suggested that the fMet remains with the proteins, because the proteins themselves act as signal peptides; thus, the amino-terminal portions are inserted into the membrane before removal of the fMet can occur. Finally, since the pVII-pIX-containing end of the phage emerges first from the infected bacteria (Lopez and Webster, 1983; see below), it can be argued that these proteins are involved in some nucleating event that starts the assembly process. Thus, since pVII and pIX are not associated with the pV/f1 ssDNA complex, they are probably at the membrane assembly site.

The product of gene *I* (pI) has been difficult to study, since it has not been detected in wildtype- or mutant-infected bacteria or minicells, presumably owing to the extremely low amounts of gene-*I* mRNA produced (Smits et al., 1984). This difficulty has recently been overcome by cloning gene *I* into a multicopy plasmid under the control of the left promoter of phage λ (Horabin and Webster, unpublished results). Moderate expression of gene *I* from this plasmid leads to the synthesis of detectable levels of pI, and this protein is found tightly associated with the bacterial envelope. Increasing amounts of pI in the bacteria appears to result in a very rapid depletion of energy, as evidenced by complete shutdown of protein and RNA synthesis. This observation suggests that pI may be capable of acting as a membrane-associated ionophore or porin in the absence of other phage proteins. Electron micrographs of bacteria containing different amounts of pI also indicate that pI is the one phage protein that must be present to form the assembly site, which resembles an adhesion zone (Lopez, 1984; Lopez and Webster, unpublished results).

Recent genetic evidence suggests that in order to promote phage morphogenesis pI must interact with thioredoxin, the product of the *trxA* gene (Russel and Model, 1983; 1984; 1985; Lim et al., 1985). This observation, together with the fact that f1 grows poorly in cells carrying a *trxB* (thioredoxin reductase) mutation suggests that reduced thioredoxin is necessary during some stage of assembly (Russel and Model, 1985).

ASSEMBLY OF THE BACTERIOPHAGE

Assembly of the bacteriophage can be broken down into three steps: nucleation, elongation, and termination. These steps will be discussed separately in this section.

Nucleation

The process of assembly of the f1 bacteriophage is essentially the extrusion of the phage DNA through the membrane as the pV from the pV/f1 ssDNA complex is exchanged for the major (pVIII) and minor (pIII, pVI, pVII, pIX) proteins of the mature phage particle. The process, as discussed in this section, is summarized schematically in Figure 7. Recent experiments in which the ends of emerging phage have been isolated from the infected bacteria and

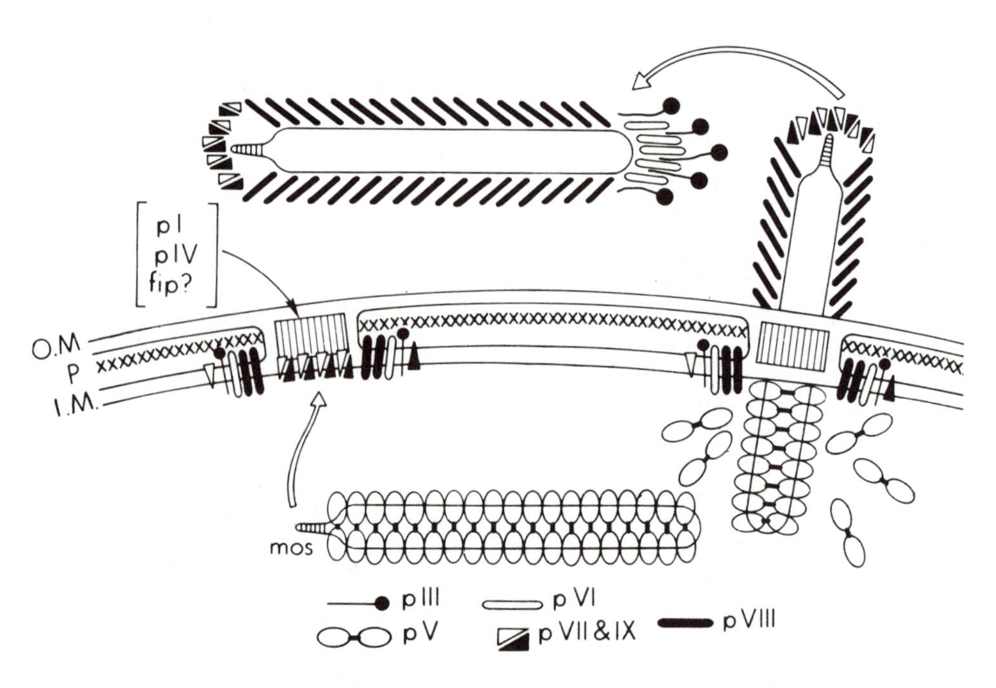

FIGURE 7 Schematic representation of the process of assembly of phage f1. OM, outer membrane; P, periplasm; IM, inner membrane; crosses; peptidoglycan layer; mos, the morphogenetic signal. The gene-*V* protein is represented as a dimer. The cytoplasmic side of the membrane containing the gV/f1-ssDNA complex is shown in the lower portion of the figure.

analyzed with respect to protein content have demonstrated that the pVII-pIX end of the phage emerges first (Lopez and Webster, 1983), and thus is the first portion of the phage to be assembled. Since the other end of the phage containing pIII and pVI is the first to enter the cell during infection, the orientation of the phage particle relative to the cell surface is the same during infection and assembly. This consistency in orientation is in accord with the observation that the major coat protein (pVIII) of infecting phage is recycled and can be found associated with the newly synthesized DNA of progeny phage (Trenkner et al., 1967; Smilowitz, 1974; Armstrong et al., 1983). Thus, in passing from the infecting particle into the cytoplasmic membrane, the parental pVIII assumes the same conformation and orientation that allows the newly synthesized pVIII to be assembled into the virion.

A specific region of the phage DNA near the junction of the intergenic region and gene *IV* (Figure 1, bottom) is always associated with the pVII-pIX-containing end of the virion (Webster et al., 1981, see section on phage particle). This region contains a sequence (nucleotides 5449-5576 in Figure 8) that appears to form a hairpin structure in the phage particle (Ikoku and Hearst, 1981). This implies that part or all of this hairpin region is involved in the nucleating event that starts the packaging of the DNA from the pV/f1 ssDNA complex into phage. Recent results using recombinant DNA techniques underscore the importance of this region in morphogenesis. The experiments involved cloning various portions of the intergenic region into plasmid pBR322. A chimeric plasmid, containing the origins of viral and complementary DNA (fragment H in Figure 8) in addition to pBR322 DNA, could enter the f1 mode of replication in the presence of f1 helper phage and produce single-stranded chimeric plasmid DNA (Cleary and Ray, 1980; Dotto et al., 1981). However, very little of this DNA was packaged into phage particles: less than 0.1% of the total number of filamentous phage particles that were produced contained chimeric DNA. When the chimeric plasmid contained the region of phage DNA located near the junction of gene IV and the intergenic region (approximately nucleotides 5400-5700 in Figure 8) in addition to the phage origins of replication, the resulting chimeric plasmid single-stranded DNA was efficiently packaged into particles: more than 50% of the total phage particles contained chimeric plasmid DNA (Dotto et al., 1981). The sequence of DNA necessary for such efficient encapsulation was termed the "morphogenetic signal" (Dotto et al., 1981 and referred to as "mos" in Figure 7). This signal appears to be necessary only for encapsulation, because single-stranded chimeric plasmid DNA that did not contain the morphogenetic signal could be isolated in a complex with pV (Grant and Webster, 1984b). Recent experiments using similar techniques have refined the location of this morphogenetic signal to the 76-nucleotide portion of DNA shown between the arrows in Figure 8 (Dotto and Zinder, 1983). Included in this 76-nucleotide fragment of DNA is a 52-nucleotide sequence of almost perfect palindromic symmetry (nucleotides 5512-5563), which constitutes the major part of the hairpin region near the junction of gene *IV* and the intergenic region shown to be always at the pVII-

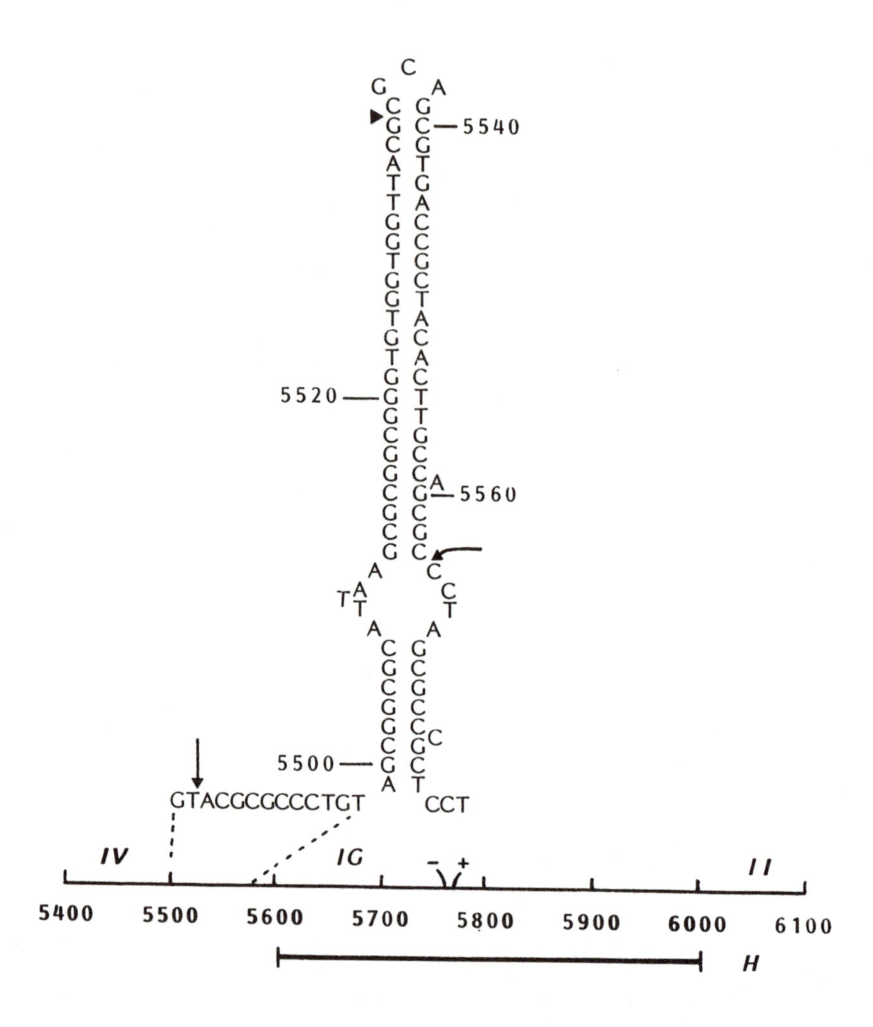

FIGURE 8 DNA sequence containing the morphogenetic signal. The upper line represents the genetic map and the nucleotide positions of the f1 genome near the intergenic region (Beck and Zink, 1981; Hill and Petersen, 1982). Genes are represented by roman numerals, the intergenic region is represented by IG, and the origins of replication of the viral and complementary strands are denoted by (+) and (-), respectively. The line marked H is the region of DNA containing the origins of replication present in all of the chimeric plasmids (see text). The nucleotide sequence of palindromic symmetry located near the border of gene *V* and the intergenic region is shown in detail. The region between the arrows is the smallest region found to contain the morphogenetic signal, as defined by the chimeric plasmid experiments. (Dotto et al, 1981, Dotto and Zinder, 1983; see text). The region from the dotted line to nucleotide 5616 is deleted in f1 pseudo-wildtype revertants from phage carrying a Tn5 insertion (Schaller, 1978). Some of the sequence 5' to the closed triangle has been shown to be necessary for efficient assembly to occur (see text).

pIX end of all phage particles (Enea et al., 1977; Webster et al., 1981; Ikoku and Hearst, 1981).

The exact sequence that constitutes the morphogenetic signal is not yet clearly defined. Infections with revertants of phage originally carrying Tn5 insertions lead to the production of f1 pseudo-wildtype phage particles missing nucleotides 5553-5616 in the intergenic region (Schaller, 1978). This would argue that nucleotides 5553-5616 are not necessary for efficient assembly. If the secondary structure of the DNA in this region is necessary for efficient packaging of the DNA, then this result will limit the morphogenetic signal to the palindromic sequence from nucleotide 5521-5553 (Figure 8). This is consistent with the observation that sequences 5' proximal to nucleotide 5535 (marked by a triangle in Figure 8) are necessary for efficient assembly (Dotto and Zinder, 1983). The GCA sequence at the tip of the hairpin may also be important, since attempts to clone DNA fragments with a GCA sequence at the end of a somewhat shorter stable hairpin into f1 cloning vectors have led to the production of mutant phage with scrambling or deletion of some or all of the inserted sequence (Rodi, 1983). Possibly, the single-stranded DNA produced from these recombinant phage has two hairpin sites that are recognized as morphogenetic signals, thereby hindering normal packaging of the DNA into phage particles.

The logical conclusion from these data is that the morphogenetic signal is composed of some portion of the DNA-hairpin structure that includes the GCA loop shown in Figure 8, and that the GCA loop is at the end of the phage particle containing pVIII-pIX. Since the pVII-pIX end of the phage emerges first from the infected bacteria, the nucleating event in the assembly process must involve some interaction of the morphogenetic signal with pVII and/or pIX. The mechanism by which this interaction occurs remains unknown. It is possible that pVII and pIX reside in the membrane (see previous section) and interact with the morphogenetic signal to bring the pV/f1 ssDNA complex in association with the site of phage assembly. This would require that the morphogenetic signal be accessible in the pV/f1 ssDNA complex. Perhaps the hairpin structure containing the GCA is located at the end of the complex represented in Figure 8, in the same manner as it is oriented in the phage particle. However, there are no data bearing on the orientation of the DNA in the pV/f1 ssDNA complex at this time.

Recognition of the morphogenetic signal by pVII and/or pIX could be mediated through interaction with other phage or bacterial proteins. One such molecule might be the *E. coli* SSB protein, since about one molecule of this protein co-purifies with the complex and may be stably associated with this structure (Grant and Webster, 1984a). Phage products pI and pIV may also be involved, since they are necessary for morphogenesis (Pratt et al., 1966; Pratt and Erdahl, 1968; Mazur and Zinder, 1975). However, little is known about their role in phage assembly other than the facts that both appear to be integral membrane proteins and that pI possibly plays an important role in the formation of the assembly site. In addition, any particular function for pI during the assembly process itself must take into account its required interaction with bacterial thioredoxin (Russel and Model, 1983).

Elongation

After the nucleation event has started the assembly of the phage, the process of elongation proceeds. The pV is displaced as the DNA is extruded through the membrane and coated with pVIII. The orientation of pVIII in the membrane and in the phage has led to the hypothesis that the DNA would first react with the group of basic amino acids located at the carboxy-terminal region of pVIII that protrudes into the cytoplasm (Marvin and Wachtel, 1975; Webster and Cashman, 1978). The gene-*VIII* protein and presumably the DNA both change their conformation as the coat protein is assembled onto the DNA, and this change would place the carboxy-terminal portion on the outside, as found in the phage particle (Figure 3). The next portion of the DNA would then react with another group of carboxy-terminal ends of pVIII, and this process would be repeated until the end of the DNA was reached.

The steady-state addition of pVIII to the DNA must involve a large conformational change in pVIII, since it contains less than 50% α helix while in the membrane but more than 80% α helix when in the phage particle (see previous section). The regions of pVIII that interact with the lipid bilayer in the membrane must change in order to interact with portions of other pVIII molecules in the phage istelf. Interaction between the four lysine side chains located at the carboxy-terminal region of pVIII and the DNA on the cytoplasmic side of the membrane may help to trigger such a conformational change. It has been suggested that the fivefold rotational symmetry of pVIII in the phage particle implies that five pVIII subunits are added at one time to the DNA during assembly (Makowski and Caspar, 1981; Makowski, 1983). This hypothesis requires that the carboxy-terminal region of each pVIII subunit being added interact both with the DNA and with a portion of the carboxy-terminal region of two subunits previously added (see Figure 3). Such a cooperative assembly of five subunits at one time would suggest that there is some interaction between pVIII molecules in the membrane assembly site to align them for addition to the assembly particle. It has been proposed that regions of β structure in pVIII might interact with each other in the hydrocarbon layer of the membrane to form a multimeric "barrel structure" (Webster and Cashman, 1978). However, there is no definitive information about the tertiary structure of the pVIII molecule in the membrane that would allow this proposal to be assessed. The incorporation of each pVIII subunit into the nascent phage must be a multistep process, because pVIII subunits (axial length 60 A) have to be added every 16 A to produce the phage structure in which these molecules are related to each other by the fivefold rotational symmetry plus a twofold screw axis of pitch 32 A, as described for Figure 3 (Makowski and Caspar, 1981; Makowski, 1983).

The structure of the DNA must also change as it goes from the pV/f1 ssDNA complex into a phage particle. However, it is difficult to describe these changes because of our limited knowledge about the conformation of the DNA in either structure. As mentioned above, one of the functions of the pV/f1 ssDNA complex may be to collapse the DNA into a conformation that is required for the next step in assembly.

However, even though the complex has been postulated to have this function, it should be pointed out that there are no definitive experiments showing that the single-stranded circular DNA need be in such a complex to be packaged into phage.

The conformational changes associated with the process of elongation appear to be quite rapid. Only high concentrations of very reactive cross-linking reagents were able to interrupt the process of phage extrusion and yield bacteria containing different lengths of emerging phage (Lopez and Webster, 1983). Attempts to "freeze" the phage during the process of extrusion by quick cooling, by the addition of energy poisons (such as arsenate, cyanide, dinitrophenol, or any combination thereof) resulted in bacteria that could not synthesize phage but were completely free of any extruding phage particles, as visualized in the electron microscope. Since it had been suggested that f1 assembly might depend on the proton motive force (Ng and Dunker, 1981), this observation would argue that the need for energy would be at the nucleation step or some earlier step, such as the synthesis and insertion of the coat protein into the membrane. In all studies regarding the possible energy requirement for phage assembly, it has been difficult to determine whether the effects observed from these metabolic poisons are direct effects on assembly or indirect effects on cellular processes such as protein synthesis.

Termination

Under normal conditions of infection, assembly of pVIII molecules onto the growing phage particle terminates when the end of the DNA is reached. This results in the production of virions whose length is determined by the size of the DNA. Miniphage containing extensive deletions of the viral genome form small virions while insertions of DNA into the viral genome lead to longer-than-normal virions whose length corresponds directly to the length of the chimeric DNA (Denhardt et al., 1978). Termination is not a passive event; it requires the active participation of at least the phage minor coat proteins pIII and pVI. Infection of nonsuppressing bacteria with phage carrying amber mutations in either gene *III* or gene *VI* results in the production, with good efficiency, of extremely long phage particles, whose length is some multiple of the normal phage length (Scott and Zinder, 1967; Salivar et al., 1967; Pratt et al., 1969; Lopez and Webster, 1983). These multiple-length particles, or polyphage, contain many separate copies of the unit-length genome, packaged end-to-end within a single virion. Thus, pIII and pVI appear to be necessary for the reaction that recognizes the end of the DNA and starts the termination process.

Proteins pIII and pVI also contribute to the structural integrity of the mature phage particle. In this respect, pVI appears to be the more important, inasmuch as polyphage produced by infection with gene-*VI* amber mutants are very unstable and highly susceptible to dissociation by salt (Lopez and Webster, 1983). Polyphage produced by infection with gene-*III* amber mutants are more stable than the gene-*VI* polyphage, suggesting that pVI has been assembled onto the ends of these particles even in the absence of any pIII. Nevertheless, these gene-*III* polyphage are much less stable than wildtype phage to heat and the nonionic detergent sarcosyl (Crissman and Smith, 1984). These data imply that pVI is assembled onto the nascent phage first during the termination step, followed by the incorporation of pIII.

The domain of pIII necessary for termination is separate from that required for infection. Removal of the amino-terminal region of pIII (the knob portion in Figure 2) with proteases results in noninfectious but stable phage particles (Armstrong et al., 1981; Grant et al., 1981a). Infection of bacteria with an f1 phage containing an internal deletion near the 5' end of gene *III* yields noninfectious particles of normal length (Nelson et al., 1981) that contain altered pIII molecules (Crissman and Smith, 1984). Expression of this mutant gene *III* would result in the synthesis of a mutant protein lacking amino acids 28-196 (of the normal 406). Presumably, it would be synthesized with its normal signal peptide, be properly inserted into the cytoplasmic membrane and then be processed like normal pIII. Thus, some portion of the carboxy-terminal half of pIII is necessary for termination and presumably interacts with pVI and possibly pVIII in the intact virion.

The gene-*VII* and *IX* proteins may also participate in termination, since polyphage can also be obtained from bacteria infected with gene-*VII* or gene-*IX* amber-mutant phage (Lopez and Webster, 1983). However, these mutants only produce 1-3 polyphage per cell, in contrast to the large numbers of polyphage produced by infection with gene-*VI* or gene-*III* amber mutants, and the polyphage are infective. The most reasonable explanation for these observations is that pVII and pIX enter the membrane from the infecting phage particles and are able to be used again in a nucleating event to start production of a new phage particle, as has been shown to occur with the pVIII protein. However, the assembly site is then depleted of the pVII or pIX that is necessary to initiate a second phage particle. In the absence of the necessary machinery to initiate a new phage particle, termination is apparently inefficient, even in the presence of pIII and pVI, and results in the production of the single polyphage per cell.

The membrane-associated assembly of the filamentous phage is probably aided, or at least influenced, by a number of host proteins. The isolation of host mutants that are temperature-sensitive for phage production is consistent with this view. These mutants have been termed *fip*, for filamentous phage production. The *fipA* mutant was isolated and cloned (Russel and Model, 1983, 1984) and later shown to be a mutation at the *trxA* locus, which codes for an altered thioredoxin (Russel and Model, 1985; Lim et al, 1985). A double mutation, *fipBC*, has been isolated that is temperature-sensitive for phage assembly (Lopez, 1984; Lopez and Webster, unpublished results). Correction of either mutation results in loss of the mutant phenotype. The *fipC* mutation maps at 73 min and has no other detectable phenotype, while *fipB* maps at 83.5 min and shows reduced binding of λ phage when grown in the presence of glucose. A fourth temperature-sensitive mutant, *fipD*, may be a mutant involved in lipopolysaccharide metabolism, as judged by its location at 65 min on the *E. coli* map (Russel and Model, pers. commun.).

CONCLUSION

In recent years the structures of the pV/f1 ssDNA and the mature phage particle have been elucidated with enough clarity to help define, in general terms, an ordered sequence of reactions that occur at the membrane during the process of assembly. First, pVII and pIX presumably interact with the morphogenetic signal of the phage DNA in the pV/ssDNA complex, thereby initiating the process of assembly and establishing the orientation of the DNA within the phage particle. Elongation of this nascent particle proceeds rapidly as pV is displaced from the DNA by pVIII, which is assembled into DNA as it is extruded through the membrane. When the end of the DNA is reached, pIII and pVI molecules are assembled onto the particle, terminating the morphogenesis reaction and releasing the completed phage into the medium. Further insights into the conformation of the DNA and its structural relation to the proteins in both the complex and phage should provide a clearer understanding of the reactions required for efficient phage formation.

A more difficult task for laboratories studying the filamentous phage will be to sort out the membrane-associated reactions involved in morphogenesis. This will require a clearer understanding of the locations and conformations of the minor capsid proteins in the membrane and of their possible interactions with each other and with bacterial components. Do conformational changes of these proteins, as they leave the membrane and interact with the nascent phage particle, contribute greatly to the mechanism of packaging? It will be necessary to determine the role of phage proteins pI and pIV, which are absolutely required for formation of the phage but are not present in the phage particle. Genetic evidence suggests an interaction of pI with thioredoxin, a host protein whose function in the uninfected bacteria is not clear. It is possible that other specific interactions between phage- and bacteria-encoded proteins are required for the particle to be put together. Hopefully, studies on these questions will lead to the establishment of an *in vitro* system in which DNA from the pV/f1 ssDNA complex can be assembled into a mature particle. Such a system would aid in determining the stages at which energy is required in the morphogenetic process.

The morphogenetic pathway of the filamentous phages differs from that of most bacteriophage, because of involvement of membranes and the lack of any effect of phage development on the viability of the host cell. Understanding this fascinating process may provide insight into how other membrane structures are assembled and yield information about mechanisms by which macromolecules such as DNA are transported across membranes.

ACKNOWLEDGEMENTS

The authors thank J. E. Coleman, L. A. Day, G. P. Dotto, L. Makowski, Z. McPherson, P. Model, M. Russel, and G. Smith for communicating results prior to publication and for helpful discussions. We are grateful to J. Horabin, P. Model, M. Russel, and G. Smith for critically reading the manuscript, and M. J. Outlaw for patiently keyboarding it. This work was supported by Public Health Service Grant GM18305.

REFERENCES

Alberts, B., L. Frey, and H. Delius. 1972. *J. Mol. Biol.*, 68, 139.

Alma, N. C. M., B. J. M. Harmsen, C. W. Helbers, G. van der Marel, and J. H. van Boom. 1981. *FEBS Lett.*, 135, 15.

Anderson, R. A., Y. Nakashima, and J. E. Coleman. 1975. *Biochemistry*, 14, 907.

Armstrong, J., J. A. Hewitt, and R. N. Perham. 1983. *EMBO J.*, 2, 1641.

Armstrong, J., R. N. Perham, and J. E. Walker. 1981. *FEBS Lett.*, 135, 167.

Banner, D. W., C. Nave, and D. A. Marvin. 1981. *Nature*, 289, 814.

Barnes, W. M. 1980. In *Genetic Engineering*. Vol. II, J. Setlow and A. Hollaender, eds. Plenum Press, p. 185.

Bayer, M. E. 1976. In *Membrane Biogenesis*. A. Tzagoloff, ed. Plenum Press. p. 393.

Beck, E., and B. Zink. 1981. *Gene*, 16, 35.

Boeke, J. D. and P. Model. 1982. *Proc. Natl. Acad. Sci. USA*, 79, 5200.

Boeke, J. D., P. Model, and N. D. Zinder. 1982. *Mol. Gen. Genet.*, 186, 185.

Boeke, J. D., M. Russel, and P. Model 1980. *J. Mol. Biol.*, 144, 103.

Bradley, D. E. 1973. *Can. J. Microbiol.*, 19, 623.

Bradley, D. E. 1981. In *Molecular Biology, Pathogenicity and Ecology of Bacterial Plasmids*. S. B. Levy, R. C. Clowes and E. L. Koenig, eds. Plenum Press. p 217.

Brayer, G. D. and A. McPherson. 1983. *J. Mol. Biol.*, 169, 565.

Brayer, G. D. and A. McPherson. 1984. *Biochemistry*, 23, 340.

Brown, L. R. and C. E. Dowell. 1968. *J. Virol.*, 2, 1290.

Burke, J. M., C. P. Novotny, and P. Fives-Taylor. 1979. *J. Bacteriol.*, 140, 525.

Caro, L. G., and M. Schnös. 1966. *Proc. Natl. Acad. Sci. USA*, 56, 126.

Casadevall, A., and L. A. Day. 1983. *Biochemistry*, 22, 4831.

Cavlieri, S. J., K. E. Neet, and D. A. Goldthwait. 1976. *J. Mol. Biol.*, 102, 697.

Chamberlain, B. K. and R. E. Webster. 1976. *J. Biol. Chem.*, 251, 7739.

Chamberlain, B. K., Y. Nozaki, C. Tanford and R. E. Webster. 1978. *Biochim. Biophys. Acta*, 510, 18.

Chang, C. N., G. Blobel and P. Model. 1978. *Proc. Natl. Acad. Sci. USA*, 75, 361.

Chang, C. N., P. Model, and G. Blobel. 1979. *Proc. Natl. Acad. Sci. USA*, 76, 1251.

Cleary, J. M. and D. S. Ray. 1980. *Proc. Natl. Acad. Sci. USA*, 77, 4638.

Coleman, J. E. and I. M. Armitage. 1978. *Biochemistry*, 17, 5038.

Coleman, J. E., R. A. Anderson, R. G. Ratcliffe, and I. M. Armitage. 1976. *Biochemistry*, 15, 5419.

Crissman, J. W. and G. P. SMith, 1984. *Virology*, 130, 252.

Cross, T. A. and S. J. Opella. 1981. *Biochemistry*, 20, 290.

Cross, T. A., P. Tsang, and S. J. Opella. 1983. *Biochemistry*, 22, 726.

Date, T., J. M. Goodman, and W. T. Wickner. 1980. *Proc. Natl. Acad. Sci. USA*, 77, 4669.

Davis, N. G., S. Boeke, and P. Model. 1985. *J. Mol. Biol.*, 181, 111.

Day, L. A. (1973) *Biochemistry*, 12, 5329.

Day, L. A., and R. L. Wiseman. 1978. In *The Single-Stranded DNA Phages*, D. T. Denhardt, D. Dressler, and D. S. Ray, eds. Cold Spring Harbor Laboratory. p. 605.

Denhardt, D. T., D. Dressler and D. S. Ray, eds. 1978. *The Single-Stranded DNA Phages*. Cold Spring Harbor Laboratory.

Dettman, H. D., J. H. Weiner and B. D. Sykes. 1982. *Biophys. J.*, 37, 243.

Dotto, G. P., V. Enea and N. D. Zinder. 1981. *Virology*, 114, 463.

Dotto, G. P. and K. Horiuchi. 1981. *J. Mol. Biol.*, 153, 169.

Dotto, G. P. and N. D. Zinder. 1983. *Virology*, 130, 252.

Dunker, A. K., and E. A. Anderson. 1975. *Biochim. Biophys. Acta*, 402, 31.

Emr, S. D., S. Hanley-Way and T. J. Silkavy. 1980. *J. Cell Biol.*, 86, 701.

Enea, V., K. Horiuchi, B. G. Turgeon and N. D. Zinder. 1977. *J. Mol. Biol.*, 111, 395.

Fodor, S. P. A., A. K. Dunker, Y. C. Ng, D. Carsten and R. W. Williams. 1981. In *Bacteriophage Assembly*, M. S. DuBow, ed., Alan R. Liss. p 441.

Fritzche, H., T. A. Cross, S. J. Opella and N. R. Kallenbach. 1981. *Biophys. Chem.*, 14, 283.

Fulford, W. and P. Model. 1984. *J. Mol. Biol.*, 178, 137.

Garssen, G. J., R. Kapein, J. G. G. Schoenmakers, and C. W. Hilbers. 1978. *Proc. Natl. Acad. Sci. USA*, 75, 5281.

Garssen, G. J., G. I. Tesser, J. G. G. Schoenmakers and C. W. Hilbers. 1980. *Biochim. Biophys. Acta*, 446. 361.

Geider, K., I. Baeumel and T. E. Meyer. 1982. *J. Biol. Chem.*, 257, 6488.

Goldsmith, M. E. and W. H. Konigsberg. 1977. *Biochemistry*, 16, 2686.

Grant, R. A., T. C. Lin, W. Konigsberg and R. E. Webster. 1981a. *J. Biol. Chem.*, 256, 539.

Grant, R. A., T. C. Lin, R. E. Webster, and W. Konigsberg. 1981b. In *Bacteriophage Assembly*. M. S. DuBow, ed. Alan R. Liss. p. 413.

Grant, R. A. and R. E. Webster, 1984a. *Virology*, 133, 350.

Grant, R. A. and R. E. Webster, 1984b. *Virology*, 133, 329.

Gray, C. W., R. S. Brown, and D. A. Marvin. 1981. *J. Mol. Biol.*, 146, 621.

Gray, D. M., C. W. Gray, and B. D. Carlson. 1982a. *Biochemistry*, 21, 2702.

Gray, C. W., G. G. Kneale, K. R. Leonard, H. Siegrist, and D. A. Marvin. 1982b. *Virology*, 116, 40.

Griffith, J., and A. Kornberg. 1974. *Virology*, 59, 139.

Griffith, J., M. Manning and K. Dunn. 1919. *Cell*, 23, 747.

Hagen, D. S., J. H. Weiner and B. D. Sykes. 1979. *Biochemistry*, 18, 2007.

Harth, G., I. Baemumel, T. F. Meyer and K. Geider. 1981. *Eur. J. Biochem.*, 119, 663.

Hill, D. F. and G. B. Petersen. 1982. *J. Virol.*, 44, 32.

Hohn, B., H. von Schutz, and D. A. Marvin. 1971. *J. Mol. Biol.*, 56, 155.

Horiuchi, K. 1980. *Proc. Natl. Acad. Sci. USA*, 77, 5526.

Horiuchi, K., J. V. Ravetch, and N. D. Zinder. 1978. *Cold Spring Harbor Symp. Quant. Biol.*, 43, 389.

Huang, C.-C. and J. E. Hearst. 1981. *Nucl. Acids Res.*, 9, 5587.

Ikoku, A. S., and J. E. Hearst. 1981. *J. Mol. Biol.*, 151, 245.

Jacobson, A. 1972. *J. Virol.*, 10, 835.

Kneale, G. G. and D. A. Marvin. 1982. *Virology*, 116, 53.

Kneale, G. G., R. Freeman and D. A. Marvin. 1982. *J. Mol. Biol.*, 156, 279.

Kornberg, A. 1980. *DNA Replication*. W. H. Freeman.

Kornberg, A. 1982. *Supplement to DNA Replication*. W. H. Freeman.

Kuo, T.-T., T.-C. Huang and T.-Y. Chow. 1969. *Virology*, 39, 548.

Lerner, T. J. and P. Model. 1981. *Virology*, 115, 282.

Lim, C.-J., B. Haller, and J. A. Fuchs. 1985. *J. Bacteriol.*, 161, 799.

Lin, N. S.-C., and D. Pratt. 1974. *Virology*, 61, 334.

Lin, T.-C., R. E. Webster, and W. Konigsberg. 1980. *J. Biol. Chem.*, 255, 10331.

Lopez, J. 1984. Ph.D. Thesis. Duke University.

Lopez, J. and R. E. Webster. 1982. *J. Virol.*, 42, 1099.

Lopez, J. and R. E. Webster. 1983. *Virology*, 127, 177.

Lyons, L. B. and N. D. Zinder. 1972. *Virology*, 49, 45.

Makowski, L. 1983. In *The Structures of Biological Macromolecules and Assemblies*. Vol. 1 The Viruses. A. McPherson and F. Jurnak, eds. Wiley.

Makowski, L., and D. L. D. Caspar. 1981. *J. Mol. Biol.*, 145, 611.

Marvin, D. A., and B. Hohn. 1969. *Bact. Rev.*, 33, 172.

Marvin, D. A., and E. J. Wachtel. 1975. *Nature*, 253, 19.

Marvin, D. A., R. L. Wiseman, and E. J. Wachtel. 1974a. *J. Mol. Biol.*, 82, 121.

Marvin, D. A., W. J. Pigram, R. L. Wiseman, E. J. Wachtel, and F. J. Marvin. 1974b. *J. Mol. Biol.*, 88, 581.

Marvin, D. A. and C. Nave. 1982. In *Structural Molecular Biology*. D. Davies, W. Saenger, and S. C. Danyluk, ed. Plenum.

Marzec, C. J., and L. A. Day. 1983. *Biophys. J.*, 42, 171.

Mazur, B. J., and P. Model. 1973. *J. Mol. Biol.*, 78, 285.

Mazur, B. J., and N. D. Zinder. 1975. *Virology*, 68, 490.

McPherson, A., F. Jurnak, A. Wang, F. Kolpak, I. Molineux, and A. Rich. 1979a. *Cold Spring Harbor Symp. Quant. Biol.*, 43, 21.

McPherson, A., F. Jurnak, A. H. J. Wang, I. Molineux, and A. Rich. 1979b. *J. Mol. Biol.*, 134, 379.

McPherson, A., F. Jurnak, A. Wang, F. Kolpak, A. Rich, I. Molineux, and P. Fitzgerald. 1980a. *Biophys. J.*, 32, 155.

McPherson, A., A. H. J. Wang, F. A. Jurnak, I. Molineux, F. Kolpak, and A. Rich. 1980b. *J. Biol. Chem.*, 255, 3174.

Merriam, M. V. 1977. *J. Virol.*, 21, 880.

Meyer, T. F. and K. Geider. 1979a. *J. Biol. Chem.*, 254, 12636.

Meyer, T. F. and K. Geider. 1979b. *J. Biol. Chem.*, 254, 12642.

Meyer, T. F. and K. Geider. 1982. *Nature*, 296, 828.

Minimishima, Y., K. Takeya, Y. Ohnishi, and K. Amako. 1968. *J. Virol.*, 2, 208.

Model, P. and N. D. Zinder. 1974. *J. Mol. Biol.*, 83, 231.

Nakashima, Y., A. K. Dunker, D. A. Marvin, and W. Konigsberg. 1974. *FEBS Lett.*, 40, 290.

Nakashima, Y. and W. Konigsberg. 1980. *J. Mol. Biol.*, 138, 493.

Nelson, F. K. S. M. Friedman, and G. P. Smith. 1981. *Virology*, 108, 338.

Newman, J., H. L. Swinney, and L. A. Day. 1977. *J. Mol. Biol.*, 116, 593.

Ng, Y. C. and A. K. Dunker. 1981. In *Bacteriophage Assembly*. M. S. DuBow, ed., Alan R. Liss. p. 467.

Nozaki, Y., B. K. Chamberlain, R. E. Webster and C. Tanford. 1976. *Nature*, 259, 335.

Nozaki, U., J. A. Reynolds and C. Tanford. 1978. *Biochemistry*, 17, 1239.

O'Connor, T. P. and J. E. Coleman. 1982. *Biochemistry*, 21, 848.

O'Connor, T. P. and J. E. Coleman. 1983. *Biochemistry*, 22, 3375.

Oey, J. L. and R. Knippers. 1972. *J. Mol. Biol.*, 68, 125.

Ohkawa, I. and R. E. Webster. 1981. *J. Biol. Chem.*, 256, 9951.

Paradiso, R. R. and W. Konigsberg. 1982. *J. Biol. Chem.*, 257, 1462.

Pluschke, G., Y. Hirota, and P. Overath. 1978. *J. Biol. Chem.*, 253, 5048.

Pratt, D., P. Laws, and J. Griffith. 1974. *J. Mol. Biol.*, 82, 425.

Pratt, D. and W. S. Erdahl. 1968. *J. Mol. Biol.*, 37, 181.

Pratt, D., H. Tzagoloff, and W. D. Erdahl. 1966. *Virology*, 30, 397.

Pratt, D., H. Tzagoloff, and J. Beaudion. 1969. *Virology*, 39, 42.

Pretorius, H. T., M. Klein, and L. A. Day. 1975. *J. Biol. Chem.*, 250, 9262.

Rasched, I. and F. M. Pohl. 1974. *FEBS Lett.*, 46, 115.

Rodi, D. J. 1983. Ph.D. Thesis, Columbia University.

Rossomando, E. F. 1970. *Virology*, 42, 681.

Russel, M. and P. Model. 1983. *J. Bacteriol.*, 154, 1064.

Russel, M. and P. Model. 1984. *J. Bacteriol.*, 157, 526.

Russel, and P. Model. 1985. *Proc. Nat. Acad. Sci. USA*, 82, 29.

Salivar, W. O., T. J. Henry, and D. Pratt. 1967. *Virology*, 32, 41.

Schaller, H. 1978. *Cold Spring Harbor Symp. Quant. Biol.*, 43, 401.

Schaller, H., E. Beck, and M. Takanami. 1978. In *The Single-Stranded DNA Phages*. D. T. Denhardt, D. Dresler and D. S. Ray, eds. Cold Spring Harbor Laboratory. p. 139.

Schwartz, F. M. and N. E. Zinder. 1968. *Virology*, 34, 352.

Scott, J. R. and N. D. Zinder. 1967. In *The Molecular Biology of Virus*. J. S. Colter and W. Paranchych, eds. Academic Press. p. 212.

Simons, G. F. M., R. N. H. Konings, and J. G. G. Shoenmakers. 1979. *FEBS Lett.*, 106, 8.

Simons, G. F. M., R. N. H. Konings, and J. G. G. Schoenmakers. 1981. *Proc. Natl. Acad. Sci. USA*, 78, 4194.

Simons, G. F. M., G. H. Veeneman, R. N. H. Konings, H. H. van Boom, and J. G. G. Schoenmakers. 1982. *Nucl. Acids Res.*, 10, 821.

Smilowitz, H. 1974. *J. Virol.*, 13, 94.

Smilowitz, H., J. Carson, and P. W. Robbins. 1972. *J. Supramol. Struct.*, 1, 8.

Smits, M. A., J. Jansen, R. N. H. Konings, and J. G. G. Schoenmakers. 1984. *Nucl. Acid Res.*, 12, 4071.

Smits, M. S., G. Simons, R. N. H. Konings, and J. G. G. Schoenmakers. 1978. *Biochim. Biophys. Acta*, 521 ,27.

Stanisich, U. A. 1974. *J. Gen. Microbiol.*, 84, 332.

Sugimoto, K., H. Sugisaki, T. Okamoto, and N. Takanami. 1977. *J. Mol. Biol.*, 111, 487.

Thomas, G. J., Jr., B. Prescott, and L. A. Day. 19983. *J. Mol. Biol.*, 165, 321.

Torbet, J., D. M. Gray, C. W. Gray, D. A. Marvin, and H. Siegrist. 1981. *J. Mol. Biol.*, 146, 305.

Torbet, J., and G. Maret. 1981. *Biopolymers*, 20, 2657.

Trenkner, E., F. Bonhoeffer, and A. Gieres. 1967. *Biochem. Biophys.*

Res. Commun., 28, 932.

Tzagoloff, H. and D. Pratt. 1964. *Virology*, 24, 372.

van Wezenbeek, P. M. G. F., T. J. M. Hulsebos, and J. G. G. Schoenmakers. 1980. *Gene*, 11, 129.

Webster, R. E. and J. S. Cashman. 1973. *Virology*, 55, 20.

Webster, R. E. and J. S. Cashman. 1978. In *The Single-Stranded DNA Phages*. D. T. Denhardt, D. Dressler and D. S. Ray, eds. Cold Spring Harbor Laboratory. p. 557.

Webster, R. E., R. A. Grant, and L. A. W. Hamilton. 1981. *J. Mol. Biol.*, 152, 357.

Webster, R. E. and M. Rementer. 1980. *J. Mol. Biol.*, 139, 393.

Wickner, W. 1975. *Proc. Natl. Acad. Sci. USA*, 72, 4749.

Wickner, W. 1976. *Proc. Natl. Acad. Sci. USA*, 73, 1159.

Woolford, J. L., J. S. Cashman and R. E. Webster. 1974. *Virology*, 58, 544.

Woolford, J. L. 1976. Ph.D. Thesis, Duke University.

Woolford, J. L., H. M. Steinman, and R. E. Webster. 1977. *Biochemistry*, 16, 2694.

Yen, T.-S. B. and R. E. Webster. 1981. *J. Biol. Chem.*, 256, 11259.

Yen, T.-S. B. and R. E. Webster. 1982. *Cell*, 29, 337.

Zinder, N. D. and J. S. Boeke. 1982. *Gene*, 19, 1.

Zwizinski, C. and W. Wickner. 1980. *J. Biol. Chem.*, 255, 7973.

8
Viral-Host Interactions in Virus Morphogenesis

Kit Tilly
Harvard University
Costa Georgopoulos
University of Utah College of Medicine

INTRODUCTION

Viruses exploit their hosts to the utmost degree. They depend on their hosts to provide energy, macromolecular building blocks, and an environment for growth. Small phages such as f1 (ssDNA) and Qβ (ssRNA) and large phages, such as T4 (dsDNA) all depend heavily on host functions throughout their growth cycles. Bacteriophage T4 has a large genome and can adapt to various differences among a wide spectrum of hosts. It protects its chromosome from degradation by various bacterial restriction enzymes by both hydroxymethylation and glucosylation of its cytosine residues. In addition, T4 codes for many "nonessential" functions that may duplicate bacterial functions not present in all bacterial strains, allowing it to have a broad host range (Mathews et al., 1983; Gauss et al., 1983). Phage Mu is of intermediate size and has an alternate strategy: it modifies a fraction of its adenine residues to protect its DNA from nucleases (Swinton et al., 1983), and it increases its host range by inversion of a particular region of its chromosome, with the two orientations coding for proteins that enable adsorption to mutually exclusive groups of bacteria to occur (Giphart-Gassler et al., 1982). Very small phages like Qβ and ϕX174 have extremely densely packed genomes. They have genes that overlap (Sanger et al., 1977) and they have bifunctional proteins; for example, the Qβ maturase also functions as a lysis protein (Winter and Gold, 1983; Karnik and Billeter, 1983). Because of their small sizes, these viruses must rely on their hosts for many of the products required for their propagation.

Phages have at least three modes of growth, which in part determine the degree of dependence on their hosts. Lytic phages infect a host cell, construct progeny, and finally lyse the cell. Since the host invariably dies in the process, its enzymatic machinery can be changed irreversibly to accommodate rapid phage growth. Temperate phages, which can either follow a lytic or lysogenic cycle, cannot afford to modify the macromolecular metabolism of their hosts irreversibly at initial stages of their growth cycles, in case growth conditions are such that the lysogenic pathway is followed. These phages may have evolved interactions that sense the physiological state of the host and gauge the best strategy. A third type of infection is exemplified by filamentous phages like f1; with these phages progeny particles are continually extruded from the host without killing the cell. Analogous variations are found with animal and plant viruses. Poxviruses are very large and self-contained; their infectious cycles are short, and the cells are lysed as the viruses are released. Retroviruses are somewhat similar to temperate phages in that they can insert copies of their genomes into the host chromosomes, where the copies are stably maintained.

Viruses and their hosts interact at many levels. In all cases, the host protein-synthetic machinery is used, though a variety of methods ensure biased translation of viral messages. Some examples are alteration of recognition of cellular messages after infection (poliovirus; Ehrenfeld, 1982), degradation of the host chromosome, modification of tRNAs, and alteration of host tRNA synthetases (phage T4; Sueoka et al., 1966; Marchin et al., 1972). Some viruses promote their own growth at a stage earlier than protein synthesis. One way this occurs is by altering the transcriptional specificity of the host RNA polymerase, so that it recognizes only viral promoters (as for T4; Horvitz, 1974; Steven, 1972) and *Bacillus subtilis* phage SP01 (Losick and Pero, 1981). Another mechanism is the inclusion of a virus-encoded polymerase as a capsid component; for example, phage N4 (Falco et al., 1980) or animal viruses such as vesicular stomatitis virus (Baltimore et al., 1970). The host also often provides necessary functions in later stages of viral infection, such as processing, modification, and translocation of proteins. For example, an *E. coli* endopeptidase, called the leader peptidase, correctly processes both the f1 (M13) major precoat protein and the precursors of some cellular proteins that are destined to be exported (Wolfe et al., 1982). Also, the infection of some animal viruses is blocked when grown in several Chinese hamster ovary cell lines that are defective in various stages of protein modification (Schmidt and Schlesinger, 1979). The host also participates in translation of the messages of the morphogenetic genes, modification and localization of the gene products, and in viral assembly itself. Host structural components (e.g., the membrane and the cytoskeleton) and specific protein-protein interactions are also required for this host function.

Techniques for studying virus-host interactions

The interplay of bacterial and phage components has been studied by two general approaches. In the first, host mutations that affect viral assembly are obtained and the effects of these mutations on phage production are analyzed. This approach has been very powerful with phages, but less so with animal and plant viruses. The difficulty in carrying out genetic experiments with cells in culture plus the diploidy of these cells has minimized the utility of genetic techniques in eukaryotic systems. In the second approach physical interactions between host and viral components occurring during assembly are studied. We will examine the mutational approach first.

Host mutations have been found in four ways. First, bacterial mutations blocking phage production have been selected by demanding that cells retain the ability to form colonies in the presence of an appropriate amount of phage (Georgopoulos, 1971). If an infected bacterium is wildtype, progeny phage will form and then infect neighboring cells, thus limiting the size of the colony. If the infected cell is a mutant in which some stage of the phage life cycle is blocked, the infected bacterium will be killed, but no phage progeny will be produced. The uninfected bacteria can multiply and form a colony. Cells surviving this test can be screened for blocks in the assembly stage of the phage growth cycle.

A second method has been to screen mutant bacteria deficient in some function that one might guess to be important in phage assembly. For example, mutations affecting localization and cleavage of membrane proteins might be expected to affect growth of filamentous and lipid-containing phages, as would mutations altering the fatty acid composition of the membrane (Cronan and Vagelos, 1971). The membrane is also important for assembly of the T4 capsid, (Simon, 1972), baseplate (Edgar and Lielausis, 1968; Kells and Haselkorn, 1974; Mason and Haselkorn, 1972; Simon, 1969), and tail fiber (King and Laemmli, 1971).

A third method has been to search directly for bacterial mutations that interfere specifically with the function of a known phage-encoded morphogenetic gene product. For example, bacterial mutants have been isolated that interfere with the growth of phage T4 carrying a temperature-sensitive mutation in gene *31* (Takahashi et al., 1975). (In the absence of a functional product of gene *31*, which is denoted gp31, the capsid protein gp23 remains as lumplike aggregates on the bacterial membrane; Laemmli et al., 1969.) The isolation was carried out under semipermissive conditions that result in limited function of gene *31*. Many of the bacterial survivors are mutants that T4 kills but that do not allow phage production. Mutations in the host *groEL* gene, which will be discussed in some detail shortly, can cause this phenotype. The advantage of this approach, in addition to its specificity, is that bacterial mutations with an otherwise subtle effect on the function of the wildtype gene can be detected.

A fourth method has been to study phage growth on natural bacterial variants (e.g., the *E. coli* strains K12 and B; Georgopoulos et al., 1977). This method takes advantage of the natural drift found among bacterial strains, which sometimes results in differences in the interactions of their components with the various phage-assembly functions.

Once an unproductive phage-bacterium combination has been identified, it is possible to isolate phage mutations that overcome the block on growth, thus identifying phage genes whose products function in a particular stage. The nature of these mutations is revealing about both phage-assembly pathways and other specific interactions with host components. So far, three ways of overcoming host mutational blocks have been identified: (1) Phage mutations may allow the complete bypass of the step at which a particular interaction is found. An example is the *byp31* mutations in phage T4, which are mutations in gene *23* that eliminate the requirement for either the phage gene-*31* function or the host *groEL* function (Doermann and Simon, 1984; Simon and Randolph, 1984.) (2) Specific phage alterations may restore protein-protein interactions disrupted by the host mutation; this identifies proteins that contact one other and often regions of proteins that interact (e.g, Katsura, 1981). (3) Control mutations that lower the level of a phage substrate may allow a partially active enzyme to complete a few structures, rather than abortively beginning many (Floor, 1970; Georgopoulos et al., 1973).

The study of phage interactions with their hosts in the assembly processes has often revealed unexpected connections. For example, bacterial mutants selected for their ability to block the assembly of one phage have been shown to have pleiotropic effects on the growth of unrelated phages (Georgopoulos et al., 1972; Sternberg, 1973a; Zweig and Cummings, 1973). In addition, although the nature of the mutant selections requires bacterial viability in the test conditions, many of the mutations cause conditional lethality and thereby identify new essential genes in *E. coli*. A related result was the discovery of an intriguing connection between genes required for phage growth and the bacterial heat shock system (to be described shortly). Finally, genes not suspected to be involved in morphogenesis (e.g, *himA* and *hip*, see below) have been shown to be essential for that process.

The second general approach, namely, studying physical interactions between host and viral components during assembly, is based on associations between proteins or with structural components; these associations are observed either after biochemical fractionation of infected cells or within completed virions (Onorato and Showe, 1975; Earnshaw and King, 1978; Murialdo, 1979; our unpublished results, see below). For example, the establishment of an *in vitro* system in which λ terminase will cut specifically at *cos* sites has led to the identification of host functions that are essential in the cutting process (Sumner-Smith et al., 1981; Gold and Becker, 1983). Further fractionation of the *in vitro* prohead-assembly system (Ferrucci and Murialdo, 1981; Kochan and Murialdo, 1983) should identify host components in that system. Although these experiments are suggestive and in some cases have corroborated genetic experiments, they are plagued by the possibility of artifactual association between components.

In the following discussion, we will detail some of the phage and bacterial interactions in assembly that have been studied and attempt to extrapolate to eukaryotic viruses. We will often draw on previous reviews of phage assembly (Hohn and Katsura, 1977; Murialdo and Becker, 1978; Wood and King, 1979; Earnshaw and Casjens, 1980; Georgopoulos et al., 1983; Feiss and Becker, 1983) and host-virus interactions (Georgopoulos and Tilly, 1981; Binkowski and Simon, 1983; Friedman et al., 1984).

HOST INVOLVEMENT IN PHAGE-HEAD ASSEMBLY

Many experiments suggest that *E. coli* proteins play roles in head assembly of double-stranded DNA (dsDNA) phages, and some results indicate that host participation is the rule. Several groups have isolated *E. coli* mutants that specifically block head construction of various phages. Some of these mutants will be described in this section.

GroE Mutants

The *groE*⁻ (Georgopoulos et al., 1972, 1973; Sternberg, 1973a,b) *mop*⁻ (Takano and Kakefuda, 1972), *tabB*⁻ (Coppo et al., 1973; Takahashi et al., 1975) and *hdh*⁻ (Revel et al., 1980) mutants are probably all mutant in one of two genes located at 94 min on the *E. coli* map, now named *groEL* (encoding a 65-kd protein; Georgopoulos and Hohn, 1978; Hendrix and Tsui, 1978) and *groES* (encoding a 15-kd protein; Tilly et al., 1981) or *mopA* and *mopB* (Bachmann, 1983). The original *groE*⁻ mutants were isolated by screening *E. coli* strains that prevented productive phage growth (isolated by the first method described above) for ones affecting head assembly. Mutants blocking T4 growth were selected directly, yet some mutants isolated by either method are nonpermissive for the growth of both phages. In addition, many of the mutations in one *groE* gene or the other interfere with bacterial growth at high temperatures. Genetic analysis has shown that in each case a single mutation in either the *groEL* or *groES* gene is responsible for all observable phenotypes (Georgopoulos et al., 1973; Georgopoulos and Eisen, 1974). Sternberg (1973a) has reported that the growth of the RNA phage R17 is also blocked in some *groE*⁻ bacteria, though the exact step affected is not known. In addition, he has reported that at least one *groE*⁻ mutation interferes with the production of stable RNA at the nonpermissive temperature (Sternberg, 1973a). It has recently been reported that some *groEL* or *groES* mutations interfere with host RNA and DNA syntheses at 42°C (Wada and Itikawa, 1984). The pleiotropic effects of the *groE* mutations make it difficult to assess the exact role of the *groE* gene products in the host.

The two *groE* genes form an operon with the order promoter-*groES*-*groEL* (Tilly, 1982). Both gene products have been purified to homogeneity. In the native state the GroEL protein forms a 25S decatetramer with sevenfold rotational symmetry and possessing weak ATPase activity (Hendrix, 1979; Hohn et al., 1979; Figure 1). The GroES protein is also oligomeric, with ca. 6 subunits in the native

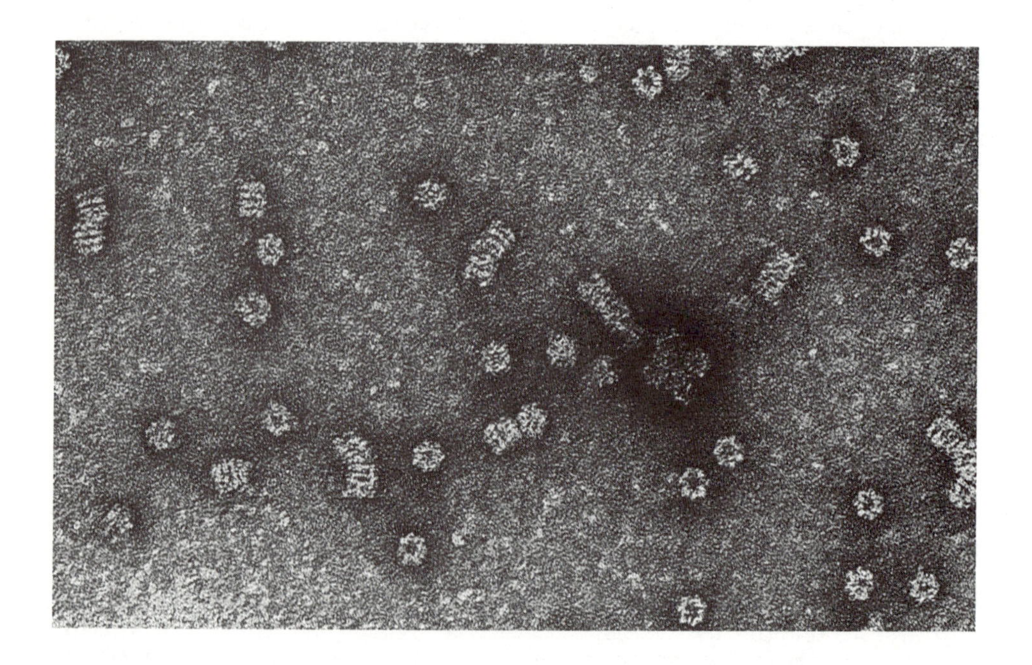

FIGURE 1 Structure of oligomers of the *groEL* gene product. Electron micrograph of purified preparation of *groEL* protein, negatively stained with 1% uranyl acetate. Magnification, ca. 340,000. Details about the structure of this protein, including rotational superposition photos can be found in Hendrix (1979). (Photo provided by R. Hendrix).

molecule (G. Chandrasekhar, pers. commun.) Three types of experiments suggest that the GroEL and GroES proteins interact: (1) *in vivo* the GroE⁻ phenotype of some *groES*⁻ mutations can be suppressed by mutations in the *groEL* gene (Tilly and Georgopoulos, 1982); (2) purified GroES protein inhibits the ATPase activity of GroEL *in vitro* (G. Chandrasekhar, pers. commun.); (3) in the presence of ATP, the GroEL and GroES proteins cosediment at 25S (G. Chandrasekhar, pers. commun.) These results suggest that the two GroE proteins act at a single step *in vivo* and help to explain why a mutation in any one of the genes results in the same phenotype with respect to phage morphogenesis and bacterial growth (see below).

All dsDNA phages assemble protein capsids, called proheads, which then encapsidate the phage chromosome (see Chapter 3, this volume). In the case of phage λ the prohead usually contains about 200 molecules of an internal protein, the product of the *Nu3* gene and called the scaffolding protein, which is removed at or before the DNA-packaging step. For λ the initial structures, containing scaffolding protein, are called immature proheads. These are converted to mature proheads by several covalent modifications of the component proteins and the cleavage and removal of the scaffolding protein (Ray and Murialdo, 1975). Normal mature λ proheads are composed of about 420 molecules of gpE (the major coat protein), gpB (2-4 molecules) and a proteolytically cleaved version pB* (10-12 molecules) (Hendrix and Casjens, 1975), and the proteins pX1 (about 6 molecules) and pX2 (about six molecules) (Hendrix and Casjens, 1974). The latter are two very similar products of cleavage and fusion of gpC and gpE (Hendrix and Casjens, 1974; Ray and Murialdo, 1975). GpB and pB* are found at the vertex at which tails will attach (Tsui and Hendrix, 1980); the positions of pX1 and pX2 in the capsids are not known. Infection of *groE*⁻ mutants leads to the accumulation of abnormal proheads (composed of gpE, gpNu3, gpC, and some gpB) and aberrantly assembled head-related structures, such as polyheads (Georgopoulos et al., 1973). This phenotype is similar to that observed in the case of infection of wildtype bacteria by λ B⁻ mutants (Kemp et al., 1968), suggesting that *groE*⁻ bacteria block λ head assembly at a very early stage in the pathway, the point at which the phage *B* product is required. Two lines of evidence suggest that the B protein interacts with at least the GroEL protein:

1. Some λ mutants, called λε, selected for their ability to form plaques on *groEL*⁻ hosts, have mutations in the *B* gene (Georgopoulos et al., 1973). Some mutations that seem to lower the levels of functional gpE, the major capsid protein also allow phage to grow on particular *groES*⁻ and *groEL*⁻ hosts (Georgopoulos et al., 1973; Sternberg, 1973b). These mutations are not allele-specific, so these phage may grow by reducing the numbers of capsid-initiation events. Mutant GroE proteins with reduced overall activity may successfully allow the completion of a few phage heads, rather than by altering a specific protein-protein interaction.

2. Murialdo (1979) showed that the *groEL* and *B* gene products cosediment in glycerol gradients. Kochan and Murialdo (1983) found that some of the cosedimentation resulted from a specific interaction between the two proteins. However, most of the cosedimentation was fortuitous, because gpB forms a dodecameric toroidal structure that also sediments with a value near 25S (Kochan and Murialdo, 1983; Kochan et al., 1984; Figure 2). Possibly, the GroEL protein, with the aid of GroES, provides a surface on which gpB can polymerize into its ring structure, which in turn may be the initiator for proper capsid assembly (Chapter 5, this volume; Murialdo and Becker, 1978).

There is no information available at present about whether B protein also interacts with the GroES protein.

It is interesting to note that similar structures to the gpB preconnector have been described for the head-tail connectors of phages ø29 (Carrascosa et al., 1982b) and T4 (Driedonks et al., 1981). The use of a preconnector for the correct assembly of phage heads may be a widespread strategy. Since an *in vitro* system for the production of biologically active λ proheads has been developed (Ferrucci and Murialdo, 1981), it may soon be feasible to determine the exact mechanism of action of each viral and host component.

(a)

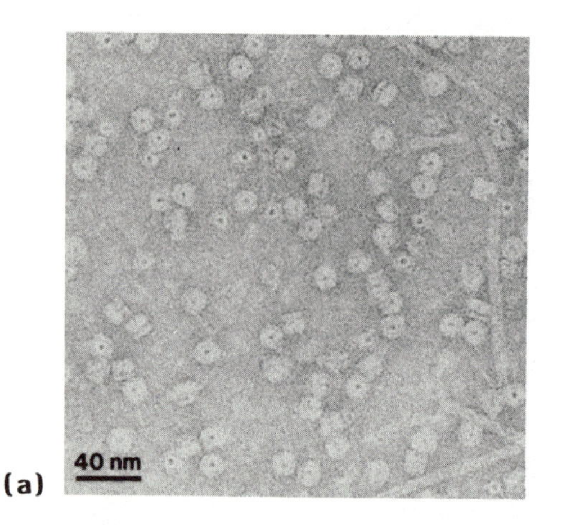

(b)

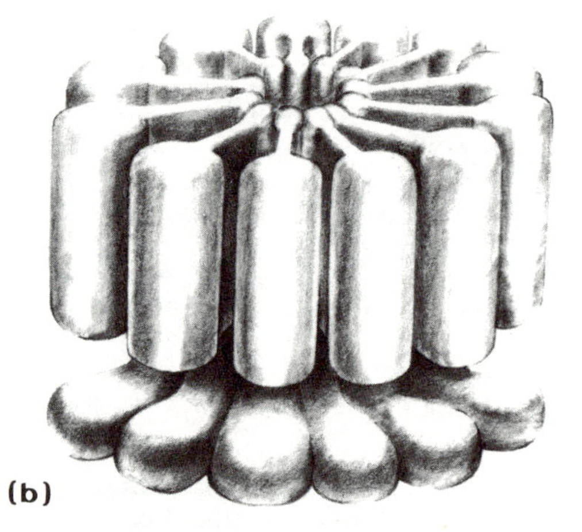

FIGURE 2 Structure of the gpB preconnector. (a) Electron micrograph of purified preparations of the gpB preconnector of phage λ, negatively stained with 1% ammonium molybdate. The bar in the lower left corner indicates the approximate scale. (b) Artist's interpretation of the gpB preconnector structure, aided by optical image processing of electron micrographs. Details about the preconnector structure are in Kochan et al., (1984). (Photos courtesy of H. Murialdo and J. Kochan.)

Head completion involves DNA packaging, with concomitant expansion of the capsids to their full sizes, and the addition of about 420 molecules of gpD and several molecules of some minor proteins to the mature λ prohead (Georgopoulos et al., 1983). GpD only adds to expanded heads *in vitro* (Imber et al., 1980) and seems to be important for head stability. The GroE proteins have not been shown to be necessary for these late steps in λ assembly, possibly because all selections used to isolate *groE* mutants required a block at an early stage in morphogenesis.

A protein similar in structure, dimensions, and antigenicity to the GroEL decatetramer has been found in *Bacillus subtilis* (Carrascosa et al., 1982a). After infection with phage ø29 this protein cosedimented with the phage gp10, which functions in the early steps of ø29 head assembly and could be analogous in function to the λ gpB (Carrascosa et al., 1982b). This result suggests that a protein similar to GroEL might also be essential for ø29 assembly. A protein with similar structure to GroEL is found in plants (Puskin et al., 1982), but its function is also not known.

It is probably significant that the rates of synthesis of the GroE proteins are increased severalfold after infection with λ (Drahos and Hendrix, 1982; Kochan and Murialdo, 1982; our unpublished results). Analysis of infection by deletion mutants has implicated functions in the left early region between and including the *red* and *ssb (Ea10)* genes. A phage that has lost this region of its chromosome is still able to grow; hence, the increased synthesis of the GroE proteins is not essential, though it may serve to augment phage production. Recently we have shown (unpublished results) that the induction of synthesis of GroE proteins is part of a phage-stimulated induction of the heat shock response (to be discussed shortly) and requires the presence of a positive bacterial regulator phage, namely, the product of the *htpR* gene (Neidhardt and VanBogelen, 1981; Yamamori and Yura, 1982).

Host Genes in Assembly of T4

The T4 gene-*31* product is also important in head assembly. Without this protein the major capsid protein, the product of gene *23*, remains in "lumps" attached to the cell membrane and is not cleaved to its mature form, gp23* (Laemmli et al., 1969; Georgopoulos et al., 1972). The bacterial mutants affecting T4 growth, including *mop*$^-$, *tabB*$^-$, *hdh*$^-$, and *groE*$^-$, all give this T4 31$^-$ phenotype. DNA replication and tail assembly proceed normally in each of these mutants. T4ε mutants, isolated by their ability to form plaques on these mutant host strains have alterations either in gene *31* (which codes for a nonstructural protein required for prohead assembly) or in gene *23* (Georgopoulos et al., 1972; Takahashi et al., 1975; Revel et al., 1980). The latter class may be analogous to the λε mutants with lowered levels of functional gpE. The existence of the former class suggests that gp31 and GroEL protein interact specifically, especially since many of the ε mutations are allele-specific (Takahashi et al., 1975; Revel et al., 1980; our unpublished results). Some mutations in gene *23* simultaneously eliminate the requirement for both the host *groEL* and the phage gene-*31* products (Doermann and Simon, 1984; Simon and Randolph, 1984), which indicates that both the GroEL and gene-*31* proteins are required for the correct assembly of gp23 and are not required during a subsequent morphogenetic step.

A gene, *fatA*, whose product affects the lipid composition of the *E. coli* inner membrane is also required for T4 prohead assembly. Mutations in gene *31* allow the phage to grow in *fatA*⁻ bacteria, suggesting that gp31 may modify the interaction between gp23 and the inner membrane (Simon et al., 1975).

Rho Mutations

Various mutations in the *E. coli rho gene, called hd590, tabC*, and *hdf*, have pleiotropic effects on T4 growth (Simon et al., 1974; Caruso et al., 1979; Simon et al., 1979; Stitt et al., 1980). Electron microscopic examination of lysates has shown the presence of empty heads, but no tails or baseplates (Simon et al., 1974). Also, the amounts of many phage proteins differ from the amounts present with a wildtype host, DNA synthesis is abnormal, and lysis is delayed, though lysozyme is produced normally (Stitt et al., 1980). It is not clear whether the effects of the *rho*⁻ mutations are a direct result of an alteration in termination of transcription or of changes in the intracellular levels of certain bacterial or phage functions. A possibly important observation is that a hyperproteolytic phenotype is exhibited by some of the *tabC* mutants (Simon et al., 1979). A single result suggests that this excessive protein-digesting activity may be a factor in the detrimental effects just described. The *lon* gene encodes an ATP-dependent protease (Chung and Goldberg, 1981). Introduction of a *lon*⁻ mutation into strain *tabC803* improves the growth of T4 on this host (Caruso et al., 1979). Some T4 phage mutants that are able to grow on *rho*⁻ hosts have mutations in gene *31*, an observation that complicates models of the action of gp31 (Simon et al., 1974).

Assembly of Filamentous Phages

A gene near the *rho* gene, named *fip*, which codes for a 12.5-kd protein required for the assembly of the filamentous phage f1, has recently been discovered (Russel and Model, 1983, 1984). Interestingly, the *fip* and *rho* genes are cotranscribed (K. Shigesada and M. Imai, pers. commun.), as are the *groES* and *groEL* morphogenetic genes. Mutants in the *fip* gene were shown to block f1 assembly at the level of the action of the product of gene *I*, but the uninfected cells exhibit no obvious growth defects. Particles are not formed by f1 gene-*I* mutants of phage f1, but the gene product has not been detected *in vivo* (Russel and Model, 1983, 1984). DNA sequencing has shown that the product coded by the *fip* gene is identical to the host protein thioredoxin (Russel and Model, 1985). Thioredoxin is required for phage T7 replication but is not essential for E. coli growth.

Tra Genes and RNA Phage Assembly

The *traD* gene of the *E. coli* F factor codes for a 77-kd protein, which is loosely bound to the membrane and necessary for successful conjugal transfer of DNA (Kennedy et al., 1977). Temperature-sensitive mutations in the *traD* gene block the propagation of group I RNA phages (Schoulaker-Schwarz and Engelberg-Kulka, 1981). Infection by phage MS2 at low temperature followed by a shift to nonpermissive temperature resulted in normal RNA replication and phage-specific protein synthesis, but no MS2-like particles or infectious units are produced. Evidence has been presented that the host membranes are altered in mutant cells, as judged by protein composition and increased sensitivity to various dyes and detergents (Schoulaker-Schwarz and Engelberg-Kulka, 1981).

It is clear that phage-head assembly processes take advantage of many pre-existing host proteins and structures. The nature of the dependence varies according to the mode of infection of the phage, a theme repeated in the cases of tail assembly and DNA packaging.

TAIL ASSEMBLY

The bacterial host participates at several levels in tail assembly. This participation has been studied by observation of the assembly process and isolation of either host or phage mutants. For example, the inner membrane seems to be the site of baseplate and tail-fiber assembly for T4, since components of those structures are membrane-associated during infection with wildtype or mutant phage (Simon, 1969; King and Laemmli, 1971; Mason and Haselkorn, 1972). These associations may be fortuitous, since *in vitro* assembly occurs without the addition of membranes (Mathews et al., 1983).

Two bacterial mutants that specifically affect T4 tail-fiber assembly have been isolated. The first, called *tabA*, causes the assembly of nonfunctional tail fibers at 42°C, and the resulting phage are not infectious (Pulitzer and Yanagida, 1971). The mutation maps between 68 and 82 min (Coppo et al., 1973), where many envelope proteins and the HtpR protein are encoded (Bachmann, 1983), but the mutation does not affect bacterial viability. The second bacterial mutation, *byp57*, was isolated for its phenotype of allowing T4 *57*⁻ phage to grow better, but it was not mapped (Revel et al., 1976). The gene-*57* product is required for the assembly of tail fibers and could serve to modify a host protein to serve the needs of T4 better. The *byp* mutations do not suppress specific *57*⁻ mutations (Revel et al., 1976), suggesting that they result in the alteration of a host protein such that a gp57-mediated modification is no longer needed for growth of T4. At least one allele causes temperature-sensitive bacterial growth, so the *byp* gene product is probably required for viability of *E. coli*.

An unexpected and unselected phenotype of some of the *groEL*⁻ and *groES*⁻ mutants is the inability of phage T5 to grow in these strains (Georgopoulos et al., 1973). T5 head assembly and head-protein processing are normal, but no functional tails form (Zweig and Cummings, 1973). The phage-protein cleavage that normally occurs during tail assembly was not observed. T5ε mutants that can grow on *groE*⁻ strains have been isolated; these have mutations in a tail-protein gene, and protein processing is restored (Zweig and Cummings, 1973). The inability of T5 to grow is a surprising observation: the effect is on T5 *tail* assembly, whereas λ tail assembly and tail-protein cleavage are normal in *groE*⁻ strains (Georgopoulos et al., 1973; unpublished results). This difference suggests that both head and tail assembly for these phages may require the *groE* gene products, and it is a result of the particular selection applied and the particular mutations obtained that there is no detectable phenotype in two of the processes. There is some precedent for this idea. For example, Hocking and Egan (1982) have recently presented data that suggests that some *E. coli groE*⁻ mutants also block tail assembly of phage 186. The evidence is that a 186ε mutant isolated as a plaque-former on a *groE⁻supE* host possesses a nonsense (Am) mutation in the tail morphogenetic gene *H* (Hocking and Egan, 1982). Growth of phage 186 *H*(Am) could result either from the lower intracellular levels of gpH owing to inefficient suppression or to the altered protein structure of gpH resulting from the insertion of glutamine at the amber codon. An alternate explanation for the participation of the *groE* gene products in some head- and tail-assembly processes is that several different morphogenetic pathways have independently evolved to use the same pre-existing host components.

Finally, host involvement in T4 baseplate assembly was also implicated by the studies of Georgopoulos et al. (1977), who isolated mutants with alterations in either gene *53* or gene *8* (whose products are necessary for baseplate assembly) that could grow on one bacterial strain, *E. coli*, K12 but not on another strain, *E. coli* B. Discrimination by these mutants may be at the level of membrane interaction, specific protein requirements, transcription, or translation.

DNA PACKAGING

Several studies have implicated host functions in maturation and packaging of phage dsDNAs.. For T4 it appears that some phage functions duplicate host functions and increase the production of normal phage. For example, Chao et al. (1974) showed that infection of *endA⁻polA⁻* bacteria by T4 leads to a decrease in the number of phage with oblate heads and a concomitant increase in petite phage particles so that 1/3 of the total number of particles are petites. Also, the effect of T4 *30*⁻ (*lig*⁻, ligase-deficient) mutations on phage growth can be modulated by the genotype of various bacterial strains. For example, bacteria with mutations in their own ligase genes are less permissive for T4 *lig*⁻ phage, whereas bacteria that overproduce ligase partially suppress the phage defect (Black et al., 1981). Endonuclease I, polymerase I, and ligase, in combination with T4 ligase and DNA-dependent ATPase-endonuclease, probably help resolve the branched DNA molecules formed during late DNA replication.

Another case in which a nonessential T4 gene is required for phage growth in some conditions is that of T4 *IPI*$^-$ infecting the host CT596. The product of the *IPI* gene is a major component of the phage, injected along with the chromosome, but is not required for growth in most bacterial strains (Black, 1974; Black and Abremski, 1974; Abremski and Black, 1979). However, the DNA of phage lacking IPI protein is degraded after injection into CT596 (Abremski and Black, 1979). A defective prophage carried in this strain may encode a nuclease that specifically recognizes either a sequence in T4 DNA or a structural feature of the molecule (Black and Abremski, 1974). A similar situation occurs with T4 gene *2*. T4 *2*$^-$ mutants will grow in *recBC*$^-$ strains but not in wildtype hosts, suggesting that the gene-*2* product makes the phage chromosomes resistant to degradation by the RecBC nuclease (Silverstein and Goldberg, 1976).

A connection has been observed between DNA cutting during packaging of lambdoid phages and a protein required for prophage integration, the integrative host factor (IHF). *In vitro* cutting at the cohesive-end site (*cos*) to generate mature λ chromosomes requires terminase (composed of the *Nu1* and *A* gene products; Sumner-Smith et al., 1981; Gold and Becker, 1983) and one of two host proteins. One is a 22-kd protein purified on the basis of its ability to participate in the reaction, and the other is IHF (Gold and Becker, 1983). Since λ can grow in bacteria with deletions of either or both of the genes encoding the two subunits of IHF (*himA* and *hip*), IHF must be nonessential for λ terminase action. However, the closely related lambdoid phage, 21, will not grow on *himA*$^-$ hosts (M. Feiss, pers. commun.). A λ mutant has been isolated that, like 21, will not grow well on strains carrying a *himA* deletion or on *himA*$^-$*hip*$^-$ double mutants (Baer et al., 1984) The responsible mutation falls in a region of dyad symmetry 154 bp from the left end of λ and may mark the binding site of another phage and/or host function involved in cutting of *cos*. These proteins might bind to DNA near a *cos* sequence and cause an alteration of its structure so that the sequence becomes a substrate for terminase.

A mutant that seems to affect an early stage of DNA maturation was isolated by Henderson and Weil (1977). This mutation, *gro256*, blocks the growth of λ carrying the *b221* deletion (which yields chromosomes about 19% shorter than wildtype) but allows wildtype phage to grow. The mutation maps near *cysC* and leads to reduced DNA cutting at *cos* sites and an increased amount of concatemeric DNA molecules. λ *b221* with additional mutations in the *B-C* region or a suppressed *D*(Am) mutation can grow on *gro256* cells. In the first case, growth may stem from facilitating the interaction between proheads and DNA during packaging, but the effect of the *D*(Am) mutation in enhancing DNA packaging is not obvious, since *in vitro* gpD only adds to expanded gpE structures (Imber et al., 1980).

STRESS RESPONSE AND INFECTION OF λ

Several observations indicate a relation between the λ infection process and the bacterial response to stress (heat shock is the stress most-often studied).

1. λ morphogenesis depends on the products of two of the genes induced by heat shock, *groES* and *groEL*. Two other heat shock genes, *dnaJ* and *dnaK*, are required for λ DNA replication (Sunshine et al., 1977; Georgopoulos, 1977). The RNA polymerase initiation factor, σ, is also required for λ transcription, and its synthesis is induced by heat shock (Gross et al., 1982).

2. λ infection causes increased synthesis of at least some of the proteins made in response to heat shock. This induction occurs via the normal cellular mechanism, since a mutant unable to induce the response to heat shock (*htpR*⁻) also fails to respond to λ infection (unpublished results). The animal viruses adenovirus (Nevins, 1982) and polyoma virus (Khandjian and Turler, 1983) also induce the synthesis of at least the host *hsp70* gene, whose analogue in *Drosophila* is homologous in amino acid sequence to the *E. coli dnaK* gene product (Bardwell and Craig, 1984). For adenovirus the function responsible for the accelerated rate of synthesis is most likely the *E1A* gene product (Nevins, 1982). The *groEL* gene product is antigenically related to proteins in *Tetrahymena* and *Drosophila* (H. Liebke, pers.commun.). It is not known whether synthesis of these proteins is induced by heat shock.

3. A host mutation called *mat*⁻, isolated by its effect on growth of λ (Fuerst et al., 1978), blocks both head assembly and the production of serum-blocking material. Waghorne and Fuerst (1985) have mapped the mutation in the *mat*⁻ strain to the *htpR* gene (Neidhardt et al., 1983).

4. Another possible connection is that mutations in the T4 *regA* gene lead to the formation of aberrant head-related structures at high temperatures (Wever et al., 1981). The *regA* gene product acts as a translational inhibitor of some T4 mRNA molecules (Karam et al., 1981). Since preincubation of the cells at the high temperature reduces the effect on head assembly, overproduction of one or more of the heat shock proteins may be involved in this phenotype.

Specific knowledge of the biochemical roles of the heat shock proteins in bacterial physiology is lacking, so it is impossible to know why the viruses have evolved to use these proteins.

HOST PROTEINS IN INFECTIONS WITH ANIMAL VIRUSES

Although the involvement of cellular functions in animal and plant viral assembly has not been studied systematically, several cases of interaction have come to light. First, host-encoded histones are required for the interaction between papovavirus DNA and coat protein

(Acheson, 1981). Second, polyoma virus assembly seems to require a second host component for the modification of VP1 that allows DNA-capsid interaction. The levels of this "permissivity factor" seem to be raised by expression of either or both middle-T and small-t antigens (Garcea and Benjamin, 1983), a situation that recalls the effect of λ infection on *groE* gene expression described above. Several types of covalent modifications of animal virus structural proteins that may affect assembly are performed by the hosts. A number of animal virus proteins are proteolytically cleaved during morphogenesis; for example, there is a hamster cell line in which the murine leukemia virus core polyprotein is not cleaved (Demsey et al., 1980). Thus, host cells are likely to participate in these processes also. The NS nucleocapsid protein of vesicular stomatitis virus must be phosphorylated by a host kinase for infectivity (Harmon et al., 1983). A portion of the Rous sarcoma virus p19 is also phosphorylated (Erikson et al., 1977), as are many other animal virus structural proteins; however, the responsible kinases have not been identified, and it is not known if phosphorylation affects assembly. All eukaryotic enveloped viruses contain glycoproteins. Host enzymes are responsible for this glycosylation, which in some cases may affect the assembly process. This host interaction is discussed in detail in Chapter 6 of this volume.

CONCLUDING REMARKS

It is clear that phages depend on their hosts for many essential functions, both structural and enzymatic. In some cases, the larger phages have duplicated host functions, for example, T4 DNA ligase and topoisomerase, and these genes are only truly essential in mutant hosts. In other cases, there is a direct interplay between phage and bacterial components. Some host genes (e.g., *tabA*, *groES*, *groEL*, etc.) were only identified because they play roles in phage assembly, whereas some host mutations in previously identified genes, such as *himA*⁻ or *hip*⁻, affect morphogenesis in unexpected ways. The requirement for IHF in *cos*-cutting by phage 21 terminase may reflect a general requirement for interaction with IHF or its alternative for any nuclease activity on lambdoid phage DNA. Miller (1981) has suggested that λ senses the HimA protein level when deciding between lysis and lysogeny. Since a high level would bias the phage toward lysogeny, it is somewhat of a paradox that HimA protein is required for phage 21 *cos*-cutting, an event found only in the lytic pathway.

The connection between λ infection and heat shock gene expression may provide insight into the physiology of *E. coli*. Phage λ not only requires expression of several of the heat shock genes for productive infection, but it accelerates their rate of expression after infection. Discovering the importance of the functions of these proteins in phage assembly, DNA replication, and transcription will help to determine their functions in uninfected cells, while finding the method of induction by λ will help to elucidate the regulation of the heat shock response. The analogous responses of animal cells to polyoma and adenovirus infections should reflect the operation of similar mechanisms. Therefore, the study of phage-host interactions may provide direction in the approach to the problem of interactions between eukaryotic cells and their viruses.

ACKNOWLEDGEMENTS

We would like to thank all of our colleagues quoted in the text for communication of results prior to publication, Sherwood Casjens for a critical reading of the manuscript, and Roger Hendrix, Helios Murialdo, and Jarko Kochan for providing the photographs shown in Figures 1 and 2.

REFERENCES

Abremski, K. and L.W. Black. 1979. *Virology*, 97, 439.

Acheson, N.W. 1981. In *DNA Tumor Viruses*. J. Tooze (ed.) p. 125. Cold Spring Harbor Laboratory.

Bachmann, B.J. 1983. *Microbiol. Rev.*, 47, 180.

Baer, S.E., D.L. Court, and D.I. Friedman. 1984. *J. Virol.*, 52, 966.

Baltimore, D., A.S. Huang, and M. Stampfer. 1970. *Proc. Nat. Acad. Sci. USA.*, 66, 572.

Bardwell, J.C.A. and E.A. Craig. 1984. *Proc. Nat. Acad. Sci. USA.*, 81, 848.

Binkowski, G. and L.D. Simon. 1983. In *Bacteriophage T4*. C.K. Mathews, E. Kutter, G. Mosig, and P. Berget (eds.). p. 342. American Society for Microbiology.

Black, L.W. 1974. *Virology*, 60, 166.

Black, L.W. and K. Abremski. 1974. *Virology*, 60, 180.

Black, L.W., A.L. Zachary, and V. Manne. 1981. In *Bacteriophage Assembly*. M. DuBow (ed.) p. 111. Alan Liss.

Carrascosa, J.L., J.A. Garcia, and M. Salas. 1982a. *J. Mol. Biol.*, 158, 731.

Carrascosa, J.L., E. Vinuela, N. Garcia, and A. Santisteban. 1982b. *J. Mol. Biol.*, 154, 311.

Caruso, M., A. Coppo, A. Manzi, and J. Pulitzer. 1979. *J. Mol. Biol.*, 135, 950.

Chao, J., L. Chao, and J.F. Speyer. 1974. *J. Mol. Biol.*, 85, 41.

Chung, C.H. and A.L. Goldberg. 1981. *Proc. Nat. Acad. Sci. USA.*, 78, 4931.

Coppo, A., A. Manzi, J. Pulitzer, and H. Takahashi. 1973. *J. Mol. Biol.*, 76, 61.

Craig, E., T. Ingolia, M. Slater, L. Manseau, and J. Bardwell. 1982. In *Heat Shock from Bacteria to Man*. M.J. Schlesinger, M. Ashburner, and A. Tisseres (eds.) p. 11. Cold Spring Harbor Laboratory.

Cronan, J.E. and P.R. Vagelos. 1971. *Virology*, 43, 412.

Demsey, A., F. Collins, and D. Kawak. 1980. *J. Virol.*, 36, 872.

Doermann, A.H. and L.D. Simon. 1984. *J. Virol.*, 51, 315.

Drahos, D.J. and R.W. Hendrix. 1982. *J. Bacteriol.*, 149, 1050.

Driedonks, R.A., A. Engel, B. ten Heggeler, and R. van Dreil. 1981. *J. Mol. Biol.*, 152, 641.

Earnshaw, W.C. and S.R. Casjens. 1980. *Cell*, 21, 319.

Earnshaw, W.C. and J. King. 1978. *J. Mol. Biol.*, 126, 721.

Edgar, R.S. and I. Lielausis. 1968. *J. Mol. Biol.*, 32, 263.

Ehrenfeld, E. 1982. *Cell*, 28, 435.

Erikson, E., J.S. Brugge, and R.L. Erikson. 1977. *Virology*, 80, 177.

Falco, S.C., W. Zehring, and L.B. Rothman-Denes. 1980. *J. Biol. Chem.*, 255, 4339.

Feiss, M. and A. Becker. 1983. In *Bacteriophage Lambda II*. R. Hendrix, J. Roberts, F.W. Stahl, and R. Weisberg (eds.). Cold Spring Harbor Laboratory.

Ferrucci, F. and H. Murialdo. In *Bacteriophage Assembly*. M. DuBow (ed.) p. 193. Alan Liss.

Floor, E. 1970. *J. Mol. Biol.*, 47, 293.

Friedman, D.I., E.R. Olson, K. Tilly, C. Georgopolous, I. Herskowitz, and F. Banuett (1984). *Microbiol. Rev.*, 48, 299.

Fuerst, C.R., H. Bingham, and J.P. Bouchard. 1978. *Virology*, 87, 416.

Garcea, R.L. and T.L. Benjamin. 1983. *Proc. Nat. Acad. Sci. USA.*, 80, 3613.

Gauss, P, D.H. Doherty, and L. Gold. 1983. *Proc. Nat. Acad. Sci. USA.*, 80, 1669.

Georgopoulos, C.P. 1971. *Proc. Nat. Acad. Sci. USA.*, 68, 2977.

Georgopoulos, C.P. 1977. *Molec. Gen. Genet.*, 151, 35.

Georgopoulos, C.P. and H. Eisen. 1974. *J. Supramol. Struct.*, 2, 349.

Georgopoulos, C.P., M. Georgiou, G. Selzer, and H. Eisen. 1977. *Experientia*, 33, 1157.

Georgopoulos, C.P., R.W. Hendrix, A.D. Kaiser, and W.B. Wood. 1972. *Nature New Biol.*, 239, 38.

Georgopoulos, C.P., R.W. Hendrix, S.R. Casjens, and A.D. Kaiser. 1973. *J. Mol. Biol.*, 76, 45.

Georgopoulos, C.P. and B. Hohn. 1978. *Proc. Nat. Acad. Sci. USA.*, 75, 131.

Georgopoulos, C. and K. Tilly. 1981. In *Bacteriophage Assembly*. M. DuBow (ed.) p. 21. Alan Liss.

Georgopoulos, C., K. Tilly, and S. Casjens. 1983. In *Bacteriophage Lambda II*. R.W. Hendrix, J. Roberts, F.W. Stahl, and R. Weisberg (eds.) p. 279. Cold Spring Harbor Laboratory.

Giphart-Gassler, M., R.H. Plasterk, and P. van der Putte. 1982. *Nature*, 297, 339.

Gold, M. and A. Becker. 1983. *J. Biol. Chem.*, 258, 14619.

Gross, C.A., Z. Burton, M. Gribskov, A. Grossman, H. Liebke, W. Taylor, W. Walter, and R.R. Burgess. 1982. In *Promoters: Structure and Function*. R. Rodriguez and M. Chamberlin (eds.), p. 252. Praeger.

Harmon, S., L. Marnell, and D. Summers. 1983. *J. Biol. Chem.*, 258, 15283.

Henderson, D. and J. Weil. 1977. *J. Mol. Biol.*, 113, 67.

Hendrix, R.W. 1979. *J. Mol. Biol.*, 129, 375.

Hendrix, R.W. and S.R. Casjens. 1974. *Proc. Nat. Acad. Sci. USA.*, 71, 1451.

Hendrix, R.W. and S.R. Casjens. 1975. *J. Mol. Biol.*, 91, 187.

Hendrix, R.W. and L. Tsui. 1978. *Proc. Nat. Acad. Sci. USA.*, 75, 136.

Hocking, S.M. and J.B. Egan. 1982. *J. Virol.*, 44, 1056.

Hohn, T., B. Hohn, A. Engel, M. Wurtz, and P.R. Smith. 1979. *J. Mol. Biol.*, 129, 359.

Hohn, T. and I. Katsura. 1977. *Curr. Top. Microbiol. Immunol.*, 78, 69.

Horvitz, H.R. 1974. *J. Mol. Biol.*, 90, 727.

Imber, R., A. Tsugita. , M. Wurtz, and T. Hohn. 1980.*J. Mol. Biol.*, 129, 359.

Karam, J., L. Gold, B.S. Singer, and M. Dawson. 1981. *Proc. Natl. Acad. Sci. USA*, 78, 4669.

Karnik, S. and M. Billeter. 1983. *EMBO J.*, 2, 1521.

Katsura, I. 1981. In *Bacteriophage Assembly*. M. DuBow, ed. p. 79. Alan Liss.

Kells, S.S. and R. Haselkorn. 1974. *J. Mol. Biol.*, 83, 473.

Kemp, C.L., A.F. Howatson, and L. Siminovitch. 1968. *Virology*, 36, 490.

Kennedy, N., L. Beutin, M. Achtman, R. Skurray, U. Rahhmsdorf, and H.P. Herrlich. 1977. *Nature*, 270, 580.

Khandjian, E.W. and M. H. Turler. 1983. *Molec. Cell Biol.*, 3, 1.

King, J. and U.K. Laemmli. 1971. *J. Mol. Biol.*, 62, 465.

Kochan, J., J.L. Carrascosa, and H. Murialdo. 1984. *J. Mol. Biol.*, 174, 433.

Kochan, J. and H. Murialdo. 1982. *J. Bacter.*, 149, 1166.

Kochan, J. and H. Murialdo. 1983. *Virology*, 131, 100.

Laemmli, U.K., F. Beguin, and G. Gujer-Kellenberger. 1969. *J. Mol. Biol.*, 47, 69.

Losick, R. and J. Pero. 1981. *Cell*, 25, 582.

Marchin, G.L., M.M. Gomer, and F.C. Neidhardt. 1972. *J. Biol. Chem.*, 247, 5132.

Mason, W.S. and R. Haselkorn. 1972. *J. Mol. Biol.*, 66, 445.

Mathews, C.K., E.M. Kutter, G. Mosig, and P.B. Berget, eds. 1983. *Bacteriophage T4*. American Society for Microbiology.

Miller, H.I. 1981. *Cell*, 25, 269.

Murialdo, H. 1979. *Virology*, 96, 341.

Murialdo, H. and A. Becker. 1978. *Microbiol. Rev.*, 42, 529.

Neidhardt, F.C. and R.A. VanBogelen. 1981. *Biochem. Biophys. Res. Commun.*, 100, 894.

Neidhardt, F.C., R.A. VanBogelen, and E.T. Lau. 1983. *J. Bacteriol.*, 153, 597.

Nevins, J.R. 1982. *Cell*, 29, 913.

Onorato, L. and M.K. Showe. 1975. *J. Mol. Biol.*, 92, 395.

Pulitzer, J.F. and M. Yanagida. 1971. *Virology*, 45, 539.

Puskin, A.V., V.L. Tsuprun, N.A. Solovjeya, V.V. Shubin, Z.G. Evstigneeva, and W.L. Kretovich. 1982. *Biochim. Biophys. Acta*, 64, 247.

Ray, P.N. and H. Murialdo. 1975. *Virology*, 64, 247.

Revel, H.R., R. Hermann, and R.J. Bishop. 1976. *Virology*, 72, 255.

Revel, H.T., B.L. Stitt, I. Lielausis, and W.B. Wood. 1980. *J. Virol.*, 33, 366.

Russel, M. and P. Model. 1983. *J. Bacteriol.*, 154, 1064.

Russel, M. and P. Model. 1984. *J. Bacteriol.*, 157, 526.

Russel, M. and P. Model. 1985. *Proc. Nat. Acad. Sci. USA*, 82, 29.

Sanger, F., G.M. Air, B.G. Barrell, N.L. Brown, A.R. Coulson, J.C. Fiddes, C.A. Hutchison III, P.M.Y. Clocombe, and M. Smith. 1977. *Nature*, 265, 687.

Schmidt, M.F.G., and M.J. Schlesinger. 1979. Cell, 17, 813.

Schoulaker-Schwarz, R. and E.B. Goldberg. 1976. *Virology*, 72, 212.

Simon, L.D. 1969. *Virology*, 38, 285.

Simon, L.D. 1972. *Proc. Nat. Acad. Sci. USA*, 69, 907.

Simon, L.D., D. Snorer, and A.H. Doermann. 1974. *Nature*, 252, 451.

Simon, L.D., J.M. McLaughlin, D. Snorer, J. Ou, C. Grisham, and M. Loeb. 1975. *Nature*, 256, 379.

Simon, L.D., M. Gottesman, K. Tomczak, and S. Gottesman. 1979. *Proc. Nat. Acad. Sci. USA*, 76, 1623.

Simon, L.D. and B. Randolph. 1984. *J. Virol.*, 51, 321.

Sternberg, N. 1973a. *J. Mol. Biol.*, 76, 1.

Sternberg, N. 1973b. *J. Mol. Biol.*, 76, 25.

Steven, A. 1972. *Proc. Nat. Acad. Sci. USA*, 69, 603.

Stitt, B.L., H.R. Revel, I. Lielausis, and W.B. Wood. 1980. *J. Virol.*, 35, 775.

Sueoka, N., T. Kano-Sueoka, and W.J. Gartland. 1966. *Cold Spring Harb. Symp. Quant. Biol.*, 31, 571.

Sumner-Smith, M., A. Becker, and M. Gold. 1981. *Virology*, 111, 642.

Sunshine, M., M. Feiss, J. Stuart, and J. Yochem. 1977. *Molec. Gen. Genet.*, 151, 27.

Swinton, D., S. Hattman, P.F. Crain, C.S. Cheng, D.L. Smith, and J.H. McCloskey. 1983. *Proc. Nat. Acad. Sci. USA*, 80, 7400.

Takahashi, H., A. Coppo, A. Manzi, G. Martire, and J.F. Pulitzer. 1975. *J. Mol. Biol.*, 96, 563.

Takano, T. and T. Kakefuda. 1972. *Nature New Biol.*, 239, 34.

Tilly, K. 1982. Ph.D. Thesis. Univ. of Utah, Salt Lake City.

Tilly, K., and C. Georgopoulos. 1982. *J. Bacteriol.*, 149, 1082.

Tilly, K., H. Murialdo, and C. Georgopoulos. 1981. *Proc. Nat. Acad. Sci. USA*, 78, 1629.

Tsui, L. and R. Hendrix. 1980. *J. Mol. Biol.*, 142, 419.

Waghorne, C. and C.R Fuerst. 1985. *Virology*, 141, 51.

Wever, G.H., B.J. Thompson, R.M. Laiken, E. Ruby, and J.S. Wiberg. 1981. In *Bacteriophage Assembly*, M. DuBow, ed. p. 167. Alan Liss.

Wada, M. and H. Itikawa. 1984. *J. Bacteriol.*, 157, 694.

Winter, R.B. and L. Gold. 1983. *Cell*, 33, 877.

Wolfe, P.B., P. Silver, and W. Wickner. 1982. *J. Biol. Chem.*, 254, 7898.

Wood, W.B. and J. King. 1979. In *Comprehensive Virology*. H.

Fraenkel-Conrat and R. Wagner (eds.). Vol. 13, p. 581. Plenum Press.

Yamamori, T. and T. Yura. 1982. *Proc. Nat. Acad. Sci. USA*, 79, 860.

Zweig, M. and D.J. Cummings. 1973. *J. Mol. Biol.*, 80, 505.

Index

A

B

C

D

E

F

G

H

I

L

M

N

O

P

Q

R

S

T

U

V

W